C++ Primer Plus

（第 6 版）中文版习题解答

[美] 史蒂芬·普拉达（Stephen Prata） 著

曹良亮 编

人 民 邮 电 出 版 社

北 京

图书在版编目（CIP）数据

C++ Primer Plus（第6版）中文版习题解答 /（美）史蒂芬·普拉达（Stephen Prata）著；曹良亮编. -- 北京：人民邮电出版社，2020.7
ISBN 978-7-115-53774-4

Ⅰ. ①C… Ⅱ. ①史… ②曹… Ⅲ. ①C语言—程序设计—题解 Ⅳ. ①TP312.8-44

中国版本图书馆CIP数据核字(2020)第058178号

◆ 著　　　[美] 史蒂芬·普拉达（Stephen Prata）
编　　　曹良亮
责任编辑　傅道坤
责任印制　王　郁　焦志炜

◆ 人民邮电出版社出版发行　　北京市丰台区成寿寺路 11 号
邮编　100164　　电子邮件　315@ptpress.com.cn
网址　https://www.ptpress.com.cn
三河市君旺印务有限公司印刷

◆ 开本：787×1092　1/16
印张：22.75　　　　2020 年 7 月第 1 版
字数：538 千字　　　2025 年 3 月河北第25次印刷
著作权合同登记号　图字：01-2012-0244 号

定价：89.00 元

读者服务热线：(010)81055410　印装质量热线：(010)81055316
反盗版热线：(010)81055315

内 容 提 要

本书是超级畅销书《C++ Primer Plus（第 6 版）中文版》的配套习题答案，针对书中的复习题和编程练习，给出了解题思路和答案。

本书共分为 18 章，每一章的主题与《C++ Primer Plus（第 6 版）中文版》完全一致。每章开篇采用思维导图的方式列出本章的知识点，然后对每章的重点内容进行了梳理总结，最后则对每章中的复习题和编程练习进行了分析并给出了解答思路，确保读者在彻底夯实理论知识的同时，进一步提升实际编程能力。

作为《C++ Primer Plus（第 6 版）中文版》的配套参考书，本书特别适合需要系统学习 C++语言的初学者阅读，也适合打算巩固 C++语言知识或者希望进一步提高编程技术的程序员阅读。

编 者 简 介

曹良亮，理学博士,高级工程师，任教于北京师范大学教育技术学院，长期从事北京师范大学本科生 C/C++语言与 Java 语言的教学工作。

前 言

作为一种功能强大的高级程序设计语言，C++语言诞生于20世纪80年代的贝尔实验室。在这之后的几十年，C++语言一直都是最重要的编程语言之一，广泛地应用于各种开发环境下。除了具备C语言高效、灵活的特性外，C++语言还引入了面向对象的设计思想和泛型编程的方法，这使得C++语言能够更好地应对现代大规模软件的设计和开发。从C语言面向过程的设计开发思想过渡到C++语言面向对象和泛型编程思想，需要学习大量新的知识和新的概念。《C++ Primer Plus（第6版）中文版》一书详细讲解和讨论了C++语言的基础知识及设计开发思想，理论讲解浅显易懂，案例设计丰富多变，是一本学习C++语言基础知识的优秀教材。

《C++ Primer Plus（第6版）中文版》一书共18章。其中第1章主要介绍了C++语言中面向对象和泛型编程的基础知识，对比和分析了C++语言和C语言在编程原理与设计思想上的差异。第2～4章主要介绍了C++语言的基本数据类型、变量、常量的使用、表达式的基本概念和常用的基本输入/输出语句。该部分的学习难点在数组、指针和new的动态存储等，读者应当从计算机的基本数据存储方式上来理解这些概念，并在实践中灵活应用。此外，该部分还简单介绍了C++的一些新数据类型，如string和模板类vector等。

第5章和第6章主要讨论了流程控制的内容。流程控制是面向过程的设计思想的核心和逻辑基础，其中主要的知识点是关系表达式、逻辑表达式、循环语句和条件语句。逻辑表达式是循环控制和条件语句的基础。应用中需要仔细分析表达式的逻辑关系和布尔值才能正确设计符合逻辑的循环和条件判断。第7章和第8章主要介绍了函数的内容。作为C++语言中最基本的程序模块，函数是实现软件开发中代码重用的基础，也是面向对象设计中数据操作的基本接口。函数设计的难点在于参数的值传递和几类特殊参数的使用，包括数组参数、指针参数、引用参数的使用。这些参数类型在程序中的区别和具体的应用场景，需要反复练习并熟练掌握。函数的重载和函数模板是C++语言泛型编程的一个具体体现，泛型编程的主要目标是创建独立于类型的通用代码。泛型函数的语法规则和设计是后续各章节的基础，也应当熟练掌握。

第9章介绍了内存模型和名称空间。读者需要牢记C++语言中变量的存储特性和作用域以及单独编译情况下的链接性，其中学习重点是静态变量的特性和使用。第10～12章主要讨论了面向对象设计思想的基础——类和对象，类是面向对象中数据的一种抽象模型，读者要理解类和对象设计原理，以及如何合理设计类的数据成员和成员函数。类是一种数据的一般性的抽象描述，只有创建了实体化的对象才能进行数据的存储和操作。第12章中的动态存储分配及对象的引用涉及的几个特殊函数和运算符是类的定义中的难点和重点，主要包括默认构造函数、默认析构函数、复制构造函数、赋值运算符和地址运算符，此外，类的类型转换函数也需要在类的设计中多加关注。

第13～15章主要介绍了类的继承和派生及相关问题。C++中的类继承是实现代码重用的重要手段。在 C++语言中，由于继承产生的类型对象的多态性和多重继承中重复基类的处理问题是这几章的难点，读者需要在此基础上理解虚函数和虚基类两个重要概念，以及由此衍生的动态联编和静态联编的问题。此外，继承过程中的数据访问权限以及数据封装产生的友元问题也需要读者重点关注。第 17 章介绍了 C++语言的标准输入/输出和文件的基本操作。输入/输出的流式操作是 C++进行数据读写的重要方式，应用中除需要熟练掌握 I/O 流的基本操作过程与相关方法外，还要注意二进制文件的读写和随机读写操作。

第 16 章和第 18 章主要介绍了 C++的部分新功能和标准模板库（Standard Template Library，STL）。STL 是 C++语言的一个标准组件，提供了大量的预定义数据结构和通用算法函数，如容器类、迭代器、智能指针和相关算法等。第 18 章的重点是移动语义和右值引用、lambda 匿名函数和可变参数模板。C++语言中 STL 的内容较多，《C++ Primer Plus（第 6 版）中文版》一书简要介绍了部分常用功能，如果在开发中需要大量用到 STL，还需要参考相关的手册。

《C++ Primer Plus（第 6 版）中文版》一书习题设置合理，基本涵盖了每章的所有重点和难点。每一位读者在学习过程中都应当参考课后的习题进行自我练习和检测，对每章重点内容的学习情况进行查漏补缺。

作为一本习题解答手册，本书首先梳理了各个章节的重要知识点，在此基础上详细分析和解答了章后的所有复习题和编程练习，供读者参考和自学使用。其中习题解析不仅给出了题目的参考答案，还强调了程序设计中分析和解决问题的基本过程。在编程分析中，首先分析题目所要求的基本功能和现有素材，然后分析与设计基本的流程和数据，最后进行代码实现。希望通过分析和实现，培养读者程序设计的基本思想和动手能力。本书提供了所有编程练习的完整代码，并且在代码中添加了详细的注释，以帮助读者理解代码的作用。

需要强调的是，本书只是提供了基本的程序设计方案，目的是易于读者理解和学习，强调代码的清晰可读，因此一些程序设计方案在逻辑设计上和算法设计上并不是最优的，编码上也远不够简洁优雅。读者可以在本书提供的解决方案和代码的基础上，不断优化和改进，实现更优的设计方案。由于程序中需要引用原书的例题，因此本书的注释采用了 C 语言风格的注释，以区别于原书的 C++风格的注释。

本书的所有代码都在 macOS Mojave 系统下使用 Xcode 验证过，但是由于部分程序用到了 C++11 的特性，因此读者可能需要升级编译器才能正确编译、运行。

由于时间仓促，编者水平有限，错误在所难免，希望读者能够谅解并提出宝贵意见。

服务与支持

本书由异步社区出品，社区（https://www.epubit.com/）为您提供后续服务。

提交勘误

作者和编辑尽最大努力来确保书中内容的准确性，但难免会存在疏漏。欢迎您将发现的问题反馈给我们，帮助我们提升图书的质量。

当您发现错误时，请登录异步社区，按书名搜索，进入本书页面，单击“提交勘误”，输入勘误信息，单击“提交”按钮即可（见下图）。本书的作者和编辑会对您提交的勘误进行审核，确认并接受后，您将获赠异步社区的100积分。积分可用于在异步社区兑换优惠券、样书或奖品。

详细信息　写书评　提交勘误

页码：　页内位置（行数）：　勘误印次：

字数统计

提交

扫码关注本书

扫描下方二维码，您将会在异步社区微信服务号中看到本书信息及相关的服务提示。

与我们联系

我们的联系邮箱是 contact@epubit.com.cn。

如果您对本书有任何疑问或建议，请您发邮件给我们，并请在邮件标题中注明本书书名，以便我们更高效地做出反馈。

如果您有兴趣出版图书、录制教学视频，或者参与图书翻译、技术审校等工作，可以发邮件给我们；有意出版图书的作者也可以到异步社区在线提交投稿（直接访问 www.epubit.com/selfpublish/submission 即可）。

如果您所在学校、培训机构或企业想批量购买本书或异步社区出版的其他图书，也可以发邮件给我们。

如果您在网上发现有针对异步社区出品图书的各种形式的盗版行为，包括对图书全部或部分内容的非授权传播，请您将怀疑有侵权行为的链接通过邮件发送给我们。您的这一举动是对作者权益的保护，也是我们持续为您提供有价值的内容的动力之源。

关于异步社区和异步图书

“异步社区”是人民邮电出版社旗下 IT 专业图书社区，致力于出版精品 IT 技术图书和相关学习产品，为作译者提供优质出版服务。异步社区创办于 2015 年 8 月，提供大量精品 IT 技术图书和电子书，以及高品质技术文章和视频课程。更多详情请访问异步社区官网 https://www.epubit.com。

“异步图书”是由异步社区编辑团队策划出版的精品 IT 专业图书的品牌，依托于人民邮电出版社近 30 年的计算机图书出版积累和专业编辑团队，相关图书在封面上印有异步图书的 LOGO。异步图书的出版领域包括软件开发、大数据、AI、测试、前端、网络技术等。

异步社区

微信服务号

目　录

第1章 预备知识

1.1 C++语言的简介

C++语言可以说是20世纪最重要的一种程序设计语言，它首先继承了C语言简洁、高效的特性；又融合了面向对象程序设计的思想及泛型编程的特性。因此 C++语言的可扩展性更高、整体性能更加强大，可用于实现复杂程度日益提高的现代软件开发任务。

C++首先继承了C语言所代表的面向过程的程序设计语言的特性，C++的语法结构与C语言基本一致。以C语言为代表的面向过程的程序设计强调软件功能和开发任务的分解，其自顶而下的开发原用于将大型程序不断分解为阶段性目标，并最终逐步设计和实现这些基本的功能模块，从而完成整个软件系统的开发。在面向过程的软件开发中，整体结构简单清晰、可操作性强，但是随着现代软件的复杂程度逐渐增加，这种面向过程的软件开发方法使得复杂结构下的功能模块剧增，各个功能模块在设计、开发、调试上互相影响和制约，各开发部门之间的协作难度日益增大，因此现代化软件开发中迫切需要一种全新的设计开发方法。

面向对象的程序设计（Object-Oriented Programming）思想与面向过程开发中强调软件的功能分析和具体过程的算法设计不同，面向对象的方法强调在软件开发中对关键性数据的描述、定义和操作。从某种角度上看，面向对象与其说是一种软件开发方法，不如说是一种世界观。面向对象的设计方法将软件看作由不同实体组成的一个整体，每一个实体有时由一定的数据及对数据的操作组成，因此软件设计的核心是将软件进行数据抽象，形成一个个抽象的数据格式——类，类中包含了具体数据的格式、内容，以及对数据的基本操作方式。将抽象的类实体化之后就可以得到具体的数据，并通过其操作方法实现数据之间的信息反馈和交流。软件由无数个这样的数据实体通过各自的操作方法实现了信息的传递，最终组合成一个整体系统。

泛型编程（Generic Programming）是一种针对大量不同类型数据的统一化操作的解决方案。由于面向过程的程序开发具有数据导向的特点，开发中需要设计大量复杂的复合类型的数据，如果针对每一个类型的数据都设计一个操作方法（操作函数），势必会导致整个软件开发的规模不断增大。泛型编程能够在繁杂的复合数据类型的基础上实现一个统一的数据操作方法，实现数据操作功能和代码的重用，从而促进面向对象的程序设计并提高开发效率。

1.2 C++语言的编译过程

和 Python 等解释型语言不同，C++语言是一种典型的编译型语言，程序员编写的 C++

源代码文件需要通过 C++的编译器的编译，才会生成在目标平台上能够直接运行的机器语言。机器语言是一种计算机能够直接识别和运行的二进制数字指令系统。计算机系统能够逐步执行序列化的机器语言指令，实现软件的计算和操作功能。作为一种高级程序设计语言，C++语言是一种更加贴近日常语言特点和规则的编码系统，需要通过相应的编译器软件，将这种高级语言的编码指令转化成当前计算机能够识别、运行的机器语言指令（即可执行文件）。一般我们将这种通过编译器将 C++语言源代码转化为机器指令的过程称为编译。

源代码的编译过程非常复杂，早期 C++的编译过程中使用了一个编码转换程序，用于将 C++代码转换为 C 语言的源代码，然后再将转换后的 C 源代码编译成目标机器语言代码。随着 C++语言的普及和发展，目前流行的编译器都直接将 C++语言的源代码直接编译成目标代码。C++的编译过程和 C 语言类似，也将整个编译工作分为编译和链接两个阶段。编译过程首先将源代码编译成目标代码文件。为了提高编译效率，编译器会将多段源代码和其他第三方代码模块分别编译，生成多段中间代码。编译的第 2 个阶段叫作链接。链接是将由编译生成的目标代码，与相应的系统标准启动代码和库代码组合起来，生成最终的可执行程序。

C++语言经过多年的发展，目前能够兼容多种软硬件平台。为了提供更好的兼容性和可移植性，ANSI 和 ISO 共同制定了 C++语言的标准化版本。目前主流的开发平台都支持 C++98 和 C++11 标准，因此我们开发的很多 C++源代码能够编译成支持不同标准的可执行文件（机器语言）。各个平台下 C++语言的编译器种类众多，在实际开发中可以选择主流的、支持新 C++标准的编译器作为工具，这样可以使用更新的 C++语言特性，也可以实现更好的兼容性。例如，Windows 平台下的 Microsoft Visual C++编译器、macOS 平台下 Xcode 中集成的 g++和 llvm 编译器以及 Linux/UNIX 平台下的 GNU C++编译器都是很好的选择。

第 2 章　开始学习 C++

本章知识点总结

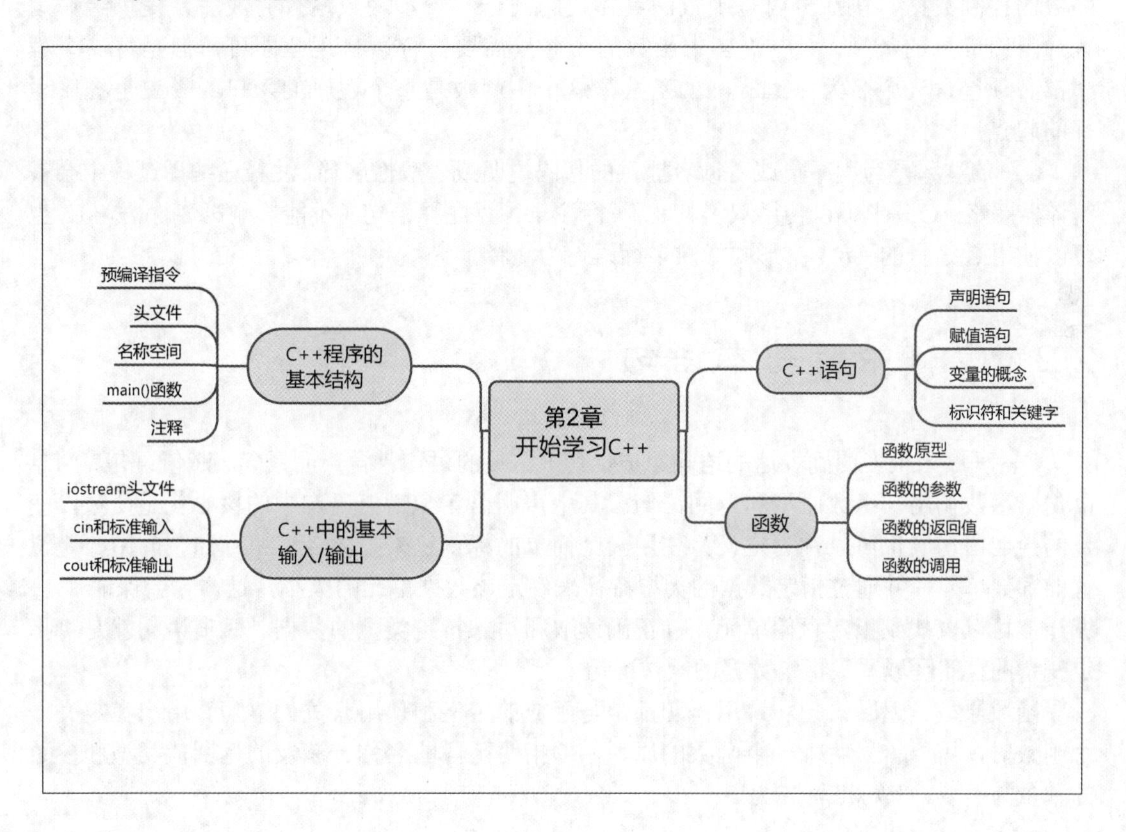

2.1　C++程序的基本结构

最简单的 C++程序“Hello World!”仅有十几行代码，但是也包含了 C++语言中一些很重要的知识点。下面我们就以 Hello World 程序为例了解 C++语言在编程中的一些基础知识。

语句`#include <iostream>`和 `using namespace std` 是一种预处理器指令，即告诉编译器在编译文件之前需要做的一些处理工作，例如，利用`#include` 语句包含 C++的标准库和第三方库的头文件，这样在编译的下一个阶段替换处理源代码文件，编译器才能正确识别和处理源代码中使用的 C++语言与系统预定义的函数和符号。`using` 编译指令明确表示以下代码中使用了哪一个名称空间定义的 C++的标准化组件，从而使编译器能够正确识别与查找这些组件并在编译过程中正确识别和编译。

cout<<"Hello World!"<<endl;语句表示 C++语言中的标准化数据输入和输出操作，即将“Hello World!”这句话输出到标准输出设备（通常默认是显示器）。其中 cout 表示一个 C++标准输出功能的对象，它实现的功能是输出运算符<<之后的字符数据，endl 是一个表示换行（end of line）的输出控制符号。

C++中的函数是可以反复调用的一整块代码的集合。int main(){}表示一个完整的函数。其中 main 是函数名，圆括号内是函数的参数列表，参数接收函数调用者发送给函数的控制数据。花括号内是函数的具体代码；int 表示函数运行完成后，提供给调用者的反馈信息的类型。C++语言中函数的命名只需要符合语言规范即可，但是每一个程序都必须包含一个名为 main()的函数，这个主函数是整个程序的入口或者说是程序运行的起点。

C++的注释是为程序员或者读者提供的代码说明或一般性解释，注释在编译过程中会被编译器忽略。C++中可以使用双斜杠（//）标注一行为注释语句（不能影响第 2 行语句），也可以使用 C 语言的（/**/）形式标注多行语句为注释。

2.2 C++中的基本语句

C++程序由很多明确表达的语句组成。其中常见的语句包含声明语句、赋值语句、消息语句、函数调用、函数原型和返回语句。其中声明语句和赋值语句中的核心就是变量，变量的声明语句（也可以称为定义）使用一个简单的标识符来描述计算机存储空间中一定位置的存储单元，并通过指定数据的类型特征来确定该数据单元的大小，这样才会保证 C++编译器可以查找到指定存储单元，并正确读取和写入特定类型的数据。赋值语句就是将指定变量的存储区域写入指定数据的操作语句。

函数原型、返回语句和调用语句都是与函数的定义和使用有关的语句。用户在定义一个函数时，需要首先明确一个函数的原型，即指明函数的名称、涉及的返回值类型和参数的数量与类型。其标准格式为

```
double sqrt(double);
```

函数的定义就是实现整个函数功能的编码，在函数完成相关功能后，函数需要使用返回语句把执行效果等具体信息反馈给函数的调用者（有时也分别称为主调函数和被调函数），返回语句的重要标志是 return 及之后的返回数据。函数的调用就是调用者通过函数名及函数参数来调用函数的具体功能并获取返回值的过程。函数调用过程中需要注意提供与函数原型定义中类型和数量相同的数据作为参数。

函数是程序设计中实现代码重用的一种重要方式。C++标准库和 STL（标准模板库）提供了大量系统函数，在程序设计过程中用户应当优先调用这些函数，这也是程序需要使用预编译指令#include 的原因。当系统预定义函数不能完成具体功能时，就需要用户自定义函数并调用。

2.3　复习题

1．C++程序的模块叫什么？

习题解析：

C++程序设计中的模块的主要形式是函数。函数是由多条语句组合而成的并且能够实现特定功能的代码模块。函数的主要作用是在程序设计过程中实现特定功能的代码的重用，提高编程的效率和可维护性。当程序员需要反复使用相同功能的代码时，可以首先声明一个函数，将这部分代码定义成函数。最后通过调用该函数就可以重复执行函数的特定功能，而不用多次重复编写相同的代码来实现类似的功能。

2．下面的预处理器编译指令的功能是什么？

```
#include <iostream>
```

习题解析：

#include 预处理器指令的主要功能是在编译器进行源代码的编译过程之前，添加或者替换相应的预编译指令，从而使得用户源代码中调用的系统预定义函数和各种标识符能够正确地被编译器识别和编译。#include <iostream>表示的含义是将 iostream 头文件添加到当前源代码中，iostream 头文件主要包含了系统的标准输入/输出函数以及数据的声明和定义。

3．下面的语句的功能是什么？

```
using namespace std;
```

习题解析：

using 预编译器指令的主要功能是表明当前源代码文件使用的名称空间 std。名称空间是 C++语言中为了解决编写大型程序时，多个厂商的独立 C++代码在标识符命名过程中可能会发生冲突的一个解决方案。不同供应商的代码模块都拥有自己的名称空间，用户在使用这个模块时也需要明确标注自己使用的是哪一个厂商的代码模块。using 预编译指令就是实现这个功能的语句。

4．什么语句可以用来输出短语“Hello，world!”，然后开始新的一行？

习题解析：

C++中输出到屏幕是通过 cout 和重定向符号 << 实现的，输出的短语是字符串，应当使用双引号将字符串括起来，这里使用以下语句。

```
cout<<"Hello, world!" ;
```

C++语言中字符串的输出是从左向右的，在当前行末尾自动换行。如果想要手动开始新的一行，需要使用换行符，换行符可以使用字符'\n'，也可以使用 C++中的控制符 endl 来表示。其中两者的用法略有区别，字符'\n'需要嵌入字符串，endl 需要单独使用重定向符来表示。如：

```
cout<<"Hello, world!\n" ;
```

等价于

```
cout<<"Hello, world!" <<endl;
```

其中 cout 和 endl 两个标识符都是定义在 std 这个名称空间中的，因此在使用时需要添加 using 预编译指令。

5．什么语句可以用来创建名为 cheeses 的整型变量?

习题解析：

变量是 C++语言中用符号标识计算机存储区域内特定单元的一种方式，通过变量程序可以进行指定类型数据的访问和存储。C++中变量的声明和定义需要首先确定变量的名称，变量的命名应当符合 C++标准中标识符的命名规范，本题中明确变量名为 cheeses。其次变量的定义需要指明变量的数据类型，C++中表示整型的关键字为 int。此外，语句末尾应当有分号。因此该语句为

```
int cheeses;
```

6．什么语句可以用来将值 32 赋给变量 cheeses?

习题解析：

对指定变量进行数据存储应当使用赋值语句，赋值语句中的核心运算符是赋值运算符（=）。赋值语句中待写入数据的变量在赋值运算符左侧，待写入数据在赋值运算符右侧，且左右两侧应当保证类型相同，本题应当保证 cheeses 是整型变量。语句末尾应当有分号。因此，该赋值语句为

```
cheeses = 32;
```

7．什么语句可以用来将从键盘输入的值读入变量 cheeses 中?

习题解析：

C++中通过系统标准输入/输出进行数据的输出和读取，读取使用 cin 对象和插入运算符（>>）符号。把标准输入数据读入变量 cheeses 中的语句为

```
cin >> cheeses;
```

8．什么语句可以用来输出“We have X varieties of cheese,”，其中 X 为变量 cheeses 的当前值。

习题解析：

C++中的标准输出操作可以通过 cout 对象来实现。cout 对象可以通过多个输出插入运

算符（<<）连接，实现输出数据的拼接，因此当需要使用多组数据统一输出时可以使用多个插入运算符组合输出语句，这里使用以下语句。

```
cout << "We have " << cheeses << " varietiers of cheese," << endl;
```

此外，我们在程序中也可以拆分多条输出信息，使用多个 cout 对象分别输出。标准输出中除非输出换行符，或者当前行信息已满，否则多个 cout 对象的输出也在同一行内，例如，下列语句等价于上一条语句。

```
cout << "We have ";
cout << cheeses;
cout << " varietiers of cheese," << endl;
```

9．下面的函数原型指出了关于函数的哪些信息?

- `int froop( double t);`
- `void rattle(int n);`
- `int prunt(void);`

习题解析：

C++中函数原型主要包含三方面的内容，分别是函数名、参数表和返回值，因此 3 个函数的主要信息如下。

- 第 1 个函数原型表明函数名为 froop；函数的返回值是整型；函数有一个参数，参数的数据类型是 double。
- 第 2 个函数原型表明函数名为 rattle；函数没有返回值；该函数有一个参数，参数的数据类型是 int。
- 第 3 个函数原型表明函数名为 prunt；该函数的返回值为整型；函数没有参数。

10．定义函数时，在什么情况下不必使用关键字 return?

习题解析：

return 关键字表示函数返回语句，通常情况下当函数执行到第 1 条 return 语句时，函数就结束运行并且将 return 后的变量值返回给函数调用者。主调函数继续执行调用函数之后的下一条语句。通常当函数返回值为空（void）时，可以不需要 return 语句，函数的语句会依次执行到函数体的最后一条语句。

11．假设你编写的 main( )函数包含如下代码:

```
cout<<" Please enter your PIN: ";
```

而编译器指出 cout 是一个未知标识符。导致这种问题的原因很可能是什么？指出 3 种修复这种问题的方法。

习题解析：

cout 是 C++预定义的一个标准输出对象，当调用该对象进行标准输入/输出操作时，可以使用预编译指令 using 将该名称空间预编译到当前源代码中。此外，也可以对 cout 对象的名称空间进行限制，即表明 cout 所在的名称空间，这样编译器就可以正确识别、查找到该对象的定义并使用它进行输入/输出操作。因此可以使用如下 3 种方法进行声明。

```
using namespace std;
//使用预编译器指令声明
using std::cout;
//使用 using 声明，仅声明 cout 的名称空间，cin、endl 等其他对象需要再次声明
std::cout<<" Please enter your PIN: ";
//直接指明使用 std::前缀表明了 cout 所在的名称空间
```

2.4 编程练习

1．编写一个 C++程序，用于显示你的姓名和地址。

编程分析：

这是一个类似于“Hello，World!”的程序，主要应用 C++的标准输出系统进行字符的输出。在编写中应当注意 C++程序的几个重要知识点——预编译指令、main()函数、cout 标准输出。完整代码如下。

```
/*第 2 章的编程练习 1*/
#include <iostream>
using namespace std;
/*预编译指令*/

int main()
{
/*main()函数*/
    cout<<"< C++ Primer Plus > author: Stephen Prata";
    cout<<endl;
    /*标准输出字符，endl 表示换行*/
    return 0;
}
/*main()函数结束，注意函数返回值和表示结束的右花括号*/
```

2．编写一个 C++程序，它要求用户输入一个以 long 为单位的距离，然后将它转换为码（1 long 等于 220 码）。

编程分析：

题目首先需要读取标准输入中的数据，并将其从 long 进制转换成码进制。进制转换可以使用乘法，用（*）运算符表示。数据的声明和定义需要注意 C++中的标识符命名规则，选择相对应的数据类型。数据输入使用 cin 和插入运算符。完整代码如下。

```
/*第 2 章的编程练习 2*/
#include <iostream>
using namespace std;
/*预编译指令*/
```

```
int main()
{
/*main()函数*/
    double distance;
    /*定义变量distance，因为数据可能有小数，所以使用 double 类型*/
    cout<<"Enter the distance (in LONG) : ";
    cin>>distance;
    /*通过 cout 输出提示，并通过 cin 读取 distance 的输入*/
    cout<<"\nThe Distance "<<distance<<" long";
    cout<<" is "<<distance*220<<" yard."<<endl;
    /*cout 用于输出，并直接在输出语句中计算和转换码的值
     *程序中可以使用 cout 连接，但是要保证所有数据都是
     *相同的输出数据，例如以上的输出语句*/
    return 0;
}
/*main()函数结束，注意函数返回值和表示结束的右花括号*/
```

3．编写一个 C++程序，它使用 3 个用户定义的函数（包括 main()），并生成下面的输出。

```
Three bline mice
Three bline mice
See how they run
See how they run
```

其中一个函数要调用两次，该函数生成前两行；另一个函数也调用两次，读函数生成其余的输出。

编程分析：

题目主要考察函数的定义和函数的调用，在函数定义中的 3 个要素是函数名、返回值和参数。在调用函数时必须在函数调用语句之前有函数的声明或者定义，保证编译器了解该函数的基本信息。通过分析可以得知本题需要定义两个自定义函数和一个 main()函数，其中自定义函数只需要完成信息输出，因此不需要参数和返回值。无参数函数可以使参数列表为空，也可以在圆括号内写 void。完整代码如下。

```
/*第 2 章的编程练习 3*/
#include <iostream>
using namespace std;
/*预编译指令*/

void print_mice(void);
void print_run(void);
/*函数的声明， 因为要保证在 main()函数内
 *调用函数时，编译器知道该函数的基本信息*/
int main()
{
/*main()函数*/
    print_mice();
    print_mice();
```

```
    print_run();
    print_run();
    /*函数调用，main()函数称为主调函数，
     *以上 4 个函数可以称为被调函数*/
    return 0;
}
/*main()函数结束，注意函数返回值和表示结束的右花括号*/

void print_mice(void)
{
    cout<<"Three bline mice"<<endl;
}
void print_run(void)
{
    cout<<"See how they run"<<endl;
}
/*函数的具体定义，无参数可以使用 void 或者使参数为空，无返回值必须写 void
 *定义也可以放置在 main()函数前，用定义替换掉声明
*/
```

4．编写一个程序，由用户输入其年龄，然后显示该年龄包含多少个月，如下所示。

```
Enter your age: 29
```

编程分析：

程序要求通过系统标准输入读取用户输入的年龄，然后计算并显示该年龄包含多少个月。计算方法是使用乘法，即年龄数值乘以 12 即可得到结果。需要注意标准输入和标准输出的使用。完整代码如下。

```
/*第 2 章的编程练习 4*/
#include <iostream>
using namespace std;
/*预编译指令*/

int main()
{
    int years;
    /*定义变量、存储读取的数据，可以使用整型数据*/
    cout<<"Enter your age: ";
    cin>>years;
    /*通过 cin 读取数据，保存至 years 内*/
    cout<<"You are "<<years<<" old, or ";
    cout<<12*years<<" months old."<<endl;
    /*输出数据，在输出语句内通过 12*years 直接计算月份并输出*/
    return  0;
}
/*main()函数结束，注意函数返回值和表示结束的右花括号*/
```

5．编写一个程序，其中的 main( )调用一个用户定义的函数（以摄氏温度值为参数，并

返回相应的华氏温度值）。该程序按下面的格式要求用户输入摄氏温度值，并显示结果。

```
Please enter a Celsius value:20
20 degrees Celsius is 68 degrees Fahrenheit.
```

下面是转换公式：

华氏温度= 1.8 × 摄氏温度+ 32.0

编程分析：

程序要求通过一个函数实现摄氏温度和华氏温度的转换，因此在函数设计中需要着重考虑函数的返回值和参数。题目要求函数输入是摄氏温度，输出是华氏温度，其含义就是函数将输入的摄氏温度值转换为华氏温度值，并返回给主调函数（这里是 main()函数），因此主调函数会获取该返回值并利用它赋值或者输出等。完整代码如下。

```
/*第 2 章的编程练习 5*/
#include <iostream>
using namespace std;
/*预编译指令*/

float convert(float f);
/*温度转换函数，将参数中的摄氏温度转换为华氏温度
 *因此返回值是华氏温度。参数和返回值都是 float 类型的数据
 */
int main()
{
    float c_degree, f_degree;
    /*声明两个变量，分别存储两种温度数据*/
    cout<<"Please enter a Celsius value:";
    cin>>c_degree;
    /*读取用户输入的摄氏温度*/
    f_degree = convert(c_degree);
    /*函数调用，并通过返回值给华氏温度赋值*/
    cout<<c_degree<<" degrees Celsius is ";
    cout<<f_degree<<" degrees Fahrenheit."<<endl;
    return 0;
}
/*main()函数结束，注意函数返回值和表示结束的右花括号*/

float convert(float f)
{
    return  f*1.8 + 32;
}
/*函数的定义，可以直接在返回语句中计算和转换，
*也可以定义一个变量，计算转换值，最后返回该变量
*例如：
*float temp = f*1.8 + 32;
*return temp;
*/
```

6．编写一个程序，其中，main()调用一个用户定义的函数（以光年值为参数，并返回

对应天文单位的值)。该程序按下面的格式要求用户输入光年值，并显示结果。

```
Enter the number of light years: 4.2
4.2 light year = 265608 astronomical units.
```

天文单位是从地球到太阳的平均距离（约 150 000 000 千米或 93 000 000 英里)，光年是光一年走的距离（约 10 万亿千米或 6 万亿英里)，除太阳外，最近的恒星大约离地球 4.2 光年。请使用 double 类型（参见程序清单 2.4)，转换公式为

1 光年=63 240 天文单位

编程分析：

本题的基本功能和编程练习 5 类似，需要将光年数转换为天文单位，因此转换函数的输入参数和输出结果分别采用光年和天文单位。由于转换后数值较大，因此需要使用 double 类型的变量来存储数据。完整代码如下。

```
/*第 2 章的编程练习 6*/
#include <iostream>
using namespace std;
/*预编译指令*/

double convert(double light);
/*数据转换函数，将光年转换为天文单位*/
int main()
{
    double astro_unit, light_year;
    /*定义两个 double 类型的变量，存储光年值和天文单位值*/
    cout<<"Enter the number of light years: ";
    cin>>light_year;
    /*读取系统标准输入 ，存储到 light_year 变量中*/
    astro_unit = convert(light_year);
    /*调用函数进行转换，并将返回值赋给 astro_unit 变量*/
    cout<<light_year<<" light year = ";
    cout<<astro_unit<<" astronomical units."<<endl;
    /*输入显示的信息*/
    return  0;
}
/*main()函数结束，注意函数返回值和表示结束的右花括号*
double convert(double light)
{
    return  63240*light;
}
/*函数的定义，直接在返回语句中计算和转换*/
```

7. 编写一个程序，要求用户输入小时数和分钟数。在 main()函数中，将这两个值传递给一个 void 函数，后者以如下格式显示这两个值。

```
Enter the number of hours: 9
Enter the number of minutes: 28
Time: 9:28
```

编程分析：

题目要求编写一个函数，对输入的时间数据进行格式化输出，并未进行数据的计算和返回数值。因此函数的参数应当是小时数和分钟数，函数的功能就是格式化输出数据，没有返回值。完整代码如下。

```
/*第 2 章的编程练习 7*/
#include <iostream>
using namespace std;
/*预编译指令*/

void format_print(int hour,int minute);
/* format_print 函数的声明*/
int main()
{
    float hours, minutes;
    /*定义 float 类型变量，用于存储时间数值*/
    cout<<"Enter the number of hours: ";
    cin>>hours;
    cout<<"Enter the number of minutes: ";
    cin>>minutes;
    /*通过标准化输入，读取数据，并存储对应变量*/
    format_print(hours,minutes);
    /*函数调用，用于输出格式化数据，函数无返回值*/
    return  0;
}
/*main()函数结束，注意函数返回值和表示结束的右花括号*/

void format_print(int hour,int minute)
{
    cout<<"Time: "<<hour<<":"<<minute<<endl;
}
/*函数的定义，无返回值的函数可以不使用 return 语句，这样
 *函数会运行到最后一句，然后自动返回*/
```

第3章　处理数据

本章知识点总结

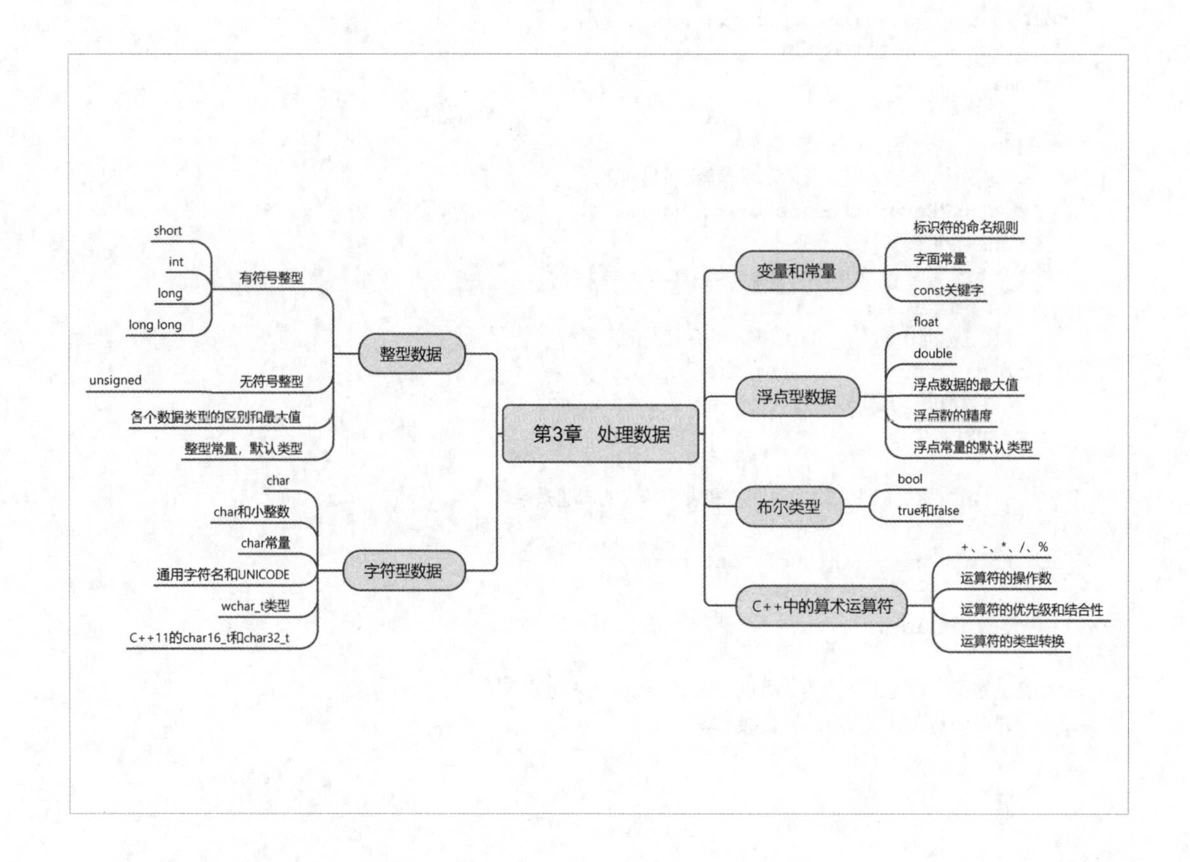

3.1　C++语言中的变量及其使用

在程序运行过程中，所有代码、数据等信息都存储在计算机系统的存储器当中，计算机按照存储设备的存储地址进行数据的查找和访问。存储器地址的表示方法相对比较复杂，因此 C++程序设计当中可以使用变量来表示一个特定存储单元，这样编程过程中对该存储区域的查找、读写等操作可以直接使用变量作为标志，从而避免使用存储器地址这种相对比较复杂的表达方式。此外，定义和使用变量还应当明确该存储区域中数据的类型，因此需要在声明该变量时标识变量的类型，以便有利于编译器进行数据访问时的类型检查。

C++语言提倡使用具有一定含义的标识符来命名变量（作为变量的名称）。程序设计中可以使用字母、数字和下划线的组合来命名变量，C++语言中命名标识符的基本规则和 C 语言相同，主要限制是首字母不能是数字。此外，以双下划线和大写字母命名的标识符保留给了编译器及其使用的资源，用户编程时应尽量避免。编程过程中可以在符合规则的前提下，使用简练易懂的标识符，以便提高程序的易读性。

3.2　C++语言中的整型数据

C++语言中的整型数据与数学中的整数类型类似，即不包括分数和小数的数字。C++中的整数类型可以依据数据大小以及是否包含负数分为很多种。整型数据中最常用的类型是 int，一个 int 类型的整数通常用 32 位二进制数表示，因此其取值范围值是$-2^{31}\sim2^{31}-1$。除 int 类型外，整型数据还包括 short、long 和 long long 以及用 unsigned 表示的无符号整数等多种类型。其中有符号类型能够表示负数，无符号类型不能表示负数，且表示的最小值为 0。C++语言规定了整型数据的最小长度，其中 short 类型至少 16 位；int 至少和 short 一样长；long 至少 32 位，且至少和 int 一样长；long long 类型至少 64 位且至少和 long 一样长。这 4 种有符号整数类型能够表示的最大整数值就存在差异。

无符号类型整型数据用 unsigned 表示，在以上 4 种类型的整型数据类型前加 unsigned 就可以表示同等字节长度的无符号整数类型，表示方式分别是 unsigned short、unsigned int、unsigned long、unsigned long long。因此，在程序设计中应当根据程序中数据运算的需要选择合适类型的整型数据类型。最后，C++语言标准明确规定了每一种整数类型的宽度，其具体数值应当由开发平台和编译器决定，这一点应当在程序设计中引起重视。

3.3　C++语言中的字符类型

char 类型是专门存储字符（字母和数字）的一种数据类型，简单看来字符类型在计算机内部和整数的存储方式相同，并且由于很多系统使用的字符都不超过 128 个，因此字符类型可以看作比 short 更小的整型数据。标准 ASCII 字符集使用 8 位作为一个字符的存储单元。在程序设计中字符使用单引号表示，并且尽量避免使用整型数据表示字符常量。对于 ASCII 码表中的非打印字符，可以使用其 ASCII 值或者转义序列来表示，通常使用反斜杠“\”来表示字符的转义。对于打印字符，%c 与%d 将会输出不同类型的数据。

C++中的字符类型在支持标准字符集的基础上还支持通用字符名。在使用中通用字符名可以使用以\u 或者\U 开头的扩展字符集。其中\u 后加 8 个十六进制数据，\U 后加 16 个十六进制数据。C++语言中使用 wchar_t 类型表示扩展字符集。其中 wchar_t 是一种整型数据，长度大于 char 类型，因此能够表示更多的字符类型。在处理 wchar_t 类型的变量时，可以使用'L'前缀来表示 wchar_t 类型，例如：

```
wchar_t bob = L'P';
```

为了解决计算机编码和字符集问题，C++11 标准又增加了 char16_t 和 char32_t 类型，长度分别为无符号 16 位和 32 位，程序设计中分别使用'u'和'U'前缀表示。

3.4 C++语言中的浮点型数据

C++语言中的浮点类型数据就是带小数部分的数字。浮点型数据在程序设计中通常需要显式地表示小数点部分，或者使用指数记数法。例如，1 和 1.0 分别表示整型与浮点型，但 3e16 可以明确表示一个浮点型数据。C++中的浮点类型有 float、double 和 long double 这 3 种类型，这些类型是按照它们可以表示的有效位数和允许的最小指数范围来描述的。C++语言规定 float 类型至少 32 位；double 类型至少 48 位且不少于 float；long double 类型至少和 double 类型的位数一样。由于二进制和十进制的转换关系，浮点型数据在计算机中不是精确数值，而是存在一定的有效位数。C++语言中浮点型数据的有效位数定义在 cfloat 头文件中，程序设计中如果需要用到有效位，可以通过头文件查询。浮点数运算过程中会产生浮点数的舍入错误。浮点型数据计算过程中可能产生上溢或者下溢，其中上溢表示数据无穷大，下溢则表示计算过程中损失了末尾有效数字。

3.5 C++语言中的常量和其他数据类型

ANSI/ISO C++标准新增了 bool 类型，用来表示逻辑运算中的真或者假。在计算中布尔变量可以是 true 或者 false 两种取值。

程序设计中的变量是指变量存储单元内数值能够反复被程序修改和重写，与此相对应的常量是指存储单元内数据值一旦初始化就不能够修改的数据。常量通常分为两类。一类是字面值，例如整型数据 365 或 24，字面值常量数据在使用过程中不能够赋值，只能参与其他表达式的运算。另一类是使用 const 关键字表示的符号常量，例如：

```
const int Months  = 12;
```

这表示 Months 是一个符号常量，常量在声明语句中被初始化为 12，在后续代码中将不能修改。

对于整数字面值，默认为 int 类型（超过 int 最大值会自动调节为 long 类型）。如果要使用 long 类型表示 365，应当使用 L 后缀，表示为 365L，其他类似后缀还有 UL 等多种。char 字面值可以直接使用一对单引号和字符表示，如'A'，也可以使用转义序列或者整型数据表示。浮点型字面值默认为 double 类型，如果要使用 float 类型需要添加 f 后缀，例如，3.14f 字面值为 float 类型。

3.6 C++中的数据类型转换和基本运算

程序设计过程中对变量的操作和运算应当与其类型相匹配。例如，在赋值以及其他运算中，应当注意运算符两侧的数据类型匹配。通常情况下 C++语言的编译器在赋值、函数参数传递、返回值处理及或其他运算中，对类型不匹配的变量自动进行数据类型的转换。数据类型的转换可能会产生潜在的问题，例如，较大的浮点型数据转换为较小的浮点型数据可能会降低精度；浮点型转换为整型会丢失小数部分等。因此用户在程序设计过程中应当明确何时会进行类型转换，并对其转换后的潜在问题有充分的了解。除自动类型转换外，也可以对数据进行强制类型转换，强制类型转换的方式有两种，分别为(long) thorn 和 long(thorn)。

C++语言中的算术运算符主要有加、减、乘、除、求模（对应的符号分别为 +、−、*、/、%）。算术运算的元素和操作数匹配，尤其整数的除法结果不包含小数部分。复杂运算中需要考虑运算符的优先级和结合性，程序设计中如果可能产生歧义，应当尽量使用圆括号明确其运算顺序。

3.7　复习题

1．为什么 C++有多种整型?

习题解析：

C++语言中包含多种整数类型，主要包括 short、int、long 和 long long 这 4 种，每一种还分别包含有符号类型和无符号类型（unsigned）。此外，char 类型也可以看作一种小整数类型。C++语言中这些整数类型的主要区别在于存储使用的数位长度不同，其中 short 类型最少 16 位长度；int 类型至少和 short 类型一样长；long 类型至少 32 位，且至少和 int 类型一样长；long long 类型至少 64 位且至少和 long 类型一样长。由于存储数据使用的数位长度不同，因此这 4 种类型能够表示的最大整数值存在差异。此外，有符号类型能够表示负数；无符号类型不能不表示负数，且表示的最小值为 0。同一种类型的有符号数和无符号数的最大值与最小值也存在不同。因此，在程序设计中应当根据数据运算的需要选择合适类型的整型数据。最后，C++语言标准并未明确规定每一种整数类型的宽度，其具体数值应当由开发平台和编译器决定，这一点应当在程序设计中引起重视。

2．声明与下述描述相符的变量。

a．short 整数，值为 80。

b．unsigned int 整数，值为 42 110。

c．值为 3 000 000 000 的整数。

习题解析：

C++中变量声明的基本格式是先标识数据类型，后标识变量名，变量的初始化可以通过声明语句中的赋值运算符实现。本题中为了声明相应的变量，需要首先考虑其数据类型是否符合题目要求。理论上 C++程序设计中变量的数据类型可以使用位数较大的类型，不能够使用位数较小的数据类型来描述较大的数值。但是为了节省存储空间通常选用与数值相匹配的数据类型。

a．题目中数据常量 80 是一个整数，在 short int 的取值范围内，因此可以使用 short int 类型。

```
short  example_a= 80;
```

或者

```
short int example_a = 80;
```

b．题目中的数值 42 110 超过了 short 的取值范围，也超过了 16 位有符号的 int 类型的最大值 32 767，因此可以使用 unsigned int 类型。声明方式如下。

```
unsigned int example_b = 42110;
```

或者

```
unsigned example_b = 42110;
```

c．题目中变量的初始值为 3 000 000 000，超过了 unsigned int 类型的取值范围，因此应当使用 unsigned long 类型或者 long long 类型表示该数值。声明方式如下。

```
unsigned long example_c = 3 000 000 000;
```

或者

```
long long example_c = 3 000 000 000;
```

3．C++提供了什么措施来防止超出整型的范围?

习题解析:

C++语言中 short、int、long 和 long long 类型的主要区别在于存储的字节长度不同，因此每种类型的最大值不同。但是 C++语言中并没有提供自动防止超出整数类型范围的功能，需要程序员预先估计数据大小与哪种数据类型匹配。同时，C++语言标准并未明确规定每一种整数类型的宽度，其具体数值应当由开发平台和编译器决定，可以使用头文件中的 climits 来确定实际的最大值。

4．33L 与 33 之间有什么区别?

习题解析:

C++语言中整型字面值具有一个默认类型，除非数值超过了 int 类型的最大值。通常整型字面值的默认类型为 int 类型，程序设计中也可以通过在常量后添加后缀来指定整型常量的具体类型，例如，L 后缀表示 long 类型，U 后缀表示 unsigned 类型。此外，也可以组合使用 U 和 L 来表示 unsigned long 类型。因此，本题目中常量 33 表示 int 型数据，33L 表示 long 型数据，两者在计算机内占用的存储空间不同。两者在存储空间上的具体差异由实现平台决定。

5．下面两条 C++语句是否等价?

```
char grade = 65;
char grade = 'A';
```

习题解析:

本题中的两条声明语句都声明了字符类型的变量 grade，并将其初始化为'A'。在基于

ASCII 码的平台下两条语句可以通用，不基于 ASCII 码的平台下则两者不能通用。此外，C++中两条语句的具体实现也稍有区别。第 1 条语句中 65 是一个 int 类型的数据常量，初始化过程中会进行类型转换，即将整型数据 65 转换成字符类型，再存储到 grade 内，数位宽度会变化。第 2 条语句直接将变量 grade 初始化为字符类型的数据常量'A'。

6．如何使用 C++来找出编码 88 表示的字符？指出至少两种方法。

习题解析：

编程过程中要查找与 ASCII 编码对应的字符，一般可以直接查询 ASCII 码表。该方法的优点是能够查找所有对应的 ASCII 字符。此外，也可以通过编写简单语句查询。通过语句查询的缺点是某些不可见字符无法显示。本题中可以通过 C++语句查询，具体方法如下。

```
char example = 88;
cout<<example<<endl;
```

以上语句通过 char 类型变量直接输出。

```
cout<<(char)88<<endl;
```

通过 C 风格的强制类型转换，将整型数据 88 转换为 char 类型并输出。

```
cout<<char(88)<<endl;
```

通过 C++类型的强制类型转换，将整型数据 88 转换为 char 类型并输出。

```
cout.put(char(88));
```

通过 cout.put()函数直接输出类型强制转换后的 char 数据。

7. 将 long 值赋给 float 变量会导致舍入误差，将 long 值赋给 double 变量呢？将 long long 值赋给 double 变量呢?

习题解析：

C++ 中的浮点类型数据表示带小数部分的数据，但是浮点型数据在很多情况下并不能精确表示所有数字，通常在很多平台下 C++中 float 类型只能够表示 6 位有效数字，double 仅能表示 15 位有效数字（浮点型数据的具体有效数字范围定义在 cfloat 头文件中）。因此在整型数据转换为浮点型数据的过程中会产生舍入误差，这种误差的产生主要是由整型数据的有效数字超过了浮点型数据的表示范围引起的。long 类型数据的最大值为 20 亿，即 10 位数，因此转换为 float 类型时会丢失精度，但 double 类型的有效数字为 15 位，因此不会产生舍入误差，但 long long 类型最大可以包含 19 位有效数字，转换为 double 类型可能会产生舍入误差。

8．下列 C++表达式的结果分别是多少?

a. `8*9+2`

b. `6*3/4`

c. `3/4*6`

d. `6.0*3/4`

e. `15%4`

习题解析：

C++的算术运算符主要有加（+）、减（−）、乘（*）、除（/）和取模（%），在进行算术表达式求值时主要需要注意运算符的优先级和结合性两个知识点，其中算术运算符的优先级是乘、除和取模高于加、减。当算术运算符的优先级相同时，需要考虑操作数（即参与运算的数值）的结合性是从左到右还是从右到左。对于同一个操作数，在 C++中从左向右运算。此外，运算过程中对于不同类型的数据，C++会进行隐式的数据类型转换。一般的转换规则是将取值范围较小的数据转换成表达式中取值范围较大的数据类型。在此基础上可以判断题目的运算结果。

a．`8*9+2` 的结果 74，先计算乘法，再做加法。

b．`6*3/4` 的结果 4，运算符优先级相同，因此操作数从左到右结合，先计算 6*3，结果为 18，由于两个操作数是整数，因此最后的除法结果为其商，无小数部分。

c．`3/4*6` 的结果 0，运算符优先级相同，因此操作数从左到右结合，先计算 3/4，结果为其商 0，最后做乘法，`0*6` 为 0。

d．`6.0*3/4` 的结果为 4.5，运算符优先级相同，因此操作数从左到右结合，先计算 `6.0*3`，结果为浮点数 18.0，后计算 18.0/4，得到浮点结果 4.5。

e．`15%4` 的结果为 3，取模运算，结果为 3。

9．假设 `x1` 和 `x2` 是两个 double 变量，要将它们作为整数相加，再将结果赋给一个整型变量。请编写一条完成这项任务的 C++语句。如果要将它们作为 double 值相加并转换为 int 呢？

习题解析：

C++中要将浮点型数据转换成整型数据，首先舍弃小数部分，其次如果原数据大于目标类型的取值范围，结果将不确定。程序中可以使用多种形式进行强制类型转换，也可以在不同时机进行类型转换，先转换成为整型数据再进行计算和先计算再转换成整型数据可能会产生不同效果，例如，对于 1.8+1.9，先计算再进行类型转换，得到结果 3；先进行类型转换再计算，得到结果 2。这些问题在程序设计中需要特别注意。

先计算再进行类型转换的语句如下所示。

`int pos = int(x1+x2);`或者`int pos = (int)(x1+x2);`

先进行类型转换再计算的语句如下所示。

```
int pos = int(x1)+int(x2);
```

或者

```
int pos = (int)x1 + (int)x2;
```

10．下面每条语句声明的变量分别是什么类型的？

a．`auto cars = 15`

b．`auto iou = 150.37f`

c．`auto level = 'B'`

d. auto crat = U'\U0002155'

e. auto fract = 8.25f/2.5

习题解析：

C++中的关键字 auto 能够使编译器根据初始值自动推断变量的类型。

a. auto cars = 15 声明了变量 cars 并初始化为整型数据 15，因此 auto 关键字使编译器推断变量 cars 的类型为 int 类型。

b. auto iou = 150.37f 声明了变量 iou 并初始化为 float 类型数据 150.37f，因此 auto 关键字使编译器推断变量 iou 的类型为 float 类型。

c. auto level = 'B'声明了变量并初始化为 char 类型的常量'B'，因此 auto 关键字使编译器推断变量 level 的类型为 char 类型。

d. 在语句 auto crat = U'\U0002155'中，常量 U'\U0002155'中前缀 U 表示该常量是 char32_t 类型的数据，因此 auto 关键字使编译器推断变量 crat 为 char32_t 类型。

e. 在语句 auto fract = 8.25f/2.5 中，变量 fract 初始化为表达式 8.25f/2.5 的值，该表达式中 8.25f 为 float 类型，2.5 为 double 类型，除法运算产生数据类型转换，其结果为 double 类型，因此通过 auto 关键字使编译器推断 fract 变量为 double 类型。

3.8 编程练习

1. 编写一个小程序，要求用户使用一个整数表示自己的身高（单位为英寸①），然后将身高转换为英尺②和英寸。该程序使用下划线字符来指示输入位置。另外，使用一个符号常量 const 来表示转换因子。

编程分析：

题目要求使用符号常量存储转换因子，结合用户的输入数据进行身高转换和计算，因此基本语句包含变量的定义、标准输入、计算转换和输出几个部分，其中常量因子使用 const 关键字在变量之前定义。完整代码如下。

```
/*第 3 章的编程练习 1*/
#include <iostream>
using namespace std;
/*预编译指令*/
const int FOOT_TO_INCH = 12;
/*定义符号常量，该常量定义后数值不会改变*/

int main()
{
    int height;
```

① 1 英寸=0.025 4 米。——编者注

② 1 英尺=0.304 8 米。——编者注

```
    cout<<"Enter your height in inchs_";
    cin>>height;
    /*定义变量，并通过标准输入读取用户输入*/
    cout<<endl<<"Your Height convert to "<<height/FOOT_TO_INCH;
    cout<<" foot and "<<height%FOOT_TO_INCH<<" inch height."<<endl;
    /*显示输出，在输出语句内直接计算转换后的数值，并显示*/
    return  0;
}
/*main()函数结束，注意函数返回值和表示结束的右花括号 */
```

2．编写一个小程序，要求以几英尺几英寸的方式输入其身高，并以磅为单位输入其体重（使用 3 个变量来存储这些信息）。该程序可以报告体重指数（Body Mass Index，BMI）。为了计算 BMI，该程序以英寸为单位指出用户的身高（1 英尺=12 英寸），并将以英寸为单位的身高转换为以米为单位的身高（1 英寸 =0.025 4 米）。然后，将以磅为单位的体重转换为以千克为单位的体重（1 千克=2.2 磅）。最后，计算相应的 BMI，即体重（单位是千克）除以身高（单位是米）的平方。用符号常量表示各种转换因子。

编程分析：

程序的功能是通过用户输入的身高、体重数据，计算用户的 BMI 数值。程序设计过程中由于单位转换较多，因此需要使用多个变量存储转换前后的数据，尽量保持程序清晰可读。计算中使用常量作为转换因子，完成计算后显示计算结果。完整代码如下。

```
/*第 3 章的编程练习 2*/
#include <iostream>
using namespace std;
/*预编译指令*/
const float KILOGRAM_TO_POUND = 2.2;
const int FOOT_TO_INCH = 12;
const float INCH_TO_METER = 0.0254;
/*定义符号常量，表示身高、体重的单位转换因子*/

int main()
{
    int height_foot, height_inch;
    /*用户输入英尺、英寸数据*/
    float weight_pound, height, weight, BMI;
    /*以磅为单位的体重和存储转换后数据的变量*/
    cout<<"Enter your height foot:";
    cin>>height_foot;
    cout<<"Enter ypour height inchs:";
    cin>>height_inch;
    cout<<"Enter ypour weight in pounds:";
    cin>>weight_pound;
    /*通过标准输入读取数据输入*/
    height = (height_foot *FOOT_TO_INCH + height_inch) *INCH_TO_METER;
    weight = weight_pound / KILOGRAM_TO_POUND;
    /*转换身高、体重数据，C++语言中可以使用括号确保正确的运算顺序*/
    BMI = weight / (height *height);
```

```
    /*计算 BMI*/
    cout<<"Your BMI is  "<<BMI<<endl;
    return 0;
}
/*main()函数结束，注意函数返回值和表示结束的右花括号 */
```

3．编写一个程序，要求用户以度、分、秒的方式输入一个纬度，然后以度为单位显示该纬度。1° 等于 60′，1′ 等于 60″，请以符号常量的方式表示这些值。对于每个输入值，应使用一个独立的变量存储它。下面是该程序运行时的输出。

```
Enter a latitude in degrees, minutes, and seconds:
First, enter the degree:37
Next, enter the minutes of arc:51
Finally, enter the seconds of arc:19
37 degrees,  51 minutes, 19 seconds = 37.8553 degrees
```

编程分析：

题目要求实现纬度数据的转换，与编程练习 2 类似，也同样需要通过常量定义转换因子，通过变量存储转换前后的数据。通过分析程序的输出，可以得出程序的输入/输出提示语句。程序的完整代码如下。

```
/*第 3 章的编程练习 3*/
#include <iostream>
using namespace std;
/*预编译指令*/
const int DEGREE_TO_MINUTE = 60;
const int MINUTE_TO_SECOND = 60;
/*定义符号常量，表示单位转换因子*/

int main()
{
    int degree, minute, second;
    float degree_style;
    /*定义数据变量，选择合适的数据类型表示度*/
    cout<<"Enter a latitude in degrees, minutes, and seconds:"<<endl;
    cout<<"First, enter the degree:";
    cin>>degree;
    cout<<"Next, enter the minutes of arc:";
    cin>>minute;
    cout<<"Finally, enter the seconds of arc:";
    cin>>second;
    /*读取用户的输入数据*/
    degree_style = degree + float(minute) / DEGREE_TO_MINUTE +
                  float(second)/(MINUTE_TO_SECOND *DEGREE_TO_MINUTE);
    /*C++语言中，可以使用括号确保正确的运算顺序*/
    cout<<degree<<" degrees,  "<<minute<<" minutes, "
        <<second<<" seconds = "<<degree_style <<" degrees"<<endl;
    /*转换数据格式，并输出*/
    return 0;
}
```

```
/*main()函数结束，注意函数返回值和表示结束的右花括号 */
```

4．编写一个程序，要求用户以整数输入秒数（使用 long 或 long long 变量存储），然后以天、小时、分钟和秒显示这段时间。使用符号常量来表示每天有多少小时、每小时有多少分钟以及每分钟有多少秒。该程序的输出应与下面类似。

```
Enter the number of  seconds:31600000
31 600 000 seconds = 365 days,  17 hours,46 minutes, 40 seconds.
```

编程分析：

题目要求实现的功能与编程练习 2、3 类似，本题需要注意的是由于数据的值较大，因此需要选择合适的数据类型存储数据，本题中在转换成秒数时使用普通 int 类型将会溢出，因此需要使用 long long 类型。完整代码如下。

```
/*第 3 章的编程练习 4*/
#include <iostream>
using namespace std;
/*预编译指令*/
const int DAY_TO_HOUR = 24;
const int HOUR_TO_MINUTE = 60;
const int MINUTE_TO_SECOND = 60;

/*定义符号常量，表示单位转换因子*/
int main()
{
    long long seconds;
    int days, hours, minutes;
    /*选择合适的数据类型 ，定义变量*/
    cout<<"Enter the number of  seconds:";
    cin>>seconds;

    cout<<seconds <<" seconds = "
    days = seconds / (DAY_TO_HOUR *HOUR_TO_MINUTE *MINUTE_TO_SECOND);
    seconds = seconds % (DAY_TO_HOUR *HOUR_TO_MINUTE *MINUTE_TO_SECOND);

    hours = seconds / (HOUR_TO_MINUTE *MINUTE_TO_SECOND);
    seconds = seconds %(HOUR_TO_MINUTE *MINUTE_TO_SECOND);
    /*C++语言中，可以使用括号确保正确的运算顺序*/
    minutes = seconds / MINUTE_TO_SECOND;
    seconds = seconds % MINUTE_TO_SECOND;
    /*读取标准输入数据*/
    cout<<days<<" days, "<<hours<<" hours,"<<minutes<< " minutes, "
        <<seconds<<" seconds."<<endl;
    /*转换数据格式，并输出*/
    return  0;
}
/*main()函数结束，注意函数返回值和表示结束的右花括号 */
```

5．编写一个程序，要求用户输入全球当前的人口和美国当前的人口（或其他国家的人口）。将这些信息存储在 long long 变量中，使程序输出显示美国（或其他国家）的人口占全球人口的百分比。该程序的输出应与下面类似。

```
Enter the world's population: 6898758899
Enter the population of US: 310783781
The population of the US is 4.50492% of the world population.
```

编程分析：

程序要求通过输入数据计算百分比，考虑到人口数量会超出 int 或 long 类型的最大值，因此使用 long long 类型，百分比应当用浮点型数据表示。完整代码如下。

```
/*第 3 章的编程练习 5*/
#include <iostream>
using namespace std;
/*预编译指令*/

int main()
{
    long long global_amount, american_amount;
    double population_percent;
    /*定义变量，型选择合适的数据类型*/
    cout<<"Enter the world's population: ";
    cin>>global_amount;
    cout<<"Enter the population of US: ";
    cin>>american_amount;
    population_percent = 100* (double)american_amount / (double) global_amount;
    /*按照百分比显示数据，因此需要先转换成浮点数据，否则除法的结果将是整数*/
    cout<<"The population of the US is "<<population_percent<<"% of the world
population.";
    return 0;
}
/*main()函数结束，注意函数返回值和表示结束的右花括号 */
```

6. 编写一个程序，要求用户输入驱车里程（单位是英里）和使用汽油量（单位是加仑），然后指出汽车耗油量为 1 加仑的里程，即油耗。如果愿意，也可以要求用户以千米为单位输入距离，并以升为单位输入汽油量，然后输出欧洲风格的结果——即每 100km 的耗油量（升）。

编程分析：

题目要求首先读取用户输入的开车的里程数和使用的汽油量，通过计算得到每加仑的里程数。也可以提示用户按照欧洲风格输入以升为单位的汽油数量，并计算欧式油耗。注意，本题并未提供单位的转换常数，要按照要求进行转换。完整代码如下。

```
/*第 3 章的编程练习 6*/
#include <iostream>
using namespace std;
/*预编译指令*/
```

```
int main()
{
    float distance_in_mile, distance_in_km;
    float fuel_in_gallon, fuel_in_litre;
    float fuel_consume;
    /*变量声明*/
    cout<<"Enter the distance in miles: ";
    cin>>distance_in_mile;
    cout<<"Enter the fuel consume in gallon: ";
    cin>>fuel_in_gallon;
    fuel_consume = distance_in_mile / fuel_in_gallon;
    /*读取数据，并计算美式油耗*/
    cout<<"The fuel consume is "<<fuel_consume<<" mpg(miles/gallon)."<<endl;
    cout<<"Enter the distance in kilometer: ";
    cin>>distance_in_km;
    cout<<"Enter the fuel consume in litre: ";
    cin>>fuel_in_litre;
    fuel_consume = (fuel_in_litre / distance_in_km)*100;
    /*计算欧式油耗*/
    cout<<"The fuel consume is "<<fuel_consume<<"L/100KM."<<endl;
    return  0;
}
/*main()函数结束，注意函数返回值和表示结束的右花括号 */
```

7. 编写一个程序，要求用户按欧洲风格输入汽车的油耗（每 100km 消耗的汽油量，单位是升），然后将其转换为美国风格的耗油量——每加仑多少英里。注意，除了使用不同的单位计量外，美式油耗（距离/燃料）与欧式油耗（燃料/距离）相反。100km=62.14mile，1 美制加仑=3.785 升。因此，19mile/gas 大约合 12.4L/100km，27mile/gas 大约合 8.7L/100km。

编程分析：

本题整体的功能要求和编程练习 6 类似，需要将用户输入的欧式油耗数据转换成美式油耗。因此本题需要对里程数和燃料数进行转换，题目已经给定转换方式和相关数值，程序内只需要做简单计算即可。完整代码如下。

```
/*第 3 章的编程练习 7*/
#include <iostream>
using namespace std;
/*预编译指令*/
const float GALLON_TO_LITER = 3.875;
const float HKM_TO_MILE = 62.14;
/*以常量作为数据转换因子*/
int main()
{
    float fuel_consume_eur, fuel_consume_us;
    /*定义变量*/
    cout<<"Enter the fuel consume in europe(l/100km): ";
    cin>>fuel_consume_eur;
```

```
    fuel_consume_us = HKM_TO_MILE / (fuel_consume_eur / GALLON_TO_LITER);
    /* 将美式油耗转换成欧式油耗*/
    cout<<"The fuel consume is "<<fuel_consume_eur<<"L/100KM."<<endl;
    cout<<"The fuel consume is "<<fuel_consume_us<<" mpg(mile/gallon)."<<endl;
    return  0;
}
/*main()函数结束，注意函数返回值和表示结束的右花括号 */
```

第 4 章 复合类型

本章知识点总结

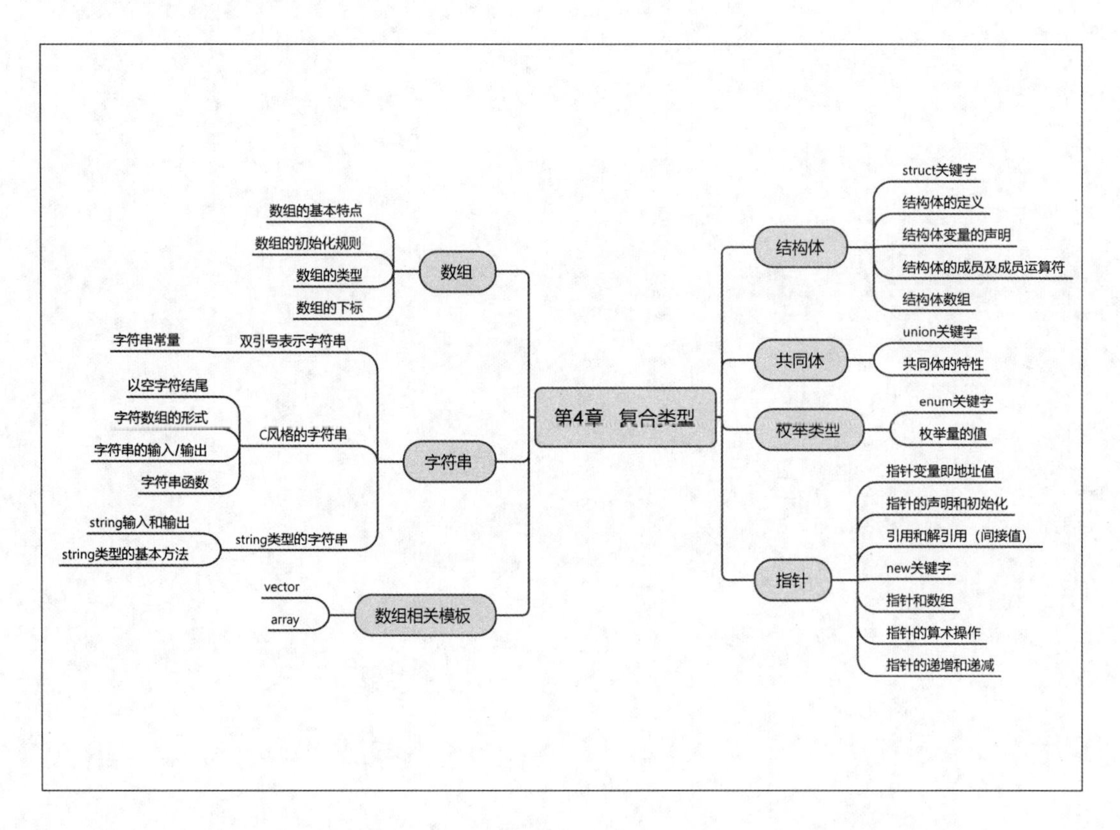

4.1 C++语言中的数组

C++语言中的数组是指一组数据类型相同的数据，按照顺序排列而成的一种数据结构形式。其基本特点是数组由多个类型相同的数据组成且排列方式为顺序结构，其中每一个数据都称为数组的元素。因为数组是顺序排列的，所以数组中的元素在存储空间中都是连续的。数组的声明中需要明确数组的名字和数组元素的个数，如：

```
short month[12];
```

这表明 month 是一个数组，其元素个数是 12。数组的元素按照顺序排列并组成了一列，因此可以使用元素在数组中的位置来表述每一个元素。表示元素位置的整数称为下标或者索引，数组的索引值从0开始计数，最后一个元素的索引值为数组的长度减1，在使用数组中应当注意不要越界访问数组元素。例如，可以使用month[0]表示month数组的第1个元素，使用month[11]表示month数组的最后一个元素。

数组元素的初始化方法主要有两种。基本初始化方式是依次对每一个元素进行赋值初始化；另一种方式是在数组声明语句中同时初始化，例如：

```
int card[4] = {3, 6, 8, 10};
```

在进行定义时也可以初始化部分元素，未初始化的元素将会自动赋值为0，因此使用{0}可以将数组元素全部初始化为0，具体的初始化方式可以根据需要选择。

4.2　C++语言中的字符串

字符串是存储在内存的连续字节中的一系列字符。C++语言中有两种风格的字符串，分别是C风格的字符串和string类型的字符串。C风格的字符串就是一个特殊的字符数组，以空字符（'\0'）表示字符串的结束。对于C风格的字符串，判断字符串的长度、复制字符串等操作都需要使用C++库中的函数，无法使用运算符。string类型的字符串需要用以下格式声明。

```
string str = "Hello World!";
```

string类型的字符串是简单变量的形式，而不是C风格字符串的字符数组形式。string类型的字符串的赋值、拼接等操作可以直接使用 + 、 = 等运算符。语句更加简洁。两种形式的字符串都可以使用cout和cin对象进行标准化输入/输出，但是在操作中需要注意空白字符的输入可能需要特殊化处理，具体处理方法可参考本章的复习题。

字符串常量一般用双引号表示，例如，"Hello world!"就是一个字符串常量。对于字符串常量和字符常量（用单引号表示）应当严格区分，两者类型不同且在存储空间中占用的位数不同。

4.3　C++语言中的结构体和结构体数组

结构体是一种可以同时存储多种不同类型数据的复合数据类型。简单来说，C++中的结构体是可以将多个不同类型的数据组合在一起，作为一个数据类型使用的一种方式。结构体使用struct定义，并需要明确标注结构内成员的类型和名称，例如：

```
struct Student{
        char name[20];
        int age;
};
```

以上定义表示结构体Student由两个成员组成，分别是字符数组name和整型数据age。定义了结构体之后就可以应用该结构体创建该结构体类型的变量。除了创建结构体变量外，还可以创建结构体数组，即该数组的每一个元素都是这个结构体的对象。

4.4 C++语言中的指针

计算机程序中的所有数据都存储在存储器中，为了更加方便和快捷地访问这些数据，我们在程序设计中通过定义变量来访问这些存储的数据。除了变量外，C++中还可以通过指针的方式来访问这些存储器中的数据单元。指针首先是一个变量，其存储单元内存储的是另一块存储单元的地址，而不是我们要访问的最终数据的值，依据这个地址我们可以间接地访问最终存储单元的数据。和指针相关的两个运算符是*运算符（间接值或解除引用运算符）和&运算符（地址运算符），例如：

```
int months = 12;
int *p = &months;
```

声明语句中首先声明了整型变量 months，其次声明了指针 p，p 是指向整型数据的指针，指针 p 最初指向 months 变量。因此程序可以通过 months 变量和*p 访问 months 数据。程序设计中也可以使用 new 运算符直接申请内存中的存储单元，然后将指针指向该地址，例如：

```
int *p = new int;
```

和上例的区别是这个存储单元只能使用指针 p 访问，且使用 new 申请的存储单元在使用完之后需要用 delete 释放。指针的使用能够大幅提高 C++程序的编译、运行效率，但是由于可以通过存储地址来访问程序内的数据，会给程序带来很大的安全隐患，因此使用中应当引起重视。

4.5 C++语言中的指针和数组

C++语言中的指针和数组是类似的，因为 C++在处理指针和数组的方式上基本相同，每次数组的索引值递增会在存储单元上向后移动一个元素所占的字节长度，而每次指针变量的递增本质上也是将指针内的地址向后移动当前类型所占字节长度。因此在很多情况下，也可以将指针和数组名按相同方式运算，即数组名是指向该数组第 1 个元素的指针。用户可以通过 new 运算符创建一个指针。

```
int*p = new int[12];
```

这样程序中可以通过`*(p+1)`或者`p[1]`来访问数组的第 2 个元素。但是，如果以数组名作为数组首元素地址，在使用中还有需要注意的问题，例如：

```
int Months[12];
```

数组名 Month 就是指向第 1 个元素的指针，但是&Months[0]和&Month 的类型又不相同，前者是 int 类型数据单元的地址，后者是 int[12]类型数据单元的地址。在程序设计中可以通过更加清晰简洁的表达方式进行数组的存取访问。

字符数组名、字符指针和字符串常量都可以视为字符串中第 1 个字符的地址，cout 对象以 char*的地址作为标准输出的起始数据单元地址。

本章还会简要介绍数据存储类别、结构指针，后面章节会进行更加详细的讨论。

4.6　复习题

1．如何声明下述数据？

a．actor 是由 30 个 char 类型的值组成的数组。

b．betsie 是由 100 个 short 类型的值组成的数组。

c．chuck 是由 13 个 float 类型的值组成的数组。

d．dipsea 是由 64 个 long double 类型的值组成的数组。

习题解析：

数组是顺序排列的一组相同类型的数据，在声明中需要表明每一个元素的数据类型和数据元素的个数，元素个数（索引值）使用方括号和括号内的整型数据表示。因此数组的声明如下。

a．char actor[30];

b．short betsie[100];

c．float chuck[13];

d．long double dipsea[64];

2．使用模板类 array 而不是数组来完成复习题 1。

习题解析：

模板类是 C++语言进行泛型编程的一种重要方式，模板类 array 可以创建指定类型和长度的数组。array 预定义在名称空间 std 内，因此为了使用该模板类需要使用以下预编译器指令。

```
#include <array>
```

添加相应的头文件。声明中使用一对尖括号标识模板类，并指定数组类型和元素数。因此本题中数组的声明如下。

a．array<char, 30> actor;

b．array<short, 100> betsie;

c．array<float, 13> chuck;

d．array<long double, 64> dipsea;

3．声明一个包含 5 个元素的 int 数组，并将它初始化为前 5 个正奇数。

习题解析：

数组元素可以在声明中使用赋值运算符进行初始化，初始化方式是使用花括号包含小于或等于元素数量并且以逗号分隔的同类型数据。本题中的初始化方式如下。

```
int arr[5] = {1, 3, 5, 7, 9};
```

4. 编写一条语句，将复习题 3 中数组的第 1 个元素与最后一个元素的和赋给变量 even。

习题解析：

数组内元素的存取和访问可以通过下标（索引）完成，需要注意的是，下标值从 0 开始，最后一个元素的下标值是元素数量减 1。因此，本题的赋值语句如下。

```
int arr[5] = {1, 3, 5, 7, 9};
int even = arr[0] + arr[4];
```

5．编写一条语句，显示 float 类型数组 ideas 中第 2 个元素的值。

习题解析：

数组的下标（索引）从 0 开始计数，因此第 2 个元素的下标是 1。语句如下。

```
float ideas[10];
…;
cout<<"The second element of array is "<<ideas[1]<<endl;
```

6．声明一个 char 的数组，并将其初始化为字符串“cheeseburger”。

习题解析：

使用字符数组和形式表示字符串有两种方式，既可以指定数组的长度，也可以不指定，而是由定义时初始化的字符串自动分配。因此可以使用以下两种方法定义，日常推荐使用第 1 种方法。

```
char st[] = "cheeseburger";
```

或者

```
char st[13] = "cheeseburger";
```

7．声明一个 string 对象，并将其初始化为字符串“Waldorf Salad”。

习题解析：

string 是 C++中预定义的一个类型，使用上更加简单，可以避免字符数组的复杂操作，像其他基本数据类型一样进行赋值操作和初始化。例如：

```
string st = "Waldorf Salad";
```

8．设计一个描述鱼的结构体声明。结构体应当包括品种、重量（整数，单位为盎司）和长度（单位为英寸，包括小数）。

习题解析：

要使用结构体进行数据类型定义，首先在语法结构上要使用关键字 struct，其次结构体的组成元素需要声明数据类型和名称，并用分号隔开。题目要求的结构体可以定义为以下形式。

```
struct fish{
```

```
char kind[20];
int weight;
float length;
};
```

9．声明复习题 8 中定义的结构体的一个变量，并对它进行初始化。

习题解析：

结构体变量的定义有两种形式。一种是类似其他基本数据类型的定义方式，另一种是直接在结构体的定义后加上变量的定义，例如：

```
struct fish{
char kind[20];
int weight;
float length;
} petes = {"BigFish", 12, 4.5 };
```

或者

```
fish petes = {"BigFish", 12, 4.5 };
```

10．用 enum 定义一个名为 Response 的类型，它包含 Yes、No 和 Maybe 等枚举量，其中 Yes 的值为 1，No 的值为 0，Maybe 的值为 2。

习题解析：

在定义枚举型变量时，其元素会自动从 0 开始赋值，用户也可以手动指定每一个元素的值，例如：

```
enum Response{No,Yes,Maybe};
```

或者

```
enum Response {No = 0,Yes = 1,Maybe = 2};
```

两者等价。

11．假设 ted 是一个 double 变量，请声明一个指向 ted 的指针，并使用该指针来显示 ted 的值。

习题解析：

要以指针作为存储数据地址的变量，也许需要标识最终指向存储单元的数据类型，因此在定义时需要使用数据类型和*表示，使用&地址运算符获得 ted 的地址并赋值给指针。

```
double ted = 0.0;
double *pd =&ted;
cout<<"The ted = "<<*pd<<endl;
cout<<"The ted = "<<ted <<endl;
```

12．假设 treacle 是一个包含 10 个元素的 float 数组，请声明一个指向 treacle 的第 1 个元素的指针，并使用该指针来显示数组的第 1 个元素和最后一个元素。

习题解析：

数组是一种特殊的数据形式，它本质上是相同类型的数据按规则排列的一系列数据，而数组的变量名本质上就是指向那一组数据的地址，因此在很多时候数组可以和指针混用。本题的声明方式如下。

```
float treacle[10];
float *pa = treacle;
```

以上直接使用数组名表示，或者可以用数组第 1 个元素的地址表示：

```
float *pa = &treacle[0];
```

13．编写一段代码，要求用户输入一个正整数，然后创建一个动态的 int 数组，其中包含的元素数目等于用户输入的值。首先使用 new 来完成这项任务，然后使用 vector 对象来完成这项任务。

习题解析：

C++内的数组（即使用 [ ] 表示的数组）需要在定义时表示其元素数量，随后在使用中该长度不会再改变，因此很多时候可以称为静态数组。动态数组是指通过指针和 new 运算符动态生成的数组，这种数组可以用 delete 删除，并重新分配长度，因此很多时候称为动态数组。代码如下。

```
unsigned int size;
cout<<"Enter a number: ";
cin>>size;
int *arr = new int[size];
```

14．下面的代码是否有效？如果有效，它将输出什么结果？

```
cout<<(int*)"Home of the jolly bytes";
```

习题解析：

"Home of the jolly bytes"在 C++语言中是一个字符串常量，采用 char 类型的首地址来记录和表示，因此 cout 能够根据该首地址查找到该字符串并输出。(int*) 使用强制类型转换将原字符串的字符类型的首地址转换成 int 类型的指针，因此 cout 语句将无法按照 char 类型的地址形式输出整个字符串，而会输出该地址的数据。作为比较，假设有如下语句，那么是否使用 (char*)进行类型转换都不会对字符串的输出产生影响。

```
cout<<(char*)"Home of the jolly bytes";
```

因为 cout 将使用 char 类型的地址进行字符串的查找，所以不需要进行强制类型转换。

15．编写一段代码，为复习题 8 中描述的结构体动态分配内存，并读取该结构体中成员的值。

习题解析：

针对用户定义的结构体，其使用方法和其他基本数据类型类似，可以直接使用结构体名，也可以按照 C 语言标准使用 struct 加名称的形式。代码如下。

```
struct fish{
char kind[20];
int weight;
float length;
};
fish *pf = new fish;
cout<<"Enter the kind of fish:";
cin>>pf->kind;
```

16. 程序清单 4.6 指出了同时输入数字和一行字符串时出现的存储问题。如果将下面的代码

```
cin.getline(address, 80);
```

替换为

```
cin>>address;
```

将对程序的运行带来什么影响?

习题解析：

cin>>address;语句将会使程序跳过输入内容中的空白字符（常用空白字符包含空格符、制表符和换行符），并且再次遇见空白字符时将会默认输入结束。而 cin.getline()函数将会读取整行数据，直到输入换行符时，将整行数据保存至 address 内。此外，该函数还对读取数据的长度进行了限制，当输入数据超过 80 个字符限制时将会自动截断，只保留前 80 个字符，而 cin>>address 则不会，并且输入数据超长时将会造成程序的错误。

17. 声明一个 vector 对象和一个 array 对象，它们都包含 10 个 string 对象。指出所需的头文件，但不要使用 using。使用 const 来指定要包含的 string 对象数。

习题解析：

vector 和 array 是 C++语言预定义的一些数据类型，如果需要使用这些类型应当添加相应的预编译指令。如果不使用 using 指令，那么需要在使用的类型和对象前添加名称空间。代码如下。

```
#include <array>
#include <string>
#include <vector>

const int size = 10;
std::vector< std::string>  vest(size);
std::array< std::string,size> arst;
```

4.7　编程练习

1. 编写一个 C++程序，如下述输出示例所示请求并显示信息。

```
What is your first name? Betty Sue
What is your last name? Yewe
```

```
What letter grade do you deserve? B
What is your age? 22
Name Yewe , Betty Sue
Grade: C
Age: 22
```

注意，该程序应该接受的名字包含多个单词。另外，程序将向下调整成绩，即向上调一个字母（即假设输入的成绩为 A，则输出为 B；输入为 B，则输出为 C，依此类推）。假设用户请求 A、B 或 C，所以不必担心 D 和 F 之间的空档。

编程分析：

程序中存储新姓名应当使用字符串，但是由于用户姓名可能不是一个单词，中间会有空白，因此不能使用 cin 和插入运算符直接读取，需要使用 getline()函数读取。当字符串使用字符数组的形式时需要注意数组长度，保证数组不会发生越界，并且 getline()函数在读取字符串会自动在字符串末尾添加空字符。完整代码如下。

```
/*第 4 章的编程练习 1*/
#include <iostream>
using namespace std;
/*预编译指令*/

int main()
{
    char first name[20], last_name[20];
    char grade;
    int age;
    /*定义程序中的变量，包括姓名、年纪等*/
    cout<<"What is your first name? ";
    cin.getline(first_name, 20);
    cout<<"What is your last name? ";
    cin.getline(last_name, 20);
    /*使用 getline()函数读取姓名，字符限制在 20 个以内*/
    cout<<"What letter grade do you deserve? ";
    cin>>grade;
    cout<<"What is your age? ";
    cin>>age;
    cout<<"Name "<<last_name<<" , "<<first_name<<endl;
    cout<<"Grade: "<<char(grade + 1)<<endl;
    cout<<"Age: "<<age<<endl;
    /*输出存储的信息*/
    return 0;
}
/*main()函数结束*/
```

2．修改程序清单 4.4，使用 C++ string 类而不是 char 数组。

编程分析：

C++语言中的 string 类与字符数组的形式相比较，优势非常明显，string 类的长度可以由 C++维护，并且可以直接使用赋值语句。本题中在使用 cin 标准输入读取数据时，需要使用 getline()函数而不是 cin 对象的 getline()函数（两者的参数有区别），且不用担心字符串长度问题。完整代码如下。

```
/*第 4 章的编程练习 2*/
#include <iostream>
#include <string>
/*若使用 string，应当修改#include 指令，添加 string 头文件*/
using namespace std;
/*预编译指令*/

int main()
{
    string name;
    string dessert;
    /*string 能够自动维护字符串长度，因此不需要长度常量*/
    cout<<"Enter your name:\n";
    getline(cin,name);
    cout<<"Enter your favorite dessert:\n";
    getline(cin,dessert);
    /*getline()函数的参数和字符数组的 cin.getline()不同*/
    cout<<"I have some delicious "<<dessert;
    cout<<" for you, "<<name<<"\n";
    return  0;
}
/*main()函数结束*/
```

3．编写一个程序，它要求用户首先输入名，然后输入姓，接着使用一个逗号和空格将姓与名组合起来，并存储和显示组合结果。请使用 char 数组和头文件 cstring 中的函数。下面是该程序运行时的情形。

```
Enter your first name: Flip
Enter your last name: Fleming
Here's the information in a single string: Fleming, Flip
```

编程分析：

在利用字符数组进行字符串处理时，相关的处理函数定义在 cstring 文件中，本题主要用到 strcpy() 函数和 strcat()函数。strcpy() 函数是复制字符串的函数，可以将一个字符串内的数据复制到另一个字符串中。strcat()函数可以将一个字符串中的内容添加到另一个字符串末尾。使用这两个函数可以完成相应的功能，完整代码如下。

```
/*第 4 章的编程练习 3*/
#include <iostream>
#include <cstring>
/*为了使用字符数组的处理函数，需要添加 cstring 头文件*/
using namespace std;
/*预编译指令*/
const int SIZE = 20;
/*使用常量表示字符数组的长度*/
```

```
int main()
{
    char first_name[SIZE], last_name[SIZE];
    char full_name[SIZE*2];
    /*分别定义姓、名和全名，注意字符数组的长度*/
    cout<<"Enter your first name: ";
    cin.getline(first_name, SIZE);
    cout<<"Enter your last name: ";
    cin.getline(last_name, SIZE);
    /*读取用户输入*/
    strcpy(full_name,last_name);
    strcat(full_name,", ");
    strcat(full_name,first_name);
    /*通过 strcpy()函数和 strcat()函数，将两个字符串复制和组合*/
    cout<<"Here's the information in a single string: ";
    cout<<full_name<<endl;
    return  0;
}
/*main()函数结束*/
```

4．编写一个程序，它要求用户首先输入名，然后输入姓，接着使用一个逗号和空格将姓与名组合起来，并存储和显示组合结果。请使用 string 对象和头文件 string 中的函数。下面是该程序运行时的情形。

```
Enter your first name: Flip
Enter your last name: Fleming
Here's the information in a single string: Fleming, Flip
```

编程分析：

题目要求使用 string 重做编程练习 3，string 的优势是在字符串的处理上能够直接使用 + 、 = 等运算符。完整代码如下。

```
/*第 4 章的编程练习 4*/
#include <iostream>
#include <string>
/*为了使用 string ，需要添加 string 头文件*/
using namespace std;
/*预编译指令*/
int main()
{
    /*char first_name[SIZE], last_name[SIZE];
    char full_name[SIZE *2 + 1];
    分别定义姓、名和全名，注意字符数组的长度*/
    string first_name, last_name, full_name;
    /*定义 3 个字符串变量*/
    cout<<"Enter your first name: ";
    getline(cin, first_name);
    cout<<"Enter your last name: ";
    getline(cin, last_name);
    full_name = last_name + ". "+ first_name;
```

```
    /*string 可以使用 + 和 = 进行字符串的合并和复制，其中 " . " 表示字符串*/
    cout<<"Here's the information in a single string: ";
    cout<<full_name<<endl;
    return  0;
}
/*main()函数结束*/
```

5．结构体 CandyBar 包含 3 个成员。第 1 个成员存储了糖块的品牌，第 2 个成员存储了糖块的重量（可以有小数），第 3 个成员存储了糖块的卡路里含量（整数）。请编写一个程序，声明这个结构体，创建一个名为 snack 的 CandyBar 变量，并将其成员分别初始化为 Mocha Munch、2.3 和 350。初始化应在声明 snack 时进行。最后，程序显示 snack 变量的内容。

编程分析：

题目要求定义 CandyBar 结构体，创建该结构体的变量并初始化。本题需要注意的是初始化结构体变量可以使用花括号，用逗号分隔成员数据。在输出数据信息时，需要使用结构体变量与成员运算符" . "限定结构体成员。完整代码如下。

```
/*第 4 章的编程练习 5*/
#include <iostream>
using namespace std;
/*预编译指令*/
struct CandyBar
{
    char brand[20];
    float weight;
    unsigned int calorie;
};
/*CandyBar 结构体的定义*/
int main()
{
    CandyBar snack = {"Mocha Munch", 2.3, 350};
    /*定义 snack 变量，并初始化*/
    cout<<"My favourite CandyBar is "<<snack.brand<<"."<<endl;
    cout<<"And its weight is "<<snack.weight<<", calorie is "<<snack.calorie;
    cout<<"."<<endl;
    /*显示 snack 的基本信息*/
    return  0;
}
```

6．结构体 CandyBar 包含 3 个成员，如编程练习 5 所示。请编写一个程序，创建一个包含 3 个元素的 CandyBar 数组，并将它们初始化为所选择的值，然后显示每个结构体的内容。

编程分析：

应用结构体创建结构体数组和其他类型数组的方法基本一致，只是在初始化数组元素的过程中，需要使用花括号的嵌套形式，才能同时初始化多个元素。成员访问需要同时使用数组下标来表示元素，使用成员运算符" . "表示成员。完整代码如下。

```
/*第 4 章的编程练习 6*/
#include <iostream>
using namespace std;
/*预编译指令*/
struct CandyBar
{
    char brand[20];
    float weight;
    unsigned int calorie;
};
/*CandyBar 结构体的定义*/
int main()
{
    CandyBar  snack[3]  =  {{"Mocha  Munch",  2.3,  350},{"Hershey  bar",  4.2,
550},{"Musketeers", 2.6, 430}};
    /*创建结构体数组，并使用花括号嵌套的方式进行初始化*/
    cout<<"My 1st CandyBar is "<<snack[0].brand<<"."<<endl;
    cout<<"And its weight is "<<snack[0].weight<<", calorie is "<<snack[0].calorie;
    cout<<"."<<endl;
    /*通过数组下标和成员运算符，可以访问数组内元素的数据成员*/
    cout<<"My 2nd CandyBar  is "<<snack[1].brand<<"."<<endl;
    cout<<"And its weight is "<<snack[1].weight<<", calorie is "<<snack[1].calorie;
    cout<<"."<<endl;

    cout<<"My 3th CandyBar  is "<<snack[2].brand<<"."<<endl;
    cout<<"And its weight is "<<snack[2].weight<<", calorie is "<<snack[2].calorie;
    cout<<"."<<endl;
    /*显示 snack 的数据信息*/
    return  0;
}
```

7. William Wingate 从事比萨饼分析服务。对于每个比萨饼，他都需要记录下列信息:

- 比萨饼公司的名称，可以由多个单词组成；
- 比萨饼的直径；
- 比萨饼的重量。

请设计一个能够存储这些信息的结构体，并编写一个使用这种结构体变量的程序。程序将请求用户输入上述信息，然后显示这些信息。请使用 cin（或它的方法）和 cout。

编程分析：

首先应当应用结构体对比萨饼的特定信息进行定义，根据题意，可以定义为

```
struct Pizza{
char company[40];
float diameter;
float weight;
};
```

应用该结构体，对比萨饼进行信息存储和输出，完整代码如下。

```
/*第 4 章的编程练习 7*/
#include <iostream>
using namespace std;
/*预编译指令*/
struct Pizza
{
    char company[40];
    float diameter;
    float weight;
};
/*Pizza 结构体的定义*/
int main()
{
    Pizza dinner;
    cout<<"Enter the Pizza's information:"<<endl;
    cout<<"Pizza's Company:";
    cin.getline(dinner.company,40);
    /*要读取用户输入的信息，可以直接使用成员运算符*/
    cout<<"Pizza's diameter(inchs): ";
    cin>>dinner.diameter;

    cout<<"CandBar's weight(pounds): ";
    cin>>dinner.weight;
    /*读取用户输入的信息，使用成员运算符表示每一个成员并赋值*/
    cout<<"The lunch pizza is "<<dinner.company<<"."<<endl;
    cout<<"And its diameter is "<<dinner.diameter<<" inch, weight is "<<dinner.weight;
    cout<<"pounds."<<endl;
    /*输出信息*/
    return  0;
}
```

8. 完成编程练习 7，使用 new 为结构分配内存，而不是声明一个结构体变量。另外，程序要在请求输入比萨饼公司名称之前输入比萨饼的直径。

编程分析：

题目要求在编程练习 7 的基础上，使用动态存储分配。因此需要将结构体变量改为结构体指针，并通过 new 运算符进行存储、分配。和结构体变量不同，成员运算符需要修改为 "->" 指针形式的成员运算符。此外，在程序结束时还必须手动调用 delete 回收存储，其他部分基本相同。完整代码如下。

```
/*第 4 章的编程练习 8*/
#include <iostream>
using namespace std;
/*预编译指令*/
struct Pizza
{
    char company[40];
    float diameter;
    float weight;
```

```
};
/*Pizza 结构体的定义*/
int main()
{
    Pizza*ppizza = new Pizza;
    cout<<"Enter the Pizza's information:"<<endl;
    cout<<"Pizza's diameter(inchs): ";
    cin>>ppizza->diameter;

    cout<<"Pizza's Company:";
    cin.getline(ppizza->company,40);

    cout<<"CandBar's weight(pounds): ";
    cin>>ppizza->weight;
    /*指针变量需要使用"->"成员运算符，而不是"."成员运算符*/
    cout<<"The lunch pizza is "<<ppizza->company<<"."<<endl;
    cout<<"And its diameter is "<<ppizza->diameter<<" inch, weight is "<<ppizza->weight;
    cout<<"pounds."<<endl;
    delete ppizza;
    /*在程序结束时，必须手动调用 delete 回收存储空间*/
    return  0;
}
```

9. 完成编程练习 6，使用 new 动态分配数组，而不是声明一个包含 3 个元素的 CandyBar 数组。

编程分析：

题目要求和编程练习 8 类似，只是对于数组形式的动态存储，需要使用 new [] 的形式，为了回收存储空间，需要使用 delete [] 形式。为了访问数组元素的成员，需要同时使用数组下标和成员运算符。两者在使用上需要引起注意，程序的注释中有详细说明。完整代码如下。

```
/*第 4 章的编程练习 9*/
#include <iostream>
using namespace std;
/*预编译指令*/
struct CandyBar
{
    char brand[20];
    float weight;
    unsigned int calorie;
};
/*CandyBar 结构体的定义*/
int main()
{
    CandyBar*pc = new CandyBar[3];
    strcpy(pc[0].brand, "Mocha Munch");
    pc[0].weight = 2.3;
    pc[0].calorie = 350;
```

```
    strcpy(pc[1].brand, "Hershey bar");
    (pc + 1)->weight = 4.2;
    pc[1].calorie = 550;
    strcpy(pc[2].brand, "Musketeers");
    pc[2].weight = 2.6;
    pc[2].calorie = 430;
    /*此处按照数组形式表示元素，也可以使用如下形式表示数组元素的成员：
     *(pc)->weight = 4.2;
     *(pc + 1)->weight = 4.2;
     *(pc + 2)->weight = 4.2;
     *下面的输出使用这种方式
     **/
    cout<<"My 1st CandyBar is "<<pc->brand<<"."<<endl;
    cout<<"And its weight is "<<pc->weight<<", calorie is "<<pc->calorie;
    cout<<"."<<endl;

    cout<<"My 2nd CandyBar  is "<<(pc+1)->brand<<"."<<endl;
    cout<<"And its weight is "<<(pc+1)->weight<<", calorie is "<<(pc+1)->calorie;
    cout<<"."<<endl;

    cout<<"My 3th CandyBar  is "<<(pc+2)->brand<<"."<<endl;
    cout<<"And its weight is "<<(pc+2)->weight<<", calorie is "<<(pc+2)->calorie;
    cout<<"."<<endl;
    delete [] pc;
    /*在程序结束时，必须手动调用 delete 回收存储空间*/
    return  0;
}
```

10．编写一个程序，要求用户输入 3 次 40 码[①]跑的成绩（也可输入 40 米跑的成绩），并显示次数和平均成绩。请使用一个 array 对象来存储数据（如果编译器不支持 array 类，请使用数组）。

编程分析：

C++语言的标准库预定义了 array 数组，需要使用模板来定义 array 对象。array 模板还定义了公用接口，用于提高编程效率，具体内容在后面章节会详细叙述。下面仅使用了 array 对象的基本功能。也可以使用下标形式访问所有的元素，但是通过直接将 array 对象赋值，就可以实现所有元素的赋值，而不用像数组那样逐一赋值。完整代码如下。

```
/*第 4 章的编程练习 10*/
#include <iostream>
#include <array>
using namespace std;
/* 预编译指令，需要添加 array 头文件*/

int main()
{
```

① 1 码=1.609 34km。——编者注

```
    array<float,3> record_list;
    /*定义 array 对象 record_list*/
    float average;
    cout<<"Please input three record of 40 miles.\n";
    cout<<"First recond:";
    cin>>record_list[0];
    cout<<"Second recond:";
    cin>>record_list[1];
    cout<<"Third recond:";
    cin>>record_list[2];
    /*依次读取数据输入*/
    cout<<"Ok, you input:\n1."<<record_list[0]<<"\n2."<<record_list[1]<<"\n3.";
    cout<<record_list[2]<<endl;
    average = (record_list[0]+record_list[1]+record_list[2])/3;
    /*计算平均值，并输出*/
    cout<<"Congratulate, your average performance is "<<average<<".";
    return  0;
}
```

第 5 章　循环和关系表达式

本章知识点总结

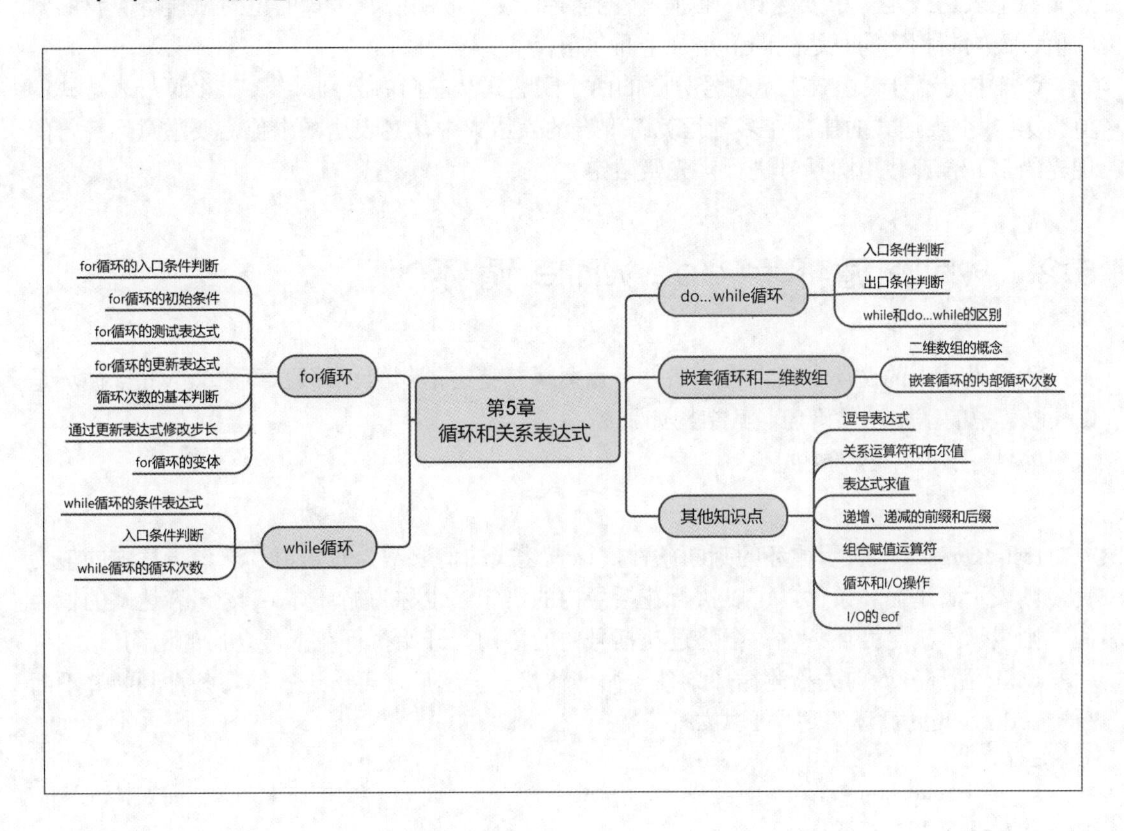

5.1　C++语言中的表达式

C++中的表达式是数值或者与数值与运算符之间的一种组合，每个表达式都有一个确定的值。在任何表达式末尾加上分号就可以得到一条 C++的语句。C++语言中的语句可以视为一个完整的操作指令，即使空语句也表示一个完整的空指令；而表达式则不是，表达式只是一个值。C++语言中的常用运算符除了加、减、乘、除和求模之外，还有递增运算符（++）和递减运算符（--）。递增与递减运算符的基本功能是对操作数进行加 1 和减 1 运算。递增和递减运算符可以作为前缀与后缀，虽然两者的运算结果相同，但是运算发生的时间不同。这个区别非常重要，也是初学者学习 C++语言的难点，递增和递减前缀是先进行递增运算再参与表达式中的其他运算，递增和递减后缀则相反。赋值运算符（=）表示赋值运算，C++语言中还有组合赋值运算运算符（+=、−=、*=、/=、%=），分别表示先算术运算再赋值。

C++中还有 6 种关系运算符（>、<、>=、 <=、==、 != ），分别表示两个操作数的大小关系。关系运算表达式的值为布尔值，即 true 或者 false。关系表达式的值为布尔型数据的特点使之大量应用于各种逻辑判断语句中，使用中需要区分判断相等的关系运算符和赋值运算符，判断相等（= =）的表达式的值为布尔值。

C++语言中赋值运算符（=）和判断相等的运算符（= =）是初学者经常混淆的一个知识点。赋值表达式是一个表达式，也有一个确定的数值，赋值表达式的值为赋值运算符右侧的值，而判断相等的表达式的值是一个布尔值。

C++中的逗号表达式表示逗号分隔的两个表达式从左向右分别计算，整个逗号表达式的值为第 2 个表达式的值。在多种运算符混合的表达式中，对表达式求值需要注意运算符的优先级，并合理使用圆括号进行优先级限定。

5.2 while 循环和 do...while 循环

程序设计中的循环是指在特定条件下反复执行特定语句的一种运算形式。while 循环是C++语言中的基本循环语句，其结构如下。

```
while(test-expression){
      statements;
}
```

test-expression 表示循环的判断条件，该表达式的值通常为布尔值。若该表达式的值为真，则表示满足循环条件，可以执行花括号内的语句（也称为循环体）；若该表达式的值为假，则表示不满足循环条件，会略过花括号内的语句，开始执行花括号之后的语句。

while 循环开始前先检测循环条件，循环体执行一遍后，再次进行循环条件判断并不断反复。do...while 循环的基本形式如下。

```
do{
       statements;
}while(test- expression);
```

do...while 循环先执行循环体的语句，再进行循环条件判断，其他运行逻辑和 while 循环相同。do...while 循环的循环体至少会执行一次，while 循环则存在不执行的可能。

5.3 for 循环

while 循环和 do...while 循环分别通过判断入口或出口条件来判定是否执行下一次循环。for 循环是另外一种循环语句，其特点在于能够清晰简洁地表示循环运行的次数，有时 for 循环也称为定次循环。for 循环的基本形式如下。

```
for(initilization; test-expression;update-expression){
```

```
        statements;
    }
```

for 循环语句中有 3 个表达式，initilization 表示循环开始的初始化条件，test-expression 表示循环入口条件，即每次循环之前都会计算该表达式的值，并判断是否进入循环；update-expression 则在每次循环体结束后执行，通常用来控制循环的次数。

由于 for 循环能够很好地控制循环初始条件，因此在处理数组数据上有很大优势。通常我们可以直接使用数组的下标作为循环控制条件，并将 update-expression 设置为数组下标的递增运算，控制和管理数组元素会更加清晰高效。实际应用中 for 循环能够通过 3 个表达式创建很多循环的变体，因此在练习过程中应当仔细分析 3 个表达式的含义、表达式的求值和判断时间，这样才能正确地控制循环的运行。

5.4　二维数组和嵌套的循环

数组是指相同类型的多个元素按规律排成一列的数据结构类型，当数组中的每一个元素都是另一个数组时，就形成了一个二维数组。二维数组可以形象地理解为行列表格的形式，每一行是一个一维数组，以多个一维数组作为元素，顺序排列形成的矩形结构就是一个二维数组。在进行二维数组的数据处理时通常要使用嵌套的循环。

嵌套的循环和二维数组在结构上类似，即以一个完整的循环作为另外一个循环的循环体，这样嵌套形成的循环形式称为嵌套的循环。在利用嵌套的循环处理二维数组时，注意，外部循环执行一遍，内部循环会执行 *m* 次，外部循环语句循环了 *n* 次，那么内部循环会执行 *n**m* 次。

5.5　标准输入/输出和循环

在通过 cin 和 cout 对象进行标准输入/输出时，C++系统经常会使用循环语句重复读写数据。在重复读写数据的过程中需要注意 cin 对象会忽略空白字符（空格、换行符和制表符）。产生这个问题的主要原因是系统的 I/O 缓冲区和字符数据的转换。因此文本输入中需要使用 cin 对象的 get()函数对空白字符进行特定处理，或者使用 cin 对象的 getline()函数读取一行数据输入。具体的处理办法在复习题中都有涉及，练习过程中需要重点关注。

5.6　复习题

1．入口条件循环和出口条件循环之间的区别是什么?各种 C++循环分别属于其中的哪一种?

习题解析：

对于入口条件循环，在进入循环体之前先进行表达式的判断。如果表达式为真，则开始执行循环体内的语句；若为假，则不执行循环体内的语句，并越过该循环，执行循环语句之后的语句。对于出口条件循环，首先执行循环体内的语句，然后再判断表达式。如果表达式为真，则再次执行循环体；否则，退出该循环，执行循环之后的语句。因此两者的差异在于入口条件循环有可能一次都不会执行循环体内语句，出口条件循环则至少会执行一次循环体。C++语言中 do...while 循环是出口条件循环，while 循环和 for 循环是入口条件循环。

2．如果下面的代码片段是有效程序的组成部分，它将输出什么内容？

```
int i;
for(i = 0;i<5;i++)
cout<<i;
cout<<endl;
```

习题解析：

代码先输出 01234，后输出换行符。由于 for 循环的循环体只有一条语句 `cout<<i;`，因此在循环完成后执行一遍 `cout<<endl;`语句。需要注意的是，使用 C++语言编写代码过程中的缩进主要是为了使代码清晰可读，并不是像 Python 语言那样通过缩进标识代码块。

3．如果下面的代码片段是有效程序的组成部分，它将输出什么内容？

```
int j;
for(j = 0;j<11;j+=3)
        cout<<j;
cout<<endl<<j<<endl;
```

习题解析：

程序的输出结果如下。

```
0369
12
```

for 循环中 `j` 从 0 开始，当 `j` 小于 11 时输出 `j` 的值，输出完成后 `j` 做加 3 的操作，然后开始下一次循环。因此循环内数据从 0 开始输出，最后输出 9，加 3 后不满足循环条件，于是退出循环。循环退出后，输出换行符和 12，最后再输出一个换行符。

4．如果下面的代码片段是有效程序的组成部分，它将输出什么内容？

```
int j = 5;
while(++j<9)
        cout<<j++<<endl;
```

习题解析：

程序的输出结果如下。

```
6
8
```

本题主要考察以递增运算符作为前缀的知识点。while 循环从 `j = 5` 开始，在进行 while 循环的条件判断时，先递增，再判断和 9 的大小关系。如果小于 9，则进入循环。进入循环后先输出 `j`，再递增。因此输出结果为 6 和 8。

5．如果下面的代码片段是有效程序的组成部分，它将输出什么内容?

```
int k = 8;
do
        cout<<" k = "<<k<<endl;
while(k++<5);
```

习题解析：

程序的输出结果如下。

```
k = 8
```

本题主要考察 do...while 循环的出口条件判断，由于 `k = 8`，因此不满足循环条件，但是该循环至少会执行一次，于是输出 `k = 8`，然后退出循环。

6．编写一个输出 1、2、4、8、16、32、64 的 for 循环，每轮循环都将计数变量的值乘以 2。

习题解析：

为了在 for 循环内控制循环计数变量，需要在更新条件处使用表达式 `i = i *2`，也可以使用表达式 `i *= 2`。输出 1、2、4、8、16、32、64 的 for 循环的基本代码如下。

```
int i;
for(i = 1;i<64;i*=2)
      cout<<i<<"、";
cout<<i<<endl;
```

7．如何在循环体中包括多条语句?

习题解析：

如果在循环体内需要执行多条语句，可以使用一对花括号{ }将多条语句组合成一个语句块或者复合语句。并且初学者应当在循环语句中尽量多使用花括号来表示循环体，即使循环体中只有一条语句。

8．下面的语句是否有效？如果无效，原因是什么？如果有效，它将完成什么工作?

```
int x = (1,024);
```

下面的语句是否有效?

```
int y;
y = 1,024;
```

习题解析：

首先应当注意到`(1,024)`中间的符号是逗号运算符，并不是浮点型数据的小数点。

在语句 `int x = (1,024);`中，赋值语句首先计算右侧圆括号内的值，`(1,024)`中逗号运算符的功能是首先计算逗号左侧表达式的值，再计算逗号运算符右侧的值，且整个逗号表达式的值是右侧表达式的值，因此赋值语句将 024 赋给变量 x，024 是八进制的常量数值，转换成十进制为 20，所以 x 被赋值为 20。

语句 `y = 1,024` 则不同，整个语句中由于赋值运算符优先级较高，因此整个表达式先进行赋值运算，y 的值为 1，然后进行逗号运算，计算整个表达式的值，即 024，但该值并未产生实际意义，语句结束后数值被丢弃，但是 y 的值不变，因此 y 为 1。

9．在查看输入方面，cin >>ch 与 cin.get(ch)和 ch=cin.get( )有什么不同?

习题解析：

C++语言中要使用输入/输出流进行字符数据的读取，通常使用 3 种方式，分别是 cin >>ch、cin.get(ch)和 ch=cin.get()。其中 cin>>ch 将会忽略所有空白字符（空格符、换行符、制表符）；cin.get(ch)和 ch=cin.get()将会读取所有 ASCII 字符并将其保存在 ch 变量内。

5.7 编程练习

1．编写一个要求用户输入两个整数的程序。该程序将计算并输出这两个整数之间（包括这两个整数） 所有整数的和。这里假设先输入较小的整数。例如，如果用户输入的是 2 和 9，则程序将输出 2～9 的所有整数的和，即 44。

编程分析：

题目要求计算两个指定整数之间所有整数的和，因此需要使用循环进行计算。通常情况下使用 while 循环、for 循环都可以实现题目要求的功能。但是由于题目的起始条件和终止条件比较明确，因此使用 for 循环更加清晰可读。下面是完整代码。

```
/*第 5 章的编程练习 1*/
#include <iostream>
using namespace std;

int main()
{
    int min, max, sum = 0;
    /*定义输入数据变量与 sum 变量*/
    cout<<"Enter the first numeral: ";
    cin>>min;
    cout<<"Enter the second numeral: ";
    cin>>max;
    /*通过标准输入读取起止范围的数据*/
```

```
    for(int i = min; i<=max; i++)
        sum += i;
    /*通过 for 循环，计算指定范围内所有整数的和*/
    cout<<"The sum of "<<min<<" +...+ "<<max<<" is ";
    cout<<sum<<endl;
    return  0;
}
```

2．使用 array 对象（而不是数组）和 long double（而不是 long long）重新编写程序清单 5.4，并计算 100!的值。

编程分析：

程序清单 5.4 实现了利用数组形式计算和存储阶乘的功能。0 的阶乘写作 0!，被定义为 1。1!= 1*0!，即 1。2!=2*1!，即 2。3!=3*2!，即 6，以此类推。每个整数的阶乘都是该整数与前一个阶乘的乘积。题目要求使用 array 对象来实现该功能， array 对象和数组在使用方法上的区别不大，只是在声明时使用 `array<long double, ArSize> factorials;`语句，循环部分的起始条件和终止条件需要修改为 100。完整代码如下。

```
/*第 5 章的编程练习 2*/
#include <iostream>
#include <array>

const int ArSize = 101;
using namespace std;

int main()
{
    array<long double, ArSize> factorials;
    factorials[1] = factorials[0] = 1;
    /*初始化阶乘中的 0! 和 1! */
    for(int i = 2;i < ArSize; i++)
        factorials[i] = i *factorials[i-1];
    /*根据阶乘定义，应用 for 循环计算 100 的阶乘，注意，此处计算了 100 以内所有整数的阶乘*/
    for(int i = 0; i < ArSize;i++)
        cout<<i<<"! = "<<factorials[i]<<endl;
    /*使用 for 循环输出求阶乘的结果*/
    return  0;
}
```

3．编写一个要求用户输入数字的程序。每次输入后，程序都将报告到目前为止，所有输入的累计和。当用户输入 0 时，程序结束。

编程分析：

本题的要求是对用户输入的数字求和，当用户输入 0 时终止。在循环语句的选择上，因为不能确定用户输入的次数，因此标准格式的 for 循环并不合适。可以选择 while 循环，因为用户至少会输入一次数据，使用 do...while 循环更加简便。当然，使用 for 循环的变体形式或者 while 循环也能解决这个问题。使用 do...while 循环的完整代码如下。

```
/*第 5 章的编程练习 3*/
#include <iostream>
using namespace std;

int main()
{
    double temp, sum = 0;
    do{
        cout<<"Input a numeral to add: ";
        cin>>temp;
        sum += temp;
        /*读取输入，并求和*/
    }while(temp != 0);
    /*当输入为 0 时退出循环*/
    cout<<"Input end.\n"<<"The sum = "<<sum<<endl;
    return 0;
}
```

4. Daphne 以 10%的单利投资了 100 美元。也就是说，每一年的利息都是投资额的 10%，即每年 10 美元，利息的计算公式如下。

利息= 0.10×原始存款

而 Cleo 以 5%的复利投资了 100 美元，利息是当前存款（包括获得的利息）的 5%，即

利息= 0.05×当前存款

Cleo 在第 1 年投资 100 美元时的利率是 5%——得到了 105 美元。下一年的利息是 105 美元的 5%，即 5.25 美元，以此类推。请编写一个程序，计算多少年后，Cleo 的总金额才能超过 Daphne 的总金额，并显示此时两个人的总金额。

编程分析：

Daphne 和 Cleo 两人通过不同方式进行投资，两人在本金相同的情况下，投资收益形式不同。题目需要计算何时 Cleo 投资价值超过 Daphne，因此基本算法是按照年度，循环计算每年两人收益并进行比较。编程过程中并不能确定何时满足条件终止循环，因此通常采用 while 循环进行判断，完整代码如下。

```
/*第 5 章的编程练习 4*/
#include <iostream>
using namespace std;
const int DEPOSIT_BASE = 100;
/*定义基准常量*/
int main()
{
    float daphne_deposit = DEPOSIT_BASE;
    float cleo_deposit = DEPOSIT_BASE;
    /*定义并初始化两人的起始金额*/
    int year = 0;
    while(daphne_deposit>=cleo_deposit)
```

```
    {
        /*若 daphone_deposit 大于或等于 cleo_deposit ，执行循环；否则，终止*/
        cout<<"In "<<year++<<" Year: Daphne = "<<daphne_deposit<<endl;
        cout<<"\tCleo = "<<cleo_deposit<<endl;
        daphne_deposit += 0.1*DEPOSIT_BASE;
        /*计算 Daphone 每年后的总金额*/
        cleo_deposit += 0.05*cleo_deposit;
        /*计算 Cleo 每年后的总金额*/
    }
    cout<<"In "<<year<<" year: Daphne = "<<daphne_deposit<<endl;
    cout<<"\tCleo = "<<cleo_deposit<<endl;
    return 0;
}
```

5. 假设要销售 *C++ For Fools* 一书。请编写一个程序，输入全年中每个月的销售量（图书数量，而不是销售额）。程序通过循环，使用初始化为月份字符串的 char *数组（或 string 对象数组）逐月进行提示，并将输入的数据存储在一个 int 数组中。然后，计算数组中各元素的总和，并报告这一年的销售情况。

编程分析：

题目要求通过数组形式存储表示月份的字符串，并将每月的图书销售量存储在另一个 int 类型字符串内，因此两个字符串可以使用下标来标注月份和图书销售量之间的对应关系，即每个数组的第 1 个元素都是第 1 个月的数据，以此类推。下面的代码通过两个 for 循环分别实现输入与求和功能，如果需要简化代码，也可以在输入数据的同时进行求和计算。完整代码如下。

```
/*第 5 章的编程练习 5*/
#include <iostream>
using namespace std;
int main()
{
    const string Month[] = {"JAN","FEB","MAR","APR","MAY","JUN","JUL","AUG","SEP","OCT",
"NOV","DEC"};
    int sale_amount[12]={};
    /*定义两个数组，分别初始化为月份数据和销售数据，销售数据通过{}初始化为 0*/
    unsigned int sum = 0;
    for(int i = 0; i < 12; i++)
    {
        cout<<"Enter the sale amount of "<<Month[i]<<" :";
        cin>>sale_amount[i];
    }
    /*通过循环，读取用户的输入*/
    cout<<"Input DONE!"<<endl;

    for(int i = 0; i < 12; i++)
    {
        cout<<Month[i]<<" SALE :"<<sale_amount[i]<<endl;
```

```
        sum += sale_amount[i];
    }
    /*通过循环再次计算总销售量*/
    cout<<"Total sale "<<sum<<" this year."<<endl;    return  0;
}
```

6．重做编程练习 5，但这一次使用一个二维数组来存储 3 年中每个月的销售量。程序将报告每年的销售量以及 3 年的总销售量。

编程分析：

题目要求和编程练习 5 基本类似，但是为了存储 3 年内每个月的销售量，需要通过二维数组，其中二维数组的主数组元素数为 3，每个元素是一个元素数是 12 的一维数组，定义方式为 `int sale_amount[3][12]={};`，花括号表示初始化数据为 0。为了实现对二维数组中每一个元素的读取访问，需要使用嵌套的循环。通常情况下，外部循环的循环次数为主数组元素数，内部循环的循环次数为下一级数组的元素数。完整代码如下。

```
/*第 5 章的编程练习 6*/
#include <iostream>
using namespace std;
int main()
{
    const string Month[] = {"JAN","FEB","MAR","APR","MAY","JUN","JUL","AUG","SEP",
"OCT","NOV","DEC"};
    int sale_amount[3][12]={};
    /*定义二维数组，表示 3 年内每个月的销售数据，可以通过行列表格来理解
     *通过{}初始化所有数据为 0*/
    unsigned int sum = 0;
    for(int i = 0; i < 3; i++)
    {
        cout<<"Starting "<<i<<" year's data."<<endl;
        for(int j = 0;j < 12 ;j++)
        {
            cout<<"Enter the sale amount of "<<Month[j]<<" :";
            cin>>sale_amount[i][j];
        }
        cout<<"End of "<<i+1<<" year's data."<<endl;
    }
    /*通过嵌套的循环实现二维数组的数据输入，内部循环对应于每年的 12 个月，外部循环对应于 3 年*/
    cout<<"Input DONE!"<<endl;

    for(int i = 0; i < 3; i++)
    {
        for(int j = 0;j < 12 ;j++)
        {
            cout<<Month[i]<<" SALE :"<<sale_amount[i][j3]<<endl;
            sum += sale_amount[i][j];
        }
        cout<<"Total sale "<<sum<<" "<<i+1<<" year."<<endl;
```

```
    }
    /*通过嵌套的循环实现二维数组的数据计算，内部循环对应于每年的 12 个月，所以每年的销售量
     *应当在外部循环中输出*/
    return 0;
}
```

7．设计一个名为 car 的结构，用它存储有关汽车的信息——生产商（存储在字符数组或 string 对象中的字符串）、生产年份（整数）。编写一个程序，询问用户有多少辆汽车。随后，使用 new 来创建一个由相应数量的 car 结构体组成的动态数组。接下来，提示用户输入每辆车的生产商（可能由多个单词组成）和年份信息。请注意，这需要特别小心，因为它将交替读取数值和字符串（参见第 4 章）。最后，显示每个结构体的内容。该程序的运行情况如下。

```
How many cars do you wish to catalog? 2
Car #1:
Please enter the maker: Hudson Hornet
Please enter the year made: 1952
Car #2:
Please enter the maker: Kaiser
Please enter the year made: 1951
Here is you collection:
1952 Hudson Hornet
1951 Kaiser
```

编程分析：

题目要求创建一个存储汽车信息的结构体数组。由于数组的长度需要用户指定，即通过标准输入读取数组长度（汽车数量），因此不能使用静态数组，而需要通过 new 关键字动态创建。汽车信息的结构体定义如下。

```
struct car_info
{
    string manufacturer;
    int date;
};
```

题目使用 string 类型的字符串存储生产商，可以简化汽车信息的输入代码。完整代码如下。

```
/*第 5 章的编程练习 7*/
#include <iostream>
using namespace std;
struct car_info
{
    string manufacturer;
    int date;
};
/*定义表示汽车信息的结构体*/
int main()
{
    int car_number;
```

```
    car_info*pcar;
    cout<<"How many cars do you wish to catalog? ";
    cin>>car_number;
    cin.get();
    pcar = new car_info[car_number];
    /*通过用户输入数量，动态创建表示汽车信息的数组*/
    for(int i = 0; i < car_number; i++)
    {
        cout<<"Car #"<<i+1<<":"<<endl;
        cout<<"Please enter the maker: ";
        getline(cin, pcar[i].manufacturer);
        cout<<"Please enter the year made: ";
        cin>>pcar[i].date;
        cin.get();
    }
    /*通过循环输入汽车信息*/
    cout<<"Here is you collection:"<<endl;

    for(int i = 0; i < car_number; i++)
    {
        cout<<pcar[i].date<<" "<<pcar[i].manufacturer<<endl;
    }
    /*输出汽车的基本信息*/
    return 0;
}
```

8．编写一个程序，它使用一个 char 数组和循环来每次读取一个单词，直到用户输入 done 为止。随后，该程序输出用户输入了多少个单词（不包括 done 在内）。下面是该程序的运行情况。

```
Enter words (to stop, type the word done):
anteater birthday category dumpster
envy finagle geometry done for sure
You entered a total of 7 words.
```

应在程序中包含头文件 cstring，并使用函数 strcmp()来进行比较。

编程分析：

程序的主要功能要求是反复读取用户输入的单词，并进行字符计数。对于不能确定循环次数的循环条件，通常使用 while 循环。循环条件判断是当用户输入 done 时退出循环，这里字符串的比较使用 strcmp()，当两个字符串完全相同时该函数返回逻辑值“真”。完整代码如下。

```
/*第 5 章的编程练习 8*/
#include <iostream>
#include <cstring>
using namespace std;

const int SIZE = 20;
const char FINISHED[] = "done";
```

```
/*定义常量*/
int main()
{
    int counter = 0;
    char words[SIZE];
    cout<<"Enter words (to stop, type the word done):"<<endl;
    while(strcmp(FINISHED,words) != 0 )
    {
        counter++;
        cin>>words;
        cin.get();
        /*题目要求读取单词，因此使用 cin，并使用 get()删除空白字符，
         *循环条件是输入的单词不是"done"
         **/
    }
    cout<<"You entered a total of "<<counter-1<<" words."<<endl;
    return 0;
}
```

9. 重做编程练习 8，但使用 string 对象而不是字符数组。请在程序中包含头文件 string，并使用关系运算符来进行比较。

编程分析：

本题的要求和编程练习 8 类似，只是要求使用 string 对象表示字符串，而不是字符数组。使用 string 对象表示字符串能够简化代码，本题中可以直接使用 == 运算符判断相等，不需要使用 strcmp()函数。程序的完整代码如下。

```
/*第 5 章的编程练习 9*/
#include <iostream>
#include <string>
using namespace std;

const char FINISHED[] = "done";
/*定义常量*/
int main()
{
    int counter = 0;
    string words;
    cout<<"Enter words (to stop, type the word done):"<<endl;
    while( words != FINISHED)
    {
        /*判断字符串相等，可以直接使用 == 运算符*/
        counter++;
        cin>>words;
        cin.get();
    }
    cout<<"You entered a total of "<<counter-1<<" words."<<endl;
    return 0;
}
```

10. 编写一个使用嵌套循环的程序，要求用户输入一个值，指定要显示的行数。然后，程序将显示相应行数的星号，其中第 1 行包括一个星号，第 2 行包括两个星号，以此类推。每一行包含的字符数等于当前行数，在星号不够的情况下，在星号前面加上句点。该程序的运行情况如下。

```
Enter the number of rows:5
....*
...**
..***
.****
*****
```

编程分析：

题目要求通过循环控制每一行输出的符号数量，通过运行结果可以看出整体形式类似于由行、列组成的表格，因此使用嵌套循环的形式输出。最基本的解决方案是句点和星号分开输出，因此内部循环又由两个并列的循环语句组成，一个递减地输出句点，一个递增地输出星号。在进行类似程序设计时，常用方法是通过循环计数的形式手动分析每一次循环的控制变量 i、j，以及循环条件的变化。这种方法可以作为一种简单的调试方法。题目的完整代码如下。

```cpp
/*第 5 章的编程练习 10*/
#include <iostream>
using namespace std;

int main()
{
    int line;
    cout<<"Enter the number of rows:";
    cin>>line;
    for(int i = 0; i < line; i++)
    {
        for(int j = 0; j < line - i - 1; j++)
        {
            cout<<".";
        }
        /*第 1 个内部循环负责输出句号，句号数量逐行递减
         *因此需要通过 j < line - i -1 来控制每一行的句号数量*/
        for(int j = 0; j <= i; j++)
        {
            cout<<"*";
        }
        /*第 2 个内部循环负责输出星号，星号数量逐行递增，
         *因此使用 j < =i 来控制每一行的星号数量*/
        cout<<endl;
    }
    return 0;
}
```

第 6 章　分支语句和逻辑运算符

本章知识点总结

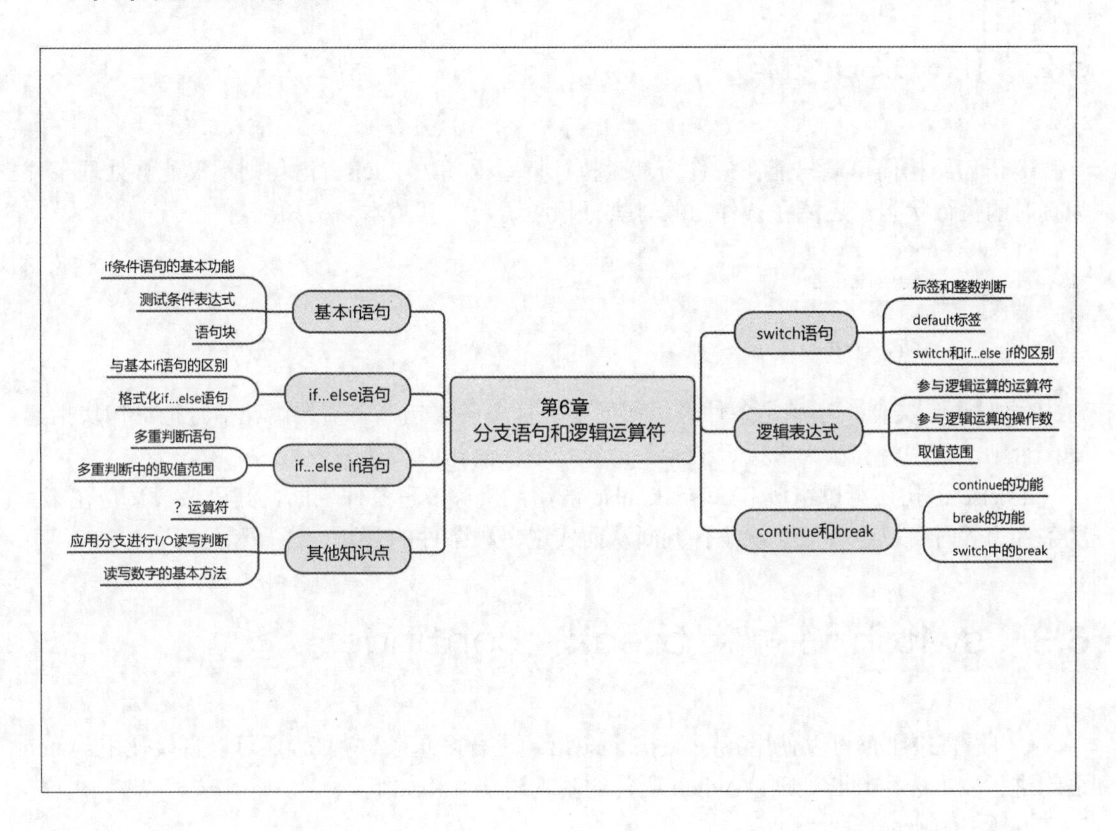

6.1　if 条件语句

C++语言中 if 条件语句是根据不同的逻辑选项执行不同分支语句的重要方式。if 语句结构比较简单，主要形式如下。

```
if(test-condition) statement;
```

或者

```
if(test-condition) {
        statements;
}
```

其中，test-condition 是一个逻辑表达式，其值应当为布尔值。当该值为 true 时，则执行 statement

语句；如果是 false，则略过 statement 语句，顺序执行后面的语句。当 statement 是一条语句时，可以不使用花括号；若是多条语句则必须使用，否则只会将最近的一条语句看作 if 语句的一部分。

C++语言中，除了关系运算表达式的结果之外，还有逻辑与、逻辑或和逻辑非（符号分别是"&&""||""！"）3 个逻辑运算，可以对布尔值进行运算并得到布尔值。程序设计中应当避免在条件判断表达式中出现非布尔值，因为编译器会将整型数据中的 0 当作 false，将非 0 当作 true，该结果在特定情况下可能会发生逻辑错误。因此，在循环的条件判断和分支语句的条件判断中，都应当重视条件判断表达式的求值。

6.2 if...else 语句

if 语句结构简单，只能在条件为真时执行特定语句。if...else 语句则不仅能够处理条件为真时的执行状态，还能在条件为假时执行相应语句，其基本结构如下。

```
if(test-condition){
        statements;
}else{
        statements;
}
```

与 if 条件语句类似，当满足条件时，如果要执行多条语句，就需要用花括号把语句块括起来；否则，程序将会认为最接近的一条语句是条件语句的一部分。

if...else 语句还可以组成 if...else if...else 语句，执行多重条件判断，其语法含义是在第 1 次条件为假时可以进行第 2 次条件判断从而实现多种条件的判断、选择和处理。

6.3 switch 语句和 break、continue

C++语言中不仅可以使用 if...else if...else 语句进行多重条件判断，还可以直接使用 switch 语句进行多重选择判断处理。switch 语句的基本语法结构如下。

```
switch(integer-expression){
case lable1: statement;
case lable2: statement;
    ...
default: statement;
}
```

为了实现多重条件判断，switch 语句中的判断条件不是布尔表达式，而是整型数据，并通过该数据和 label 进行比较。integer-expression 是一个整型表达式，它与 label 值相等时则执行该标签下的语句；否则，进行下一个标签的比较。在 switch 语句中，当表达式的值与某个 labler 的值相等后，程序会执行之后该 lable 之下的所有语句，因此必须使用 break 语句停止当前的 switch 语句。break 和 continue 语句是控制循环执行的重要语句，两者的差异在于 break 会跳出当前循环或者条件语句，continue 语句会终止当前循环，进入下一次循环。

综合应用循环语句和条件语句能够较好地实现 C++程序中的流程控制。在具体应用中应当选择合适的语句，控制程序的运行。

6.4 复习题

1．请看下面两个计算空格和换行符数目的代码片段。

```
//版本 1
while(cin.getch())   //在遇到 EOF 时退出
{
    if(ch == ' ')
        spaces++;
    if(ch == '\n')
        newlines++
}
//版本 2
while(cin.getch())   //在遇到 EOF 时退出
{
    if(ch == ' ')
        spaces++;
    else if(ch == '\n')
        newlines++
}
```

第 2 个版本比第 1 个版本好在哪里呢?

习题解析：

题目中的两种分支语句都能够正确统计空格符和换行符。两者的差异在于版本 1 使用了两次 if 语句，针对所有输入的字符变量 ch，都要进行两次判断，分别判断是否是空格符和是否是换行符；版本 2 的代码使用 if … else 语句，针对所有输入字符 ch，都首先判断是否是空格符，针对所有非空格符 ch 再次判断是否是换行符。原理上版本 2 的执行效率会略高于版本 1。但是对于现代编译器来说，编译器的优化可能会消除这两种版本之间的差异。对于非大规模的分支语句，可以选择更加清晰易懂的语句来实现。

2．在程序清单 6.2 中，用 ch+1 替换++ch 将发生什么情况呢?

习题解析：

程序清单 6.2 中关于++ch 的代码如下。

```
while(ch != '.')
{
    if(ch == '\n')
        std::cout<< ch;
    else
        std::cout<<++ch;
    std::cin.get(ch);
```

```
}
```

该程序清单的主要功能是将非换行符的 ASCII 字符加 1，例如，若输入的 ch 为字符'A'，则输出的++ch 为'B'。通常情况下 C++语言中的++运算符表示变量的递增，但是++操作符和 ch+1 的区别在于，对于后者，加法运算符会在左右两侧数据参与运算时进行数据类型的转换。ch 为字符型变量，常量 1 为 int 类型常量，因此其运算结果会转换为 int 类型，所以会输出 ASCII 字符的十进制整型数值。而++运算符则不会发生这种情况，它仅保证了原变量的递增。输出仍为字符型数据。

3．请认真考虑下面的程序。

```
#include <iostream>

using namespace std;
int main() {
    char ch;
    int ct1, ct2;
    ct1 = ct2 = 0;
    while((ch = cin.get()) != '$')
    {
        cout << ch;
        ct1++;
        if(ch = '$')
             ct2++;
        cout << ch;
    }
    cout<<"ct1 = "<<ct1<<", ct2 = "<<ct2<<"\n";
    return 0;
}
```

假设输入如下（请在每行末尾按 Enter 键）。

```
Hi!
Send $10 or $20 now!
```

则输出将是什么?

习题解析：

程序的输出结果如下。

```
Hi!
H$i$!$
$Send $10 or $20 now!
S$e$n$d$ $ct1 = 9, ct2 = 9
```

在程序中，if(ch = '$')语句使用了赋值运算符而不是==逻辑运算符，因此会造成对于输入的非'$'运算符都会首先执行 ct1++，随后执行 ch = '$' 赋值语句，由于赋值语句右侧的表达式的值总为真，因此总是执行 ct2++。例如，当输入 `Hi!`时，循环语句块会首先输出 ch（第 1 次输出的 ch 为字符'H'），随后 ct1++，接着 ch 被赋值为'$'，`ct2++`，最后输出 ch（第 2 次输出 ch 时，ch 为字符'$'）。然后进入下一次循环。最终形成 H$i$!$的输出结果，且 ct1 与 ct2 为 3。重要的是，此时还应该注意空白字符在程序内的处理方式，这里主要是空格符和换行符。由于这两个字符在显示上容易被忽略，但是依然会被重新赋值为'$'并显示，因此最终 ct1 和 ct2 同为 9。这一点可以根据回显的'$'字符计算。

4．创建表示下述条件的逻辑表达式。

a．weight 大于或等于 115，但小于 125。

b．ch 为 q 或 Q。

c．x 为偶数，但不是 26。

d．x 为偶数，但不是 26 的倍数。

e．donation 介于 1 000～2 000 或 guest 为 1。

f. ch 是小写字母或大写字母（假设小写字母是依次编码的，大写字母也是依次编码的，但在大小写字母间编码不是连续的）。

习题解析：

```
a. weight >=115 && weight <125
b. ch == 'q'||ch == 'Q'
c. x%2 == 0 && x != 26
d. x%2 ==0 && x%26 != 0
e. (donation>=1000 && donation<2000)[1]|| guest == 1
f. (ch>='a' && ch<='z')||(ch>='A' && ch<='Z')
```

5．在英语中，"I will not not speak（我不会不说）" 的意思与 "I will speak（我要说）" 相同。在 C++中，`!!x` 是否与 x 相同呢?

习题解析：

逻辑运算符 "!" 表示逻辑非，作用是将其右侧的表达式的值取反，因此其运算结果是 true 或者 false。`!!x` 的运算顺序可以表示为`!(!x)`，即先执行右侧的逻辑非运算，再运算左侧的逻辑非运算。当 x 是一个布尔值时，`!!x` 的运算结果等于 x；当 x 是其他类型时，则 x 和`!!x` 在类型上并不相同，运算符会将 x 转换成合适的布尔类型值之后，再进行逻辑运算。因此，并不能简单认为两次取反（!!）就等于原值。

6．创建一个条件表达式，其值为变量的绝对值。也就是说，如果变量 x 为正，则表达式的值为 x；如果 x 为负，则表达式的值为-x——这是一个正值。

习题解析：

首先本题的逻辑规则可以使用 if 条件语句判断，即 x 和 0 的大小关系。如果使用条件表达式，则可以表示如下。

```
(x <0) ? -x : x
```

或者

```
(x>=0) ? x : -x
```

---

① 这里不加括号&&运算符仍可运行，但加上括号后更容易理解。——编者注

7．用 switch 改写下面的代码片段。

```
if(ch == 'A')
        a_grade++;
else if(ch =='B')
        b_grade++;
else if(ch =='C')
        c_grade++;
else if(ch =='D')
        d_grade++;
else
        f_grade++;
```

习题解析：

代码中的多重条件语句依次判断输入数据 ch 分别是'A''B''C''D'的情况，首先字符数据也可以当作整型数据进行比较判断，因此该代码可以改写为 switch 语句。由于 switch 语句在处理多种条件判断上格式更加清晰可读，因此在多重条件语句中应当优先选择使用 switch 语句。本题中 switch 语句末尾应当添加 break 语句。完整代码如下。

```
switch(ch)
{
      case 'A': a_grade++;
              break;
      case 'B': b_grade++;
              break;
      case 'C': c_grade++;
              break;
      case 'D': d_grade++;
              break;
      default: f_grade++;
              break;
}
```

8．对于程序清单 6.10，与使用数字相比，使用字符（如 a 和 c）表示菜单选项和 case 标签有何优点呢?（提示：思考用户输入 q 和输入 5 的情况。）

习题解析：

程序清单 6.10 中相关代码如下。

```
int choice;
cin>>choice;
while(choice != 5){
      switch(choice){
           case 1:
                cout<<"\a\n";
                break;
            ⋮
           default:
```

```
            cout<<"That's not a choice.\n";
        }
        cin>>choice;
    }
}
```

由于多重选择语句中标签需要使用整数常量，因此 choice 使用 int 类型或者 char 类型都可以正确编译和执行这些语句，但是用户的输入可以有很多种情况。当 choice 类型为 int 类型时（标签为整型数据），如果用户输入整型数据（阿拉伯数字 0～9），程序可以正常运行，但是当用户输入字符（A～Z 等英文字母或者其他特殊符号）时，输入读取语句 `cin>>choice;`将会产生读取异常，最终导致程序无法处理 switch 分支语句而崩溃。但是如果 choice 是一个字符型变量，所有的键盘输入都可以以 ASCII 字符的形式正确读取（例如，用户输入 5，既可以正确转换为整数 5，也可以正确转换为字符'5'），从而保证了 switch 分支语句的正确运行。也就是说，使用字符类型的标签能够在一定程度上提高程序的健壮性。

9．请看下面的代码片段。

```
int line = 0;
char ch;
while (cin.get(ch))
{
        if(ch == 'Q')
                break;
        if(ch != '\n')
                continue;
        line++;
}
```

请重写该代码片段，不要使用 break 和 continue 语句。

习题解析：

首先应当对代码的逻辑关系进行分析，代码首先读取系统标准输入。当输入数据为'Q'时，利用 break 语句退出整个循环；如果输入为换行符，则暂停当前循环（即 line 不计数）并开始下一次读取。为了避免使用 break 和 continue，可以在循环入口条件中添加退出循环的条件，在循环体内添件计数的判断条件。修改后的代码如下。

```
int line = 0;
char ch;
while (cin.get(ch) && ch != 'Q')
{
        if(ch == '\n')
                line++;
}
```

6.5　编程练习

1．编写一个程序，读取键盘输入，直到遇到@符号为止，并回显输入（数字除外），同时将大写字符转换为小写，将小写字符转换为大写（别忘了 cctype 函数系列）。

编程分析：

题目要求处理键盘输入的字符数据，分别转换字母大小写、删除（不回显）数字。为了实现这个功能，需要对输入数据应用多重条件语句进行比较判断。由于字母在一定区间内且数量较大，因此使用 switch 语句比较麻烦，推荐使用 if...else 语句。完整代码如下。

```
/*第 6 章的编程练习 1*/
#include <iostream>
#include <cctype>
/*为了使用 toupper()函数，需要添加 cctype 头文件*/
using namespace std;

int main()
{
    char input;
    cout<<"Enter the character: ";
    cin.get(input);
    while(input != '@')
    {
        /*循环入口条件是输入字符不等于 @*/
        if(isdigit(input))
        {
            /*输入数据是数字时的处理方法*/
            cin.get(input);
            continue;
        }else if(islower(input))
        {
            input = toupper(input);
            /*小写字母的处理方法*/
        }else if(isupper(input))
        {
            input = tolower(input);
            /*大写字母的处理方法*/
        }
        /*通过多重条件语句处理输入数据*/
        cout<<input;
        cin.get(input);
    }
    return 0;
}
```

2．编写一个程序，最多将 10 个 donation 值读入一个 double 数组中（也可使用模板类 array）。在程序遇到非数字输入时将结束输入，并报告这些数字的平均值以及数组中有多少个数字大于平均值。

编程分析：

在数组形式的数据结构中，计算平均数以及高于平均值的元素数量的基本方法是首先计算平均数，随后再通过循环依次比较每一个元素值与平均值的大小关系。由于题目要求最多输入 10 个元素，因此需要单独记录元素个数，但数组长度可以使用固定长度 10。完整代码如下。

```
/*第 6 章的编程练习 2*/
#include <iostream>
#include <array>
using namespace std;

int main()
{
    array<double ,10> donation;
    /*使用 array 模板定义数组，长度为 10*/
    double input;
    int counter = 0;
    double average, sum = 0;
    int bigger = 0;
    /*sum、average、counter、bigger 分别记录和、平均值、元素数与大于平均值的元素个数*/
    cout<<"Enter the double numerial: ";
    cin>>input;
    while(input != 0 && counter<10)
    {
        donation[counter++] = input;
        cout<<"No."<<counter<<" Data input to Array."<<endl;
        cout<<"Enter the double numerial: ";
        cin>>input;
    }
    /*通过 while 循环输入数据，当输入非数字时或超过 10 个元素时退出循环*/
    for(int i = 0;i < counter; i++)
    {
        sum +=  donation[i];
    }
    average = sum / counter;
    /*求和并计算平均值*/
    for(int i = 0;i < counter; i++)
    {
        if(donation[i] > average)
            bigger++;
    }
    /*通过遍历计算大于平均值的元素个数*/
    cout<<"The Average is "<<average<<" and "<<bigger;
    cout<<" data bigger than average."<<endl;
    return 0;
}
```

3．编写一个菜单驱动程序的雏形。该程序显示一个提供 4 个选项的菜单——每个选项用一个字母标记。如果用户使用有效选项之外的字母进行响应，程序将提示用户输入一个有效的字母，直到用户这样做为止。然后，该程序使用一条 switch 语句，根据用户的选择执行一个简单操作。该程序的运行情况如下。

```
Please enter one of the following choices:
c) carnivore          p) pianist
t) tree               g) game
f
Please enter a c, p, t, or g: q
Please enter a c, p, t, or g: t
A maple is a tree.
```

编程分析：

题目要求实现文本格式的菜单，基本功能是对用户输入的不同字符，执行相应的功能模块。这是计算机文本形式下一种常用的用户交互模式。这类程序的主要难点在于对用户输入字符的读取和判断，考虑到用户输入的字符（菜单）能够直接与字符类型相对应，因此可以采用 switch 语句。此外，在处理数据输入时，还要考虑对非菜单数据的检测和错误信息的反馈。题目并未要求菜单的反复选择和使用，如果有这方面的需求，即在用户选择输入后，会退回主菜单。可以设计菜单显示函数，专门负责主菜单的显示功能，这样可以重复利用代码，并简化程序。完整代码如下。

```
/*第 6 章的编程练习 3*/
#include <iostream>
using namespace std;

void showmenu();
/* showmenu 函数的声明*/
int main()
{
    char choice;
    showmenu();
    cin.get(choice);
    /*显示菜单，并读取用户输入，保存至 choice 变量中*/
    while(choice != 'c' && choice != 'p'&& choice != 't'&& choice != 'g' )
    {
        cin.get();
        cout<<"Please enter a c, p, t, or g: ";
        cin.get(choice);
    }
    /*判断用户的输入是否符合菜单选项，如果不符合，则要求下一次输入*/
    switch(choice)
    {
        case 'c':
            break;
        case 'p':
```

```
            break;
        case 't':
            cout<<"A maple is a tree.";
            break;
        case 'g':
            break;

    }
    /*针对输入的菜单做出多重选择和反馈*/
    return 0;
}

void showmenu()
{
    cout<<"Please enter one of the following choices:\n";
    cout<<"c) carnivore\t\t\tp) pianist\n";
    cout<<"t) tree\t\t\t\tg) game\n";
}
/* showmenu 函数只负责菜单信息的输出*/
```

4．加入 Benevolent Order of Programmer 后，在 BOP 大会上，人们便可以通过加入者的真实姓名、头衔或秘密 BOP 姓名来了解他（她）。请编写一个程序，可以使用真实姓名、头衔、秘密姓名或成员偏好来列出成员。在编写该程序时，请使用下面的结构体。

```
//Benevolent Order of Programmer 姓名结构体
struct bop{
        char fullname[srtsize];//真实姓名
        char title[strsize];   //头衔
        char bopname[strsize]; //秘密 BOP 姓名
        int preference;        //0 = fullname,1 = title, 2 = bopname
};
```

该程序创建一个由上述结构体组成的小型数组，并将其初始化为适当的值。另外，该程序使用一个循环，让用户在下面的选项中进行选择。

a．display by name　　　　b．display by title
c．display by bopname　　　d．display by preference
q．quit

注意，“display by preference”并不意味着显示成员的偏好，而是意味着根据成员的偏好来列出成员。 例如，如果偏好号为 1，则选择 d 将显示程序员的头衔。该程序的运行情况如下。

```
a. display by name      b. display by title
c. display by bopname   d. display by preference
q. quit
a
Wimp Macho
Raki Rhodes
Celia Laiter
Hoppy Hipman
```

```
Pat Hand
Next choice:d
Wimp Macho
Junior Programmer
MIPS
Analyst Trainee
LOOPY
Next choice:q
Bye!
```

编程分析：

题目首先需要定义结构体数组，并通过菜单形式和用户进行交互性输入，因此首先结合编程练习 3，创建显示菜单。当用户选择菜单进行输出后，需要在 switch 语句的标签下进行数组的显示，为了维持主程序和 switch 语句的简短性与可读性，可以将输出功能定义成函数，在 switch 语句内通过函数调用来实现对应的功能。

```
/*第 6 章的编程练习 4*/
#include <iostream>
#include <cstring>
using namespace std;

const int strsize = 40;
const int usersize = 40;
//Benevolent Order of Programmer 姓名结构体
struct bop{
    char fullname[strsize];    //真实姓名
    char title[strsize];       //头衔
    char bopname[strsize];     //秘密 BOP 姓名
    int preference;            //0 = fullname,1 = title, 2 = bopname
};
bop bop_user[usersize] =
        {{"Wimp Macho","Programmer","MIPS",0},
         {"Raki Rhodes","Junior Programmer","",1},
         {"Celia Laiter","","MIPS",2},
         {"Hoppy Hipman","Analyst Trainee","",1},
         {"Pat Hand","","LOOPY",2},};
/*定义常量，定义结构体，初始化 bop 数组信息*/
void showmenu();
void print_by_name();
void print_by_pref();
void print_by_title();
void print_by_bopname();
void create_info();
/*为了保持主程序清晰，将部分功能代码定义成函数形式*/
int main()
{
    char choice;
   //create_info();
   /*此处调用 create_info()函数自定义创建数组的信息*/
```

```
    showmenu();
    cin.get(choice);
    /*显示菜单，读取用户的输入*/
    while(choice != 'q')
    {
        switch(choice)
        {
            case 'a':
                print_by_name();
                break;
            case 'b':
                print_by_title();
                break;
            case 'c':
                print_by_bopname();
                break;
            case 'd':
                print_by_pref();
                break;
            default:
                cout<<"Please enter character a, b, c, d, or q: ";
        }
        /*通过 switch 语句对应的函数进行显示*/
        cin.get();
        cout<<"Next choice:";
        cin.get(choice);
    }
    /*将 switch 语句放置在 while 循环内，可以反复选择各项功能*/
    cout<<"Bye!"<<endl;
    return 0;
}

void showmenu()
{
    cout<<"a. display by name \t\tb. display by title\n";
    cout<<"c. display by bopname\t\td. display by preference\n";
    cout<<"q. quit\n";
}
/*显示菜单*/
void print_by_name()
{
    for(int i = 0; i < usersize; i++){
        if(strlen(bop_user[i].fullname) == 0)
            break;
        else
            cout<<bop_user[i].fullname<<endl;
    }
}
/*通过循环，按名字输出信息*/
void print_by_pref()
```

```
{
    for(int i = 0; i < usersize; i++)
    {
        if(strlen(bop_user[i].fullname) == 0)
            break;
        else{
            switch(bop_user[i].preference)
            {
                case 0 :
                    cout<<bop_user[i].fullname<<endl;
                    break;
                case 1:
                    cout<<bop_user[i].title<<endl;
                    break;
                case 2:
                    cout<<bop_user[i].bopname<<endl;
                    break;
            }
        }
    }
}
void print_by_title()
{
    for(int i = 0; i < usersize; i++)
    {
        if(strlen(bop_user[i].fullname) == 0)
            break;
        else
            cout<<bop_user[i].title<<endl;
    }
}
/*按头衔输出数组信息*/
void print_by_bopname()
{
    for(int i = 0; i < usersize; i++)
    {
        if(strlen(bop_user[i].fullname) == 0)
            break;
        else
            cout<<bop_user[i].bopname<<endl;
    }
}
/*按 BOP 姓名输出数组信息*/
void create_info()
{
    for(int i = 0; i < usersize; i++)
    {
        cout<<"Enter the user's full name: ";
        cin.getline(bop_user[i].fullname, strsize);
        cout<<"Enter the user's title: ";
```

```
        cin.getline(bop_user[i].title, strsize);
        cout<<"Enter the user's bopname: ";
        cin.getline(bop_user[i].bopname, strsize);
        cout<<"Enter the user's preference: ";
        cin>>bop_user[i].preference;
        cout<<"Next...(f for finished):";
        cin.get();
        if(cin.get() == 'f') break;
    }
}
/*向数组添加 BOP 成员信息*/
```

5．在 Neutronia 王国，货币单位是 tvarp，收入所得税的计算方式如下。

- 收入不超过 5 000 tvarp：不纳税。
- 收入介于 5 001～15 000 tvarp：税率为 10%。
- 收入介于 1 5001～35 000 tvarp：税率为 15%。
- 收入在 35 000 tvarp 以上：税率为 20%。

例如，当收入为 38 000 tvarp 时，所得税为 5 000 × 0.00 + 10 000 × 0.10 + 20 000 × 0.15 + 3 000 × 0.20，即 4 600tvarp。请编写一个程序，使用循环来要求用户输入工资，并报告所得税。当用户输入负数或非数字时，循环将结束。

编程分析：

按照数据的区间进行数据处理是多重选择的常见应用场景，但是这里需要实现的是一个较大数据区间的判断，而不是整型数据的多重选择的判断，因此并不适合使用 switch 的多重选择语句，而需要通过 if...else 语句实现。多重判断内的条件语句还有一些值得注意的地方，可以参考程序注释。完整代码如下。

```
/*第 6 章的编程练习 5*/
#include <iostream>

using namespace std;
int main()
{
    float tax, salary = 0.0;
    cout<<"Hello, enter your salary to calculate tax:";
    cin>>salary;
    /*读取用户输入的工资*/
    while(salary > 0)
    {
        if(salary <= 5000)
        {
            tax = 0;
        }else if(salary <= 15000)
        {
            tax = (salary - 5000)*0.10;
        }else if(salary <= 35000)
```

```
        {
            tax = 10000*0.10 + (35000-15000)*0.15;
        }else if(salary > 35000)
        {
            tax = 10000*0.10 + 20000*0.15 + (salary - 35000)*0.20;
        }
        /*通过多重选择进行判断，这里需要注意的是条件表达式 并未使用与、或、非逻辑运算符
         *例如 salary <= 35000 的条件，当能够进行改条件语句判断时，salary 必然已经大于 15000
         **/
        cout<<"Your salary is "<<salary<<" tvarps, and you should pay ";
        cout<<tax<<" tvarps."<<endl;
        cout<<"enter your salary to calculate tax:";
        cin>>salary;
    }
    cout<<"Bye!"<<endl;
    return 0;
}
```

6．编写一个程序，记录捐助给“维护合法权利团体”的资金。该程序首先要求用户输入捐献者数目，然后要求用户输入每一个捐献者的姓名和款项。这些信息存储在一个动态分配的结构数组中。每个结构有两个成员——用来存储姓名的字符数组（或 string 对象）和用来存储款项的 double 成员。读取所有的数据后，程序将显示所有捐款超过 10 000 的捐款者的姓名及其捐款金额。该列表前应包含一个标题，指出下面的捐款者是重要捐款人（Grand Patrons）。然后，程序将列出其他的捐款者，该列表要以 Patrons 开头。如果某种类别没有捐款者，则程序将输出单词“NONE”。该程序只显示这两种类别，而不进行排序。

编程分析：

题目要求显示一个捐款人名单，因此首先需要定义一个结构体，再通过结构体定义一个动态数组。程序通过标准输入读取捐款人的基本信息（姓名和捐款金额），最后按照捐款金额分别显示重要捐款人和一般捐款人。整体来看，程序需要通过多个循环来进行数据的输入、判断和显示。在显示信息的循环内需要通过 if 条件语句判断捐款人属于哪一个类别。完整代码如下。

```
/*第 6 章的编程练习 6*/
#include <iostream>
#include <string>
using namespace std;

struct patrons{
    string full_name;
    double fund;
};
/*定义表示捐款人基本信息的结构体*/
int main()
```

```
{
    int patrons_number;
    patrons*ppatrons;
    cout<<"How many patrons? ";
    cin>>patrons_number;
    cin.get();
    /*读取捐款人名单的长度，多用一个 get()的函数目的是删除缓冲区中的换行符*/
    ppatrons = new patrons[patrons_number];
    /*建立动态数组*/
    int id = 0;
    bool empty = true;
    cout<<"Starting to input patrons' info:"<<endl;
    while(id < patrons_number)
    {
        cout<<"Enter the full name of patrons: ";
        getline(cin, ppatrons[id].full_name);
        cout<<"Enter the fund of "<<ppatrons[id].full_name<<" :";
        cin>>ppatrons[id].fund;
        cin.get();
        id++;
        cout<<"Continue to input, or press (f) to finished.";
        if(cin.get() == 'f') break;
    }
    /*建立捐款人名单*/
    cout<<"Grand Patrons"<<endl;
    for(int i = 0; i < patrons_number; i++)
    {
        if(ppatrons[i].fund >= 1000){
            cout<<ppatrons[i].full_name<<" : "<<ppatrons[i].fund<<endl;
            empty = false;
        }
    }
    /*查询 Grand Patrons 名单，如果 empty 为 true，输出 NONE*/
    if(empty) cout<<"NONE"<<endl;
    empty = false;
    cout<<"Patrons"<<endl;
    for(int i = 0; i < patrons_number; i++)
    {
        if(ppatrons[i].fund < 1000){
            cout<<ppatrons[i].full_name<<" : "<<ppatrons[i].fund<<endl;
        }
    }
    /*查询 Patrons 名单，如果 empty 为 true，输出 NONE*/
    if(empty) cout<<"NONE"<<endl;
    return 0;
}
```

7. 编写一个程序，它每次读取一个单词，直到用户只输入 q。然后，该程序指出有多少个单词以元音开头，有多少个单词以辅音开头，还有多少个单词不属于这两类。为此，方法之一是，使用 isalpha()来区分以字母和其他字符开头的单词，然后对于通过了 isalpha()测试的单词，使用 if 或 switch 语句来确定哪些单词以元音开头。该程序的运行情况如下。

```
Enter words (q to quit):
The 12 awesome oxen ambled
quietly across 15 meters of lawn. q
5 words beginning with vowels
4 words beginning with consonants
2 others
```

编程分析：

对于给定单词的基本信息分析，可以使用 if 语句或者 switch 语句。对于单词的字母分析，可以使用单个字符数据的比较判断，且元音字母较少，因此本题可以优先考虑使用 switch 语句进行分析。switch 语句中所有元音字母可以使用合并标签的方式进行分析。完整代码如下。

```
/*第 6 章的编程练习 7*/
#include <iostream>
#include <cctype>

using namespace std;

int main()
{
    char words[40];
    int vowel, consonant, others;
    vowel = consonant = others = 0;
    cout<<"Enter words (q to quit):"<<endl;
    cin>>words;
    /*设置变量，使用字符数组来实现单词输入*/
    while(strcmp(words,"q") != 0)
    {
        if(!isalpha(words[0]))
        {
            others ++;
        }
        /*统计以非字母开头的单词*/
        else{
            switch(words[0])
            {
                case 'a':
                case 'e':
                case 'i':
                case 'o':
                case 'u':
                    vowel++;
                    /*统计以元音字母开头的单词*/
                    break;
```

```
            default:
                consonant++;
                /*统计以非元音字母开头的单词*/
        }
      }
      cin>>words;
    }
    cout<<vowel<<" words beginning with vowels"<<endl;
    cout<<consonant<<" words beginning with consonants"<<endl;
    cout<<others<<" others"<<endl;
    return 0;
}
```

8．编写一个程序，它打开一个文件，逐个字符读取该文件，直到到达文件末尾，然后指出该文件中包含多少个字符。

编程分析：

题目要求从文件读取字符，并统计文件中的字符个数。和标准输入不同，文件输入需要从文件流读取数据，主要代码如下。

```
ifstream fin;//定义文件流对象
fin.open("file_name");//打开文件
```

随后就可以使用 `fin>>ch;`的形式读取文件内的字符，在使用时可以把 fin 与 cin 进行简单对比。字符计数等部分较简单，完整代码如下。

```
/*第 6 章的编程练习 8*/
#include <iostream>
#include <fstream>
using namespace std;

int main()
{
    ifstream fin;
    string file_name;
    cout<<"Enter the file name: ";
    getline(cin, file_name);
    /*等待用户输入文件名*/
    fin.open(file_name);
    /*通过文件流对象打开文件*/
    if(!fin.is_open()){
        cout<<"Error to open file."<<endl;
        exit(EXIT_FAILURE);
    }
    /*如果在打开文件时出现错误，则终止程序*/
    char read_char;
    int char_counter = 0;
    while(!fin.eof()){
        fin>>read_char;
        char_counter++;
```

```
    }
    /*通过 eof()函数判断是否到达文件末尾*/
    cout<<"The file "<<file_name<<" contains "<<char_counter<<" characters."<<endl;
    fin.close();
    /*关闭文件*/
    return 0;
}
```

9．重做编程练习 6，但从文件中读取所需的信息。该文件的第 1 项应为捐款人数，余下的内容应为成对的行。在每一对中，第 1 行为捐款人姓名，第 2 行为捐款数额，即该文件形式如下。

```
4
Sam Stone
2000
Freida Flass
100500
Tammy Tubbs
5000
Rich RTaptor
5500
```

编程分析：

本题在编程练习 6 的基础上将捐款人信息存储在文件内。根据文件样例可知，文件第 1 行为捐款人数——4。从第 2 行开始是姓名和捐款金额的信息，共 4 人，8 条信息。在文件数据的读取过程中，循环次数应当以第 1 行数据为参考，在循环内读取两行数据，分别是一行字符串、一行 double 型数据。完整代码如下。

```
/*第 6 章的编程练习 9*/
#include <iostream>
#include <fstream>
#include <string>
using namespace std;

struct patrons{
    string full_name;
    double fund;
};
/*表示捐款人信息的结构体*/
int main()
{
    ifstream fin;
    string file_name;
    cout<<"Enter the file name: ";
    getline(cin, file_name);
    fin.open(file_name);
    if(!fin.is_open())
    {
        cout<<"Error to open file."<<endl;
        exit(EXIT_FAILURE);
```

```
    }
    /*定义文件对象，打开文件*/
    int patrons_number;
    patrons*ppatrons;
    int id = 0;
    bool empty = true;

    fin>>patrons_number;
    if(patrons_number <= 0)
    {
        exit(EXIT_FAILURE);
    }
    ppatrons = new patrons[patrons_number];
    fin.get();
    /*读取人数，创建动态数组*/
    while(!fin.eof() && id < patrons_number)
    {
        getline(fin,ppatrons[id].full_name);
        cout<<"Read Name: "<<ppatrons[id].full_name<<endl;
        fin>>ppatrons[id].fund;
        cout<<"Read fund: "<<ppatrons[id].fund<<endl;
        fin.get();
        id++;
    }
    /*循环读取捐款人信息，也可以使用 for 循环*/
    fin.close();
    /*关闭文件*/
    cout<<"Grand Patrons"<<endl;
    for(int i = 0; i < patrons_number; i++)
    {
        if(ppatrons[i].fund >= 10000){
            cout<<ppatrons[i].full_name<<" : "<<ppatrons[i].fund<<endl;
            empty = false;
        }
    }
    if(empty) cout<<"NONE"<<endl;
    empty = false;
    cout<<"Patrons"<<endl;
    for(int i = 0; i < patrons_number; i++)
    {
        if(ppatrons[i].fund < 10000)
        {
            cout<<ppatrons[i].full_name<<" : "<<ppatrons[i].fund<<endl;
        }
    }
    if(empty) cout<<"NONE"<<endl;
    return 0;
}
```

第 7 章 函数——C++的编程模块

本章知识点总结

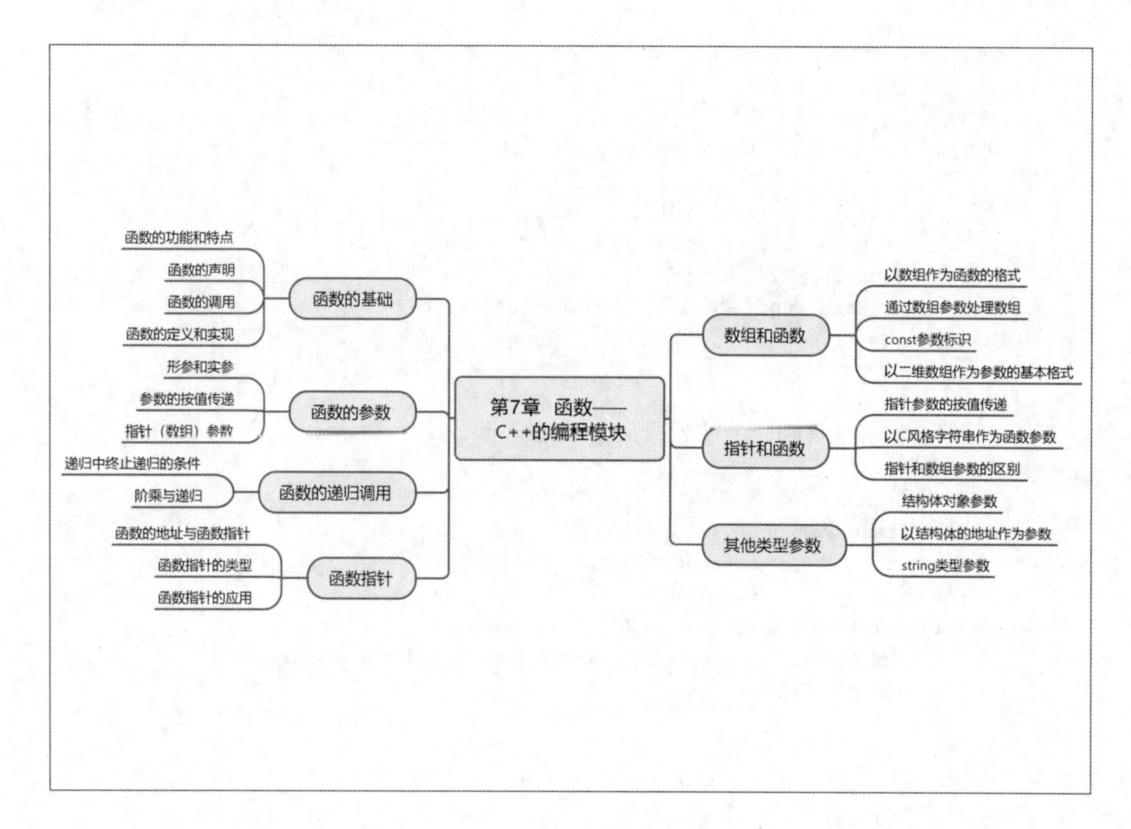

7.1 函数的原型和定义

C++语言中函数是一种可重用的标准化模块，程序设计中可以通过对函数的调用，反复使用函数模块的基本功能。C++语言中的函数有系统预定义的函数和第三方库中的函数，例如对于常用的字符串处理函数 strcpy()、strcmp()等，系统预先提供了函数的功能代码模块，用户通过对函数的调用来重复应用这些功能模块。第三方函数库通常是 C++语言标准之外的公司或组织提供的功能性函数库。程序员在程序设计中也可以自行定义一个函数，并通过调用来重复使用这个功能模块。

函数的定义中首先应当声明函数的原型，函数的原型是标识函数基本信息的一条语句，也称为函数的声明。函数的原型中需要提供函数的名称、返回值类型、参数个数和参数类型，可以不提供参数的变量名，并且语句以分号结束。C++语言要求程序设计中必须提供函数的原型，以确保编译器在源代码的编译阶段能够检查被调函数的参数和返回值是否正确，从而保证多文件编译过程中语法检查的正确性并提高编译效率。

函数的定义是指整个函数的功能代码的实现。函数定义中的参数称作函数形参。函数的调用是指在程序中通过函数名和提供函数的实参调用该函数的功能并利用其返回值的C++语句。

函数的调用过程可以简单理解为，主调函数会在调用函数时首先暂停当前的代码运行，等待被调函数的功能模块运行完；被调函数运行过程中会将返回值复制到计算机指定的存储单元内，并结束被调函数的运行；主调函数在被调函数返回后会从指定存储单元获取被调函数的返回值，随后继续执行被暂停的代码。

7.2　函数调用中的按值传递

C++语言中函数调用过程中使用的参数称为实参。函数调用的实参类型应当与函数定义中的形参在数量和类型上匹配。函数的调用过程是将实参的值通过副本的形式传递给形参，被调函数在运行中会得到一个新的临时变量，其值与实参相同。在程序运行过程中，函数体内语句中参与运算的变量是函数内的临时变量。完成函数调用后，函数内使用的临时变量都会被系统回收，返回值会存储在指定单元并通过返回语句传递给主调函数。函数的按值传递能够有效保证主调函数的数据安全。因此，C++语言中函数的参数传递也称为按值传递。

7.3　以数组和指针作为函数的参数

数组可以作为函数的参数，函数的定义中通常需要使用两个参数来表示一个完整的数组参数。函数声明中一般把数组的类型和数组名放在前面，把数组的长度放在后面。C++中由于数组名表示数组第 1 个元素的地址，按照前面所说的参数的按值传递，函数获得了该数组的地址的副本，但是该副本仍然是指向原数组的地址。也就是说，值传递只完成了数组地址的复制、传递，并没有把整个数组内容复制和传递一遍，这样在函数内通过值传递获得的数组地址可以直接操作原数组。因此在实际应用中使用数据进行函数参数传递时，需要明确函数内部可以修改数组的内容。当我们需要确保该数组不被修改时，函数的声明中应当使用 const 关键字。例如：

```
void show_array(const double ar[], int size);
```

对于 int 类型的简单数据变量 size，由于按值传递参数，因此不需要使用 const。数组的参数传递同样适用于指针，实际应用中可以将数组和指针在参数传递中等价使用。值传递的特性同样会涉及指针指向单元的内容是否需要保护以免被修改的问题，一般建议尽可能使用 const。

7.4 字符串、二维数组和函数

C++语言中要表示字符串，可以使用 char 类型的数组或者指向 char 类型的指针。此外，双引号引起的字符串字面值作为一种常量也经常用于函数的调用。本质上说，把字符串作为参数传递，同样传递的是字符串的首地址，即 char*类型的指针。此外，C++中也可以使用 string 对象作为函数的参数，其基本用法和 C 风格字符串基本类似。二维数组在参数传递中可以使用特定的数组形式传递，格式如下。

```
int sum(int ar[][4], int size);
```

由于二维数组可以看作一个由行和列组成的数据单元格，因此可以通过参数 size 传递主数组的元素数量，列数必须通过数值标识。此外，由于二维数组的特性，也可以通过指针来表示该数组，但是这过于麻烦和含混，因此推荐尽量使用数组表示法。

7.5 参数传递中的结构体

结构体作为函数的参数可以通过两种方式进行传递。一是使用结构体的对象，这样可以直接按值把该结构的副本传递给函数并参与运算。这种形式能够确保主调函数内数据的安全，但是由于创建结构体副本需要消耗较多资源，为了提高效率可以使用第 2 种方式，即传递结构体的指针并根据需要使用 const 关键字，以指针形式传递结构体。为了使用地址来传递结构体，需要使用“&”地址运算符获得结构体的地址。当使用指针形式表示结构体内的成员需要时，使用“->”间接成员运算符，而不是使用结构体对象时使用的成员运算符“.”。

7.6 递归函数与函数指针

函数的递归调用是指在一个函数的函数体内部调用了它本身。因此递归调用中必须明确递归调用的终止条件，并能够满足终止条件，否则函数将会陷入无限的循环调用当中。递归函数的这种特性使它在很多领域有非常重要的应用，例如，各种数学应用、算法设计等。

C++语言中的函数也有自己的地址，函数的地址是存储机器代码的内存的起始地址。因此，与其他数据类似，程序设计中可以定义一个指针指向一个函数的机器代码的起始地址，这样的指针就称为函数指针。此外，也可以将一个函数的指针作为参数传递到另外一个函数当中，这样可以通过函数的调用参数运行不同的函数。函数指针的声明中需要注意的是表示函数的圆括号，例如：

```
double pam(int);
double (*pf)(int);
pf = pam;
```

第 1 行声明了一个函数 pam。第 2 行声明了一个指向函数的指针，指向的函数的返回值为 double 类型，参数为圆括号内的 int，函数指针名为 pf。第 3 行代码为指向函数的指针和函数建立指向关系，因此可以通过 pf 调用函数 pam()。C++语言可以使用两种形式调用函数指针，一种是直接使用指针名 pf(5)，另一种是使用间接运算符(*pf)(5)。

7.7　复习题

1．使用函数的 3 个步骤是什么?

习题解析：

函数为程序设计中的编程模块，可以通过对函数的调用实现函数的代码重用功能。通常系统预定义函数可以在添加对应预编译指令的头文件后直接调用。用户自定义的函数需要首先声明函数原型、定义函数并实现函数的功能性代码，然后才能调用函数。

2．请创建与下面的描述匹配的函数原型。

a．igor()没有参数，且没有返回值。

b．tofu()接受一个 int 参数，并返回一个 float 值。

c．mpg()接受两个 double 参数，并返回一个 double 值。

d．summation()以 long 数组名和数组长度作为参数，并返回一个 long 值。

e．doctor()接受一个字符串参数（不能修改该字符串），并返回一个 double 值。

f．ofcourse()以 boss 结构作为参数，不返回值。

g. plot()以 map 结构体的指针作为参数，并返回一个字符串。

习题解析：

函数的原型声明主要标识一个函数的基本信息，从而提高编译器的工作效率。函数原型需要提供函数名、返回值类型、参数数量以及参数类型，其中无返回值可以标识为 void，无参数可以使参数列表空白，或者标注为 void。

a．`void igor();` 或者 `void igor(void);`

b．`float tofu(int n);` 或者 `float tofu(int)`

c．`double mpg(double d1, double d2);`或者 `double mpg(double double)`

d．`long summation(long arr[], int size);`

e．`double doctor(const char*str);`

f．`void ofcourse(boss bs);`

g．`char*plot(map*pm );`

3．编写一个接受 3 个参数的函数，3 个参数分别为 int 数组名、数组长度和一个 int 值，并将数组的所有元素设置为该 int 值。

习题解析：

当以数组作为函数的参数时，需要将数组的类型和数组的元素个数作为两个参数分别表示，通常数组类型在前，数组长度参数在后并且使用整型变量。此外，本题函数不需要返回值，因此代码如下。

```
void set_value(int arr[], int size, int value){
        for(int i = 0;i<size;i++)
                arr[i]=value;
}
```

4．编写一个接受 3 个参数的函数，3 个参数分别为指向数组区间中第 1 个元素的指针、指向数组区间中最后一个元素后面的指针以及一个 int 值，并将数组中的每个元素都设置为该 int 值。

习题解析：

在处理数组数据时，函数可以直接使用数组作为参数，也可以使用指针作为参数来表示相关数据。为了使用指针来表示数组区间，需要使用数组的头指针和数组的尾指针两个参数，这样才能在函数体内正确查找和表示数组的元素。此外，函数不需要返回值，因此本题的代码如下。

```
void set_value(int* begin; int* end; int value){
       for(int*p = begin; p!=end;p++)
       *p = value;
}
```

5．编写以 double 类型的数组名和数组长度作为参数并返回该数组中最大值的函数。该函数不应修改数组的内容。

习题解析：

将本题和复习题 3 对比，可以发现本题的数组元素在函数体内不需要修改，而复习题 3 需要对数组进行赋值操作，因此对于本题中的参数需要添加 const 关键字，这样才能更好地维护函数代码的功能和运行效率，此外，函数返回值应当为 double 类型。完整代码如下。

```
double max(const double arr[], int size){
       int max_value = arr[0];
       for(int i = 0; i< size; i++){
               if(max_value<arr[i])
               max_value = arr[i]
      }
}
```

6．为什么不对类型为基本类型的函数参数使用 const 限定符？

习题解析：

C++语言中的参数传递是以值传递的形式实现的，即函数的实参表示以值的副本的形式传递给函数的形参，因此，在实参传递过程中主调函数的实参并不会被修改。但是以指针形式传递或者其他复合数据类型，则可以通过地址的指向间接访问和存取主调函数中的数据，因此需要使用 const 关键字进行数据的保护。

7．C++程序可使用哪 3 种 C 风格字符串?

习题解析：

C++程序中使用的 C 风格字符串主要有 3 种形式，其中最常用的是字符数组的形式，其次是以双引号标识的字符串常量形式，最后也可以使用指向字符串首字符的指针的形式来表示。其中使用字符数组形式表示和使用指针形式表示的字符串在程序中类似，使用双引号表示的是常量形式的字符串，程序中不能修改字符串的数据。

8．编写一个函数，其原型如下。

```
int replace(char* str, char c1, char c2);
```

该函数将字符串中所有的 c1 都替换为 c2，并返回替换次数。

习题解析：

本题的函数有 3 个参数，分别是字符串的指针以及表示查找和替换的两个字符数据。题目要求将字符串内的数据进行替换，基本算法就是从头开始遍历整个字符串，进行字符比对。其中循环代码的判断条件为是否到了字符串的末尾，因此使用 while(*str)或者while(*str != '\0')表示均可。完整代码如下。

```
int replace(char* str, char c1,char c2){
        int count;
        while(*str){
                if(*str == c1){
                        *str = c2
                        count++;
                }
                str++;
        }
        return count;
}
```

9．表达式*"pizza"的含义是什么？"taco" [2]呢？

习题解析：

C++语言中使用双引号标识字符串常量，字符串常量使用其第 1 个元素的地址进行表示，C++中通过这个首元素的地址实现了字符串常量的存储位置的查询，因此通过*运算符，*"pizza"能够直接找到字符'p'。同理，"taco"[2]则表示通过字符串中首字符的地址查找到该字符串的第 3 个元素'c'。

10．C++允许按值传递结构体，也允许传递结构体的地址。如果 glitz 是一个结构体变量，如何按值传递它？如何传递它的地址？这两种方法有何利弊？

习题解析：

和其他基本数据类型类似，C++中的值传递只需要将变量直接传递给函数即可，若使用地址传递，对于变量需要使用&取地址，因此按值传递使用结构体名 glitz 即可。传递地址使用地址运算符，传递形式为&glitz。按值传递通过数据副本形式保护原始数据，但效率低下；按地址传递可节省时间和内存，但不能保护原始数据，除非对函数参数使用了 const 限定符。

11．函数 judge( )的返回类型为 int，它将以一个函数的地址作为参数，以 const char 指针作为参数，并返回一个 int 值。请编写 judge( )函数的原型。

习题解析：

judge()函数的参数是一个函数指针，函数指针通常可以理解为一个指向函数代码起始位置的地址，通过这个函数的指针可以调用函数。函数指针的声明中需要注意的是表示函数的圆括号和函数的返回值、函数的参数等。本题中需要一个函数指针，返回 int 类型，参数是 const char*，因此函数指针可以表示为 int (*pf)(const char*)，其中 pf 表示函数指针变量名，因此 judge()函数的定义如下。

```
int judge(int (*pf)(const char*))
```

12．假设有如下结构体声明。

```
struct application{
        char name[30];
        int credit_ratings[3]
};
```

a．编写一个函数，它以 application 结构体作为参数，并显示该结构体的内容。

b．编写一个函数，它以 application 结构体的地址作为参数，并显示该参数指向的结构体的内容。

习题解析：

以结构体作为参数和以结构体的地址作为参数，在实现上最大的差异是前一种方案中直接将变量作为参数，通过值传递，函数内不会修改实参的内容，但是通过地址传递参数，函数内可以修改实参的内容，因此必要时要使用 const 关键字。此外，指针和变量形式的成员运算符略有不同。代码如下。

```
void display_struct(application app){
        cout<<app.name<<endl;
        for(int i = 0;i<3;i++)
               cout<<app.credict_ratings[i]<<endl;
}
```

```
void display_struct(const application* app){
        cout<<app->name<<endl;
        for(int i = 0;i<3;i++)
                cout<<app->credict_ratings[i]<<endl;
}
```

13．假设函数 f1()和 f2()的原型如下。

```
void f1(applicant *a);
const char *f2(const applicant *a1, const applicant *a2);
```

请将 p1 与 p2 分别声明为指向 f1 和 f2 的指针；将 ap 声明为一个数组，它包含 5 个类型与 p1 相同的指针；将 pa 声明为一个指针，它指向的数组包含 10 个类型与 p2 相同的指针。使用 typedef 来帮助完成这项工作。

习题解析：

函数指针首先是一种特殊的指针变量，它的定义只描述了它指向的函数的类型（参数、返回值），对比其他类型的指针，例如，int*p；定义了指针 p 是指向整型数据的指针，该指针具体指向哪一个整型变量还需要通过赋值确定。例如，以下代码才能使 p 指向变量 i 所表示的整型数据。

```
int i = 0;
int *p;
p = &i;
```

本题中的函数指针也是这样，需要首先定义函数指针的指向类型，然后将 f1()、f2()函数与两个指针建立联系。完整代码如下。

```
void f1(applicant *a);
const char *f2(const applicant *a1, const applicant *a2);
//声明了两个函数 f1()、f2()
trpedef void (*p_f1)(applicant*);
//通过 typedef 简化代码
p_f1 p1 = f1;
//p1 是 p_f1 类型的函数指针，现在指向 f1
trpedef const char*(*p_f2)(const applicant*, const applicant*);
p_f2 p2 = f2;
//p2 是 p_f2 类型的函数指针，现在指向 f2
p_f1 ap[5];
//ap 是一个由函数指针组成的数组
p_f2 (*pa)[10];
//pa 是一个指针，它指向一个数组，该数组由 10 个函数指针组成
```

7.8 编程练习

1．编写一个程序，不断要求用户输入两个数，直到其中的一个为 0。对于每两个数，程序将使用一个函数来计算它们的调和平均数，并将结果返回给 main( )，而后者将报告结果。调和平均数指的是倒数平均值的倒数，计算公式如下。

调和平均数=2.0 * x* y / (x + y)

编程分析：

题目要求设计一个计算调和平均数的函数。由题目要求可以判断函数的参数是两个 double 类型数据，函数计算通过参数输入的两个非零数的调和平均数，并返回调和平均数的计算结果，即一个 double 类型数据。函数模块由计算调和平均数的公式实现。完整代码如下。

```
/*第 7 章的编程练习 1*/
#include <iostream>
using namespace std;

double harmonic(double, double);
/*函数的原型*/
int main()
{
    double input_num1, input_num2;
    cout<<"Enter the two number( seperate by blank):";
    cin>>input_num1>>input_num2;
    /*读取系统输入的两个浮点数据*/
    while(input_num1 != 0 || input_num2 != 0){
        /*函数入口条件是两个数不同时为 0*/
      cout<<"The "<<input_num1<<" and "<<input_num2;
      cout<<" harmonic mean is "<<harmonic(input_num1,input_num2)<<endl;
        cout<<"Enter the operand( seperate by blank):";
        cin>>input_num1>>input_num2;
    }
    return 0;
}
double harmonic(double x, double y){
   double result = 2.0 *x *y / (x + y);
   return result;
   /*返回值在算式比较简单时可以直接
    *返回 2.0 *x *y / (x + y)的值，
    *定义变量、计算再返回变量较麻烦*/
}
```

2．编写一个程序，要求用户输入最多 10 个高尔夫成绩，并将其存储在一个数组中。程序允许用户提早结束输入，并在一行上显示所有成绩，然后报告平均成绩。请使用 3 个数组处理函数来分别输入、显示和计算平均成绩。

编程分析：

用户输入的高尔夫球成绩是少于或等于 10 个数的一组数据，因此使用数组存储时可能会出现数组未完全使用的情况。在设计数据分析函数时，必须要通过函数的参数指定当前数组的有效数据下标，不能直接使用数组的长度 10。题目要求使用 3 个函数分别输入、显示和计算平均成绩，因此输入数据的函数必须返回输入数据的个数，否则可能会影响计算平均数并显示错误数据。完整代码如下。

```
/*第 7 章的编程练习 2*/
#include <iostream>
using namespace std;

const int SIZE = 10;
int set_mark(int[],int);
void display_mark(int[],int);
double average_mark(int[],int);
/*函数原型*/
int main()
{
    int size, golf_mark[SIZE];
    size = set_mark(golf_mark, SIZE);
    /*调用输入成绩的函数，并通过返回值获得成绩数量*/
    display_mark(golf_mark, size);
    cout<<"The average marks is : "<<average_mark(golf_mark, size)<<endl;
    /*调用函数，输出并计算平均分*/
    return 0;
}
int set_mark(int arr[],int size)
{
    int i = 0;
    do{
        cout<<"Enter the No."<<i+1<<" golf marks: ";
        cin>>arr[i++];
        cin.get();
        cout<<"press enter to continue, or 's' for STOP input : ";
        if(cin.get() == 's'){
            for(;i<size;i++) arr[i] = 0;
            break;
        }
    }while(i < size);
    /*通过 while 循环获取输入，但是使用 do...while 循环会至少需要输入一次成绩，
     *通常可以使用 while 循环来实现 0~10 次输入*/
    return i;
}
void display_mark(int arr[],int size)
{
    cout<<"The marks is below:"<<endl;
    for(int i = 0; i < size ;i++)
        cout<<arr[i]<<"\t";
    cout<<endl;
 /*循环输出数组内数据*/
}
double average_mark(int arr[],int size)
{
    int sum = 0;
    for(int i = 0; i < size; i++)
        sum += arr[i];
```

```
    return 1.0 *sum / size;
/*计算数组元素和、取平均值，乘以 1.0 转换为浮点数据*/
}
```

3．下面是一个结构体声明。

```
struct box{
        char maker[40];
        float height;
        float width;
        float length;
        float volume;
};
```

a．编写一个函数，按值传递 box 结构体，并显示每个成员的值。

b．编写一个函数，传递 box 结构体的地址，并将 volume 成员设置为其他三维长度的乘积。

c．编写一个使用这两个函数的简单程序。

编程分析：

题目要求通过函数分别实现数据输出和计算 box 体积的功能，分别使用传值和传地址两种方式作为参数传递结构体值。其中计算体积的功能需要修改结构体变量内的 volume 数据，因此如果使用按值传递，那么需要通过返回值将计算结果返回给主调函数；如果使用传地址的方式，则可以直接在实参的结构体内修改，不需要返回值并通过返回值赋值，能够提高运行效率。下面是完整代码。

```
/*第 7 章的编程练习 3*/
#include <iostream>
using namespace std;

struct box{
    char maker[40];
    float height;
    float width;
    float length;
    float volume;
};
/*结构体定义*/
void display(box);
void calc_volume(box *);
/*函数的原型声明*/
int main()
{
    box Orange = {"China",12,12,12,0};
    calc_volume(&Orange);
    display(Orange);
    /*创建结构体变量并初始化，简单应用函数计算体积并显示*/
    return 0;
}
```

```
void display(box b)
{
    cout<<"This box made by "<<b.maker<<".\nAnd height = "<<b.height;
    cout<<", width = "<<b.width<<", length = "<<b.length<<", volume = ";
    cout<<b.volume<<".";
/*输入box结构体的基本数据信息*/
}
void calc_volume(box *pb)
{
    pb->volume = pb->width *pb->height *pb->length;
    cout<<"Calculate box's volume done."<<endl;
    /*计算体积，并将其直接存放在参数 pb 指向的数据对象内，
     *因此可以不使用返回值返回计算结果*/
}
```

4．许多州的彩票发行机构使用如程序清单 7.4 所示的简单彩票玩法的变体。在这些玩法中，玩家从一组称为域号码（field number）的号码中选择几个。例如，可以从域号码 1～47 中选择 5 个号码，还可以从第 2 个区间（如 1～27）中选择一个号码（称为特选号码）。要赢得头奖，必须正确猜中所有的号码。中头奖的概率是选中所有域号码的概率与选中特选号码概率的乘积。例如，在这个例子中，中头奖的概率是从 47 个号码中正确选取 5 个号码的概率与从 27 个号码中正确选择 1 个号码的概率的乘积。请修改程序清单 7.4，以计算中彩票头奖的概率。

编程分析：

程序清单 7.4 通过公式 1/*R* 计算中奖概率，因此需要使用一个循环来计算排列组合的乘积。题目要求在程序清单 7.4 的基础上，计算中头奖的概率，中头奖的概率是 47 选 5 的正确概率和 27 选 1 的正确概率的乘积。完整代码如下。

```
/*第 7 章的编程练习 4*/
#include<iostream>
using namespace std;

long double probability(double fnumbers, double snumber, double picks);
/*修改程序中计算多选多的中奖概率，添加表示特别号码的数字 */
int main()
{
   cout << "Field number is 45 , and special number is 27 ."<<endl;
   cout << "the probability is one of the: "<<probability(45,27,5)<<endl;
   return 0;
}
long double probability(double fnumbers, double snumber, double picks)
{
   long double result = 1.0;
   long double n;
   unsigned p;
   for(n = fnumbers, p = picks ; p > 0; n--, p--)
      result = result *n / p;
```

```
    /*计算域号码的选中概率*/
    return result /=  snumber ;
    /*正确选中域号码的概率乘以正确选中特选号码的概率*/
}
```

5．定义一个递归函数，该函数接受一个整数参数，并返回该参数的阶乘。前面讲过，3 的阶乘写作 3!，等于 3×2!，以此类推；而 0!被定义为 1。通用的计算公式是，如果 n 大于零，则 $n!=n\cdot(n-1)!$。在程序中对该函数进行判断，程序使用循环要求用户输入不同的值，程序将报告这些值的阶乘。

编程分析：

阶乘的计算是程序设计中函数递归算法的典型应用。阶乘的计算公式可以表示为 $n! = n\cdot(n-1)!$。在使用递归算法时必须要注意设计递归函数的返回点，阶乘中使用 0! == 1 返回，程序设计中必须将返回点判断放置在递归函数的前面。完整代码如下。

```
/*第 7 章的编程练习 5*/
#include <iostream>
using namespace std;

long long factorial(int);
/*阶乘函数的原型*/
int main()
{
    int n;
    cout<<"Enter a number to calc factorial: ";
    cin>>n;
    while(n > 0)
    {
        cout<<n<<"! = "<<factorial(n)<<endl;
        cout<<"Enter a number to calc factorial: ";
        cin>>n;
    }
    cout<<"Done!"<<endl;
    return 0;
}
long long factorial(int n)
{
    if(n == 0)
    {
        return 1;
        /*函数的第一条语句，0! 是递归返回点*/
    }else{
        return n *factorial(n-1);
    }
}
```

6．编写一个程序，它使用下列函数。

- Fill_array()以一个 double 型数组的名称和长度作为参数。它提示用户输入 double 值，并将这些值存储到数组中。当数组被填满或用户输入非数字时，输入将停止，并返回实际输入了多少个数字。
- Show_array()以一个 double 型数组的名称和长度作为参数，并显示该数组的内容。
- Reverse_array()以一个 double 型数组的名称和长度作为参数，并将存储在数组中的值的顺序反转。

程序将使用这些函数来填充数组，然后显示数组；反转数组，然后显示数组；反转数组中除第 1 个和最后一个元素之外的所有元素，然后显示数组。

编程分析：

本题要求使用 3 个函数，分别实现数组的数据填充、数据显示和数据反转。为了填充和显示数组，需要使用数组和数组长度作为参数，实现较简单。数组的反转中，题目要求实现部分数据反转，当使用数组和长度作为参数时，可以从实际参数输入来考虑进行部分反转，即实际参数的数组首元素后移，数组长度缩减。完整代码如下。

```
/*第 7 章的编程练习 6*/
#include <iostream>
using namespace std;

int Fill_array(double[], int);
void Show_array(double[], int);
void Reverse_array(double[], int);
/*函数的原型声明*/
const int SIZE = 20;
/*定义数组最大长度*/
int main()
{
    double Array[SIZE];
    int size = Fill_array(Array, SIZE);
    Show_array(Array, size);
    Reverse_array(Array, size);
    Show_array(Array, size);
    Reverse_array(&Array[1], size - 2);
        /*通过控制函数参数的形式实现部分数据的反转*/
    Show_array(Array, size);
    return 0;
}

int Fill_array(double arr[], int size)
{
    int count = 0;
    double temp;
    cout<<"Enter the number seperate by blank, 's' to stop : ";
    cin>>temp;
    while(count < size)
    {
        if(cin.get() == 's')
```

```
            {
                return count;
            }else{
                arr[count++] = temp;
                cin>>temp;
            }
        }
        /*读取数据并输入数组，在输入数据时计数，并返回计数值，作为数组长度*/
        return count;
    }
    void Show_array(double arr[], int size)
    {
         cout<<"The array's data: "<<endl;
         for(int i = 0; i < size; i++)
         {
             cout<<arr[i]<<"\t";
        }
        cout<<endl;
        /*循环输出数组内容，数组长度使用参数 size 表示*/
    }
    void Reverse_array(double arr[], int size){
        double temp;
        for(int i = 0; i < size/2; i++)
        {
            temp = arr[i];
            arr[i] = arr[size - i - 1];
            arr[size - i - 1] = temp;
        }
        /*反转数组，分别从头和尾互换数据，直到数组的中间*/
    }
```

7．修改程序清单 7.7 中的 3 个数组处理函数，使之使用两个指针参数来表示区间。fill_array()函数不返回实际读取了多少个数字，而返回一个指针，该指针指向最后填充的位置；其他的函数以该指针作为第 2 个参数，以标识数据结尾。

编程分析：

程序清单 7.7 使用 3 个函数来处理数据的填充、显示和重新赋值。要修改函数参数为指针，需要使用两个参数分别表示数组的头和尾，函数内部通过与头尾指针的比较实现相关的循环操作，指针的自增操作能够实现移动数组下标的功能。完整代码如下。

```
    /*第 7 章的编程练习 7*/
    #include <iostream>
    const int Max = 5;
    //function prototypes
    double* fill_array(double* begin, double* end);
    void show_array(double* begin, double* end);
    void revalue(double r, double* begin, double* end);
    /*修改函数原型，把参数修改为指针类型*/
    int main(int argc, char *argv[])
```

```
{
    using namespace std;
    double properties[Max];

    double* pend = fill_array(properties, properties+Max);
    show_array(properties, pend);
    if(pend - properties > 0)
    {
        cout<<"Enter revalue factor: ";
        double factor;
        while(!(cin>>factor))
        {
            cin.clear();
            while(cin.get() != '\n')
                continue;
            cout<<"bad input; Please input a number: ";
        }
        revalue(factor, properties, pend);
        show_array(properties, pend);
    }
    cout<<"Done.\n";
    cin.get();
    cin.get();
    return 0;
}

double* fill_array(double*begin, double*end)
{
    using namespace std;
    double temp;
    double* p;
    for(p = begin; p != end; p++){
        cout<<"Enter value #"<< (p - begin) / sizeof(double) + 1 <<":";
        /*用指针数据的差除以 double 类型数据的长度，可以得到当前数据的排序*/
        cin>>temp;
        if(!cin)
        {
            cin.clear();
            while(cin.get() != '\n')
                continue;
            cout<<"bad input; input process terminated.\n";
            break;
        }else if(temp < 0)
            break;
        *p = temp;
    }
    return p;
}
void show_array(double* begin, double* end)
{
```

```
    using namespace std;
    for(double* p = begin; p != end; p++)
    {
        cout<<"Property #"<< (p - begin) / sizeof(double) + 1<<":$";
        cout<<*p<<endl;
    }
    /*编号显示通过地址差显示*/
}
void revalue(double r, double* begin, double* end){
    double* p = begin;
    for(double* p = begin; p != end; p++)
    {
        *p = r;
    }
    /*通过首尾指针判断循环是否完成*/
}
```

8．在不使用 array 类的情况下完成程序清单 7.15 所做的工作。编写两个这样的版本。

a．使用 const char *数组存储表示季度名称的字符串，并使用 double 型数组存储开支。

b．使用 const char*数组存储表示季度名称的字符串，并使用一个结构体，该结构体只有一个成员，即一个用于存储开支的 double 型数组。这种设计与使用 array 类的设计基本类似。

a 部分的编程分析：

程序清单 7.15 应用 array 模板进行数据填充和显示，功能较简单。题目要求使用 const char *数组存储表示季度名称的字符串，并使用 double 型数组存储开支。完整代码如下。

```
/*第 7 章的编程练习 8a*/

#include <iostream>
#include <string>
//常量数据
const int Season = 4;
const char* Sname[] = {"Spring","Summer","Fall","Winter"};
//修改数组对象的函数

void fill(double arr[], int size);
void show(const double arr[], int size);

int main()
{

    double expenses[Season];
    fill(expenses, Season);
    show(expenses, Season);
    return 0;
}
```

```
void fill(double arr[], int size)
{
    using namespace std;
    for(int i = 0; i < size; i++)
    {
       cout<<"Enter "<< Sname[i] <<" expenses: ";
       cin>>arr[i];
    }
}
void show(const double arr[], int size)
{
    using namespace std;
    double total = 0.0;
    cout<<"\nEXPENSES\n";

    for(int i = 0; i < size; i++)
    {
       cout<< Sname[i] <<":$ "<<arr[i]<<endl;
       total += arr[i];
    }
    cout<<"Total Expenses:$ "<<total<<endl;
}
```

b 部分的编程分析：

在 a 部分的基础上，设计和使用一个结构体来存储开支，因此结构体的定义如下。

```
struct Spend{
    double money[Season];
};
#
```

完整代码如下。

```
/*第 7 章的编程练习 8b*/
#include <iostream>
#include <string>
//常量数据
const int Season = 4;
const char* Sname[] = {"Spring","Summer","Fall","Winter"};
struct Spend{
    double money[Season];
};
//修改数组对象的函数

void fill(double arr[], int size);
void show(const double arr[], int size);

int main()
{

    Spend expenses;
    fill(expenses.money, Season);
```

```
    show(expenses.money, Season);
    return 0;
}

void fill(double arr[], int size)
{
    using namespace std;
    for(int i = 0; i < size; i++)
    {
        cout<<"Enter "<< Sname[i] <<" expenses: ";
        cin>>arr[i];
    }
/*通过循环，读取标准输入，填充数据*/
}
void show(const double arr[], int size)
{
    using namespace std;
    double total = 0.0;
    cout<<"\nEXPENSES\n";

    for(int i = 0; i < size; i++)
    {
        cout<< Sname[i] <<":$ "<<arr[i]<<endl;
        total += arr[i];
    }
    cout<<"Total Expenses:$ "<<total<<endl;
/*通过循环，输出数组信息*/
}
```

9．编写处理数组和结构体的函数。下面是程序的框架，请提供其中描述的函数，以完成该程序。

```
#include <iostream>
using namespace std;
const int SLEN = 30;
struct student{
    char fullname[SLEN];
    char hobby[SLEN];
    int ooplevel;
};
//getinfo() 有两个参数，一个是 student 结构体中第 1 个元素的指针，
//另一个是表示数组中元素个数的整数。
//该函数存储学生的数据，
//在填充数据之后或者遇到空行时，该函数终止输入
//该函数返回实际填充的元素个数
int getinfo(student pa[], int n);

//display1()以 student 结构体作为参数并显示其内容
void display1(student st);
```

```
//display2()以 student 结构体的地址作为参数并显示其内容
void display2(const student*ps);

//display3()以 student 结构体中第 1 个元素的地址和数组中的元素个数作为参数，
 //并显示其内容
void display3(const student pa[], int n);

int main() {

    cout<<"Enter the class size: ";
    int class_size;
    cin>>class_size;
    while(cin.get() != '\n')
       continue;
    student *ptr_stu = new student[class_size];
    int  entered = getinfo(ptr_stu,class_size);
    for(int i = 0 ; i < class_size; i++)
    {
       display1(ptr_stu[i]);
       display2(&ptr_stu[i]);
    }
    display3(ptr_stu, entered);
    delete[] ptr_stu;
    cout<<"Done\n";
    return 0;
}
```

编程分析：

题目提供了程序的基本框架，需要在此基础上实现相关函数的定义。函数的主要功能是信息输入以及信息显示。其中信息显示功能涉及 3 个函数，主要区别在于函数的参数不同。3 个函数的参数分别是结构体变量、结构体指针和结构体数组。以结构体指针作为函数参数，应重点考虑实参对象的数据安全问题，以上代码中使用了 const 关键字，因此不能修改参数的对象信息。完整代码如下。

```
/*第 7 章的编程练习 9*/
#include <iostream>
using namespace std;
const int SLEN = 30;
struct student{
    char fullname[SLEN];
    char hobby[SLEN];
    int ooplevel;
};
/*
getinfo() 有两个参数，一个是 student 结构体中第 1 个元素的指针，另一个是表示数组中元素个数的整数。
该函数存储学生的数据，在填充数据之后或者遇到空行时，该函数终止输入。该函数返回实际填充的元素个数
*/
int getinfo(student pa[], int n);
```

```
//display1()以 student 结构体作为参数并显示其内容
void display1(student st);

//display2()以 student 结构体的地址作为参数并显示其内容
void display2(const student* ps);

//display3()以 student 结构体中第 1 个元素的地址和数组中的元素个数作为参数，
//并显示其内容
void display3(const student pa[], int n);

int main() {

    cout<<"Enter the class size: ";
    int class_size;
    cin>>class_size;
    while(cin.get() != '\n')
       continue;
    student *ptr_stu = new student[class_size];
    int  entered = getinfo(ptr_stu,class_size);
    for(int i = 0 ; i < class_size; i++)
    {
       display1(ptr_stu[i]);
       display2(&ptr_stu[i]);
    }
    display3(ptr_stu, entered);
    delete[] ptr_stu;
    cout<<"Done\n";
    return 0;
}

int getinfo(student pa[], int n)
{
    int i = 0;
    for( i = 0 ;i< n ; i++)
    {
       cout<<"Enter the info of student name: ";
       cin>>pa[i].fullname;
       cout<<"Enter the info of student hobby: ";
       cin>>pa[i].hobby;
       cout<<"Enter the info of student level: ";
       cin>>pa[i].ooplevel;
       if(!cin)
       {
          cin.clear();
          while(cin.get() != '\n')
             continue;
          cout<<"Bad input. procerss terminated\n";
          break;
       }
    }
```

```
    return i;
}
/*getinfo()函数实现学生信息录入功能，以数组及其长度作为参数，返回值为录入的信息数量*/

void display1(student st)
{
    cout<<"Student Name: "<<st.fullname<<endl;
    cout<<"Student hobby: "<<st.hobby<<endl;
    cout<<"Stuent level: "<<st.ooplevel<<endl<<endl;
}
/*以结构体变量作为函数参数，输出相关信息*/

void display2(const student* ps)
{
    cout<<"Student Name: "<<ps->fullname<<endl;
    cout<<"Student hobby: "<<ps->hobby<<endl;
    cout<<"Stuent level: "<<ps->ooplevel<<endl<<endl;
}
/*以指针作为函数参数，输出相关信息*/
void display3(const student pa[], int n)
{
    for(int i = 0; i < n; i++)
    {
        cout<<"Student Name: "<<pa[i].fullname<<endl;
        cout<<"Student hobby: "<<pa[i].hobby<<endl;
        cout<<"Stuent level: "<<pa[i].ooplevel<<endl<<endl;
    }
}
/*以数组作为函数参数，输出整个数组的信息*/
```

10．设计一个名为 calculate()的函数，它接受两个 double 值和一个指向函数的指针，而被指向的函数接受两个 double 参数，并返回一个 double 值。calculate()函数的类型也是 double，并返回被指向的函数。使用 calculate()的两个 double 参数计算得到的值。例如，假设 add()函数的定义如下。

```
double add(double x, double y)
{
    return x + y;
}
```

则下述代码中的函数调用将导致 calculate()把 2.5 和 10.4 传递给 add()函数，并返回 add()的返回值（12.9）。

```
double q = calculate(2.5, 10.4, add);
```

请编写一个程序，它调用上述两个函数和至少另一个与 add()类似的函数。该程序使用循环要求用户成对地输入数字。对于每对数字，程序都使用 calculate()来调用 add()和至少一个其他的函数。如果读者爱冒险，可以尝试创建一个指针数组，其中的指针指向 add()样式的函数，并编写一个循环，使用这些指针连续让 calculate()调用这些函数。提示：下面是声明这种指针数组的方式，其中包含 3 个指针。

```
double (*pf[3]) (double, double)
```

可以采用数组初始化语法，并以函数名作为地址来初始化这样的数组。

编程分析：

C++中的函数指针是表示函数机器代码的起始地址的变量。将一个函数的指针作为参数传递到另外一个函数当中，通过调用参数不同、运行方式不同的函数，可以实现相对比较灵活的多功能函数调用。题目要求通过函数指针实现一个 calculate()函数，该函数以两个double 类型的数据和一个函数指针作为参数。函数指针的类型是指向两个 double 类型参数，返回 double 类型参数的函数，因此，该函数指针可以定义为

```
double (*pf ) (double, double)
```

calculate()函数通过函数指针 pf 调用对应的实参指定的函数，计算其他两个参数的结果并返回，因此 calculate()函数的定义如下。

```
double calculate(double x, double y, double (*pf)(double x1, double x2))
```

程序的完整代码如下。

```
/*第 7 章的编程练习 10*/
#include <iostream>
using namespace std;
double add(double, double);
double subtract(double ,double);
double calculate(double, double, double (*)(double ,double));

int main(int argc, char *argv[])
{
    double q = calculate(2.5, 10.4, add);
    cout<<"The Answer of add is "<<q<<endl;
    double t = calculate(2.5, 10.4, subtract);
    cout<<"The Answer of substract is "<<t<<endl;
    return 0;
}

double add(double x, double y)
{
    return x + y;
}
double subtract(double x,double y){
    return x - y;
}

double calculate(double x, double y, double (*pf)(double x1, double x2)){
    return pf(x,y);
}
```

第 8 章　函数探幽

本章知识点总结

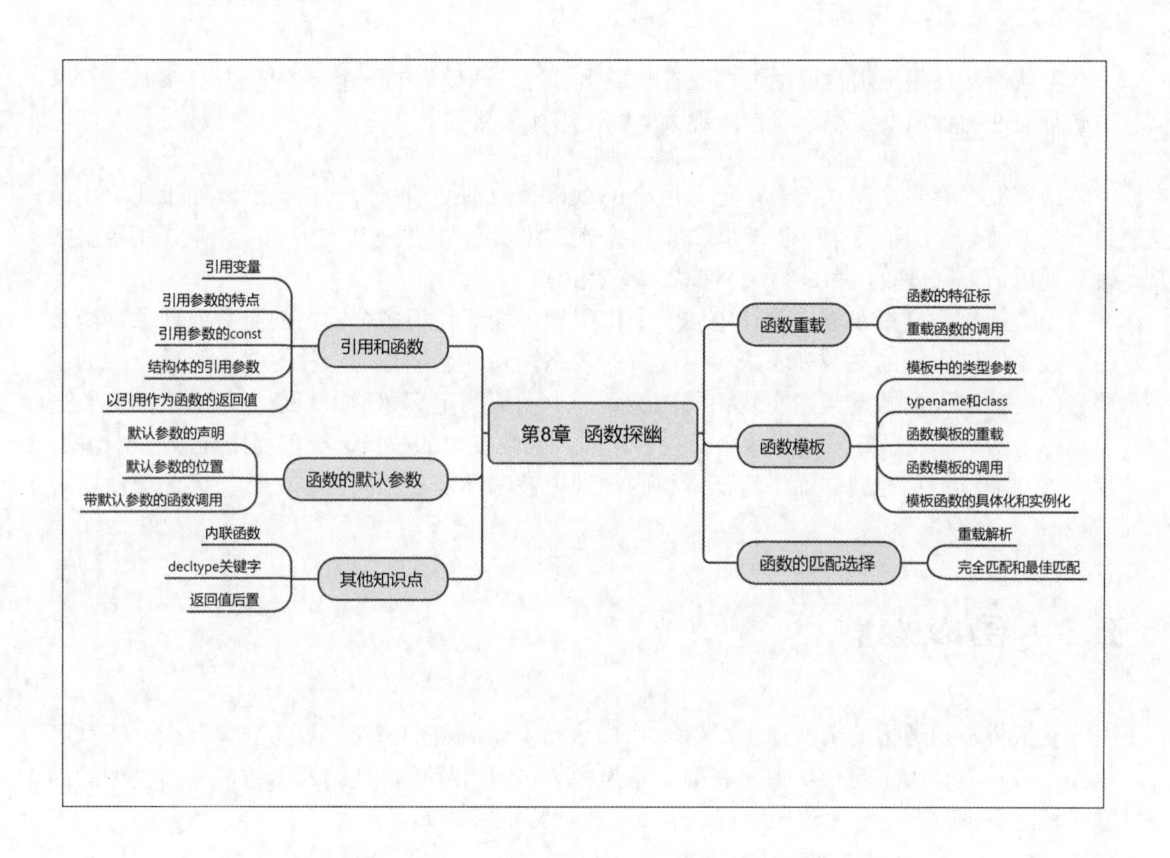

8.1　引用变量和引用参数

C++语言中的引用类型是一种新增的复合类型，本质上，引用就是其他已定义变量的别名。当引用变量和其他类型的变量赋值连接后，引用变量和原变量具备相同的功能，可以交替使用。引用变量最重要的功能是作为函数的形参。通过引用变量作为函数的形参，函数在运行中将使用实参的原始数据而不是副本。使用引用参数的这种特性在大型结构数据中作为参数传递，会大幅提高函数的运行效率，但是由于引用参数会直接修改实参，因此需要尽可能使用 const 关键字避免实参被误修改。当使用的 const 关键字或者实参和引用参数类型不匹配时，C++会创建临时变量从而影响到某些函数功能的实现，因此在程序设计中还应当注意函数调用时的参数匹配问题。

引用变量也可以作为函数的返回值使用。传统的返回值与按值传递的参数类似，也是一个临时数据被复制的过程。而当使用引用作为返回值时，实际上返回的是被引用变量的别名，这同样避免了数据的复制过程，能够提高程序的运行效率。实际应用中不能使用函数内定义的局部变量作为返回值返回。

8.2　函数的默认参数与重载

默认参数是指当函数调用过程中省略了实参时自动使用的一个参数值。默认参数在C++中必须通过函数原型实现，即在函数原型中使用如下格式。

```
char* left(const char* str, int n = 0);
```

函数能够有多个默认参数，但是所有的默认参数都必须从右向左排列。如果在调用函数时主调函数使用了两个实参，那么第 2 个实参将覆盖 0 并向函数传递；如果在调用函数时只使用了一个参数，那么第 2 个参数将使用 0。

函数的重载是指在程序设计中同一个作用域内能够使用多个同名的函数的现象。函数的重载是通过不同的参数列表来实现函数的多态性和泛型编程的一种方式。函数重载的主要目的是针对不同数据类型实现相同功能而减少函数的定义和调用复杂度。函数的参数列表也称为函数的特征标。如果两个函数名相同，函数参数的数目、类型和排列顺序也相同，那么就称为特征标相同。如果特征标相同，则会被编译器视为同一个函数，因此无法实现重载。

8.3　函数模板

函数模板是通用的函数描述，是 C++中进行泛型编程的一个重要方式。其中，泛型可以使用多种具体类型替换，通过将类型作为参数传递给模板，使编译器生成具体类型的函数。泛型编程有时也称为通用编程。函数模板的标准格式如下。

```
template <typename AnyType> void Swap(AnyType& a, AnyType& y);
```

其中，template 是表示模板的关键字，<typename AnyType>是类型定义，因为模板函数是一种泛型编程，具体类型需要通过函数调用生成，所以在定义函数时需要使用 typename。函数内部如果需要用到参数的具体类型，也应统一使用 AnyType 来表示。

在泛型编程中，由于简单数据类型和复杂数据类型在数据处理方面存在差异，因此函数模板有时需要针对特定的结构体参数或者类参数进行函数模板的具体化。具体化是在原有的模板基础上，新添加一个具体化模板函数定义的过程。例如，在以上原有模板的基础上，具体化如下。

```
template <> void Swap(job& a, job& y);
```

通过具体化定义函数模板，可以在函数内部特殊化处理复杂类型 Swap。

8.4　函数的重载解析

对于函数重载、函数模板和函数模板重载，C++需要一个定义良好的策略，来决定在函数调用中具体使用哪一个函数定义，这个过程称为重载解析（overloading resolution）。其基本过程如下。

（1）创建候选函数列表。其中包含与被调函数的名称相同的函数和模板函数。

（2）使用候选函数列表创建可行函数列表。这些都是参数数目正确的函数，为此有一个隐式转换序列，其中包括实参类型与相应的形参类型完全匹配的情况。

（3）确定是否有最佳的可行函数。如果有，则使用它；否则，该函数调用出错。

编译器必须确定哪个可行函数是最佳的。为了使函数调用的参数与可行的候选函数的参数相匹配，需要进行转换。通常，从最佳到最差的顺序如下所述。

（1）完全匹配，但常规函数优先于模板。

（2）提升转换（例如，char 和 short 类型自动转换为 int 类型，float 类型自动转换为 double 类型）。

（3）标准转换（例如，int 类型转换为 char 类型，long 类型转换为 double 类型）。

（4）用户定义的转换，如类声明中定义的转换。

通过函数的重载解析，C++编译器按照这些基本规则实现函数的匹配调用，因此程序设计过程中，应当明确规则，确保函数调用过程总能够满足程序设计者的意图。

此外，本章还介绍了内联函数的基本特点和使用，其主要功能在复习题内有详细描述。

8.5　复习题

1．哪种函数适合定义为内联函数?

习题解析：

内联函数是一种特殊的函数，其主要特点是在进行源代码编译时，C++编译器使用内联函数的代码替换函数的调用语句，这样程序在运行中将不需要执行类似于常规函数的复杂操作，因此可以节省大量的系统资源，提高程序的执行效率。对于常规函数，在程序进行函数调用时，需要进行一系列复杂的操作，包括函数地址查询、地址保存、函数参数复制、程序代码跳转、执行函数代码、函数代码执行完毕后返回原始代码中的地址等。

通常情况下，程序设计中代码较简单的非递归函数可以使用内联函数。内联函数使用关键字 inline。但是考虑到现代编译器的具体优化特性，原始代码中的内联函数会在编译器过程中有选择地进行内联优化。

2．假设 song()函数的原型如下。

```
void song(const char* name, int times);
```

a．如何修改原型，使 times 的默认值为 1?

b．函数定义需要做哪些修改?

c．能否为 name 提供默认值“O. My Papa”？

习题解析：

a．函数默认值需要在函数的原型声明处添加，必须从右向左排列，因此可以修改声明为 `void song(const char* name, int times = 1);`。

b．函数默认值只需要在声明处表示，在函数的定义处不需要标注参数默认值。

c．可以为 name 提供默认值，但是在为 name 提供默认值时，必须也要保持 times 的默认值，因为默认值必须从右向左排列，不能 name 在参数列表左侧有默认值，而 times 在右侧但是没有默认值。

3．编写 iquote()的重载版本——显示它用双引号括起来的参数。编写 3 个版本。其中，一个用于 int 参数，另一个用于 double 参数，最后一个用于 string 参数。

习题解析：

在函数的重载中函数名相同但是函数的参数列表不同，因此本题中的重载如下。

```
void iquote(int n)
  {
        cout<<"\""<<n<<"\""<<endl;
  }

void iquote(double d)
  {
        cout<<"\""<<d<<"\""<<endl;
  }

void iquote(string s)
  {
        cout<<"\""<<s<<"\""<<endl;
  }
```

4．下面是一个结构体模板。

```
struct box
{
      char maker[40];
      float height;
      float width;
      float length;
      float volume;
};
```

a．请编写一个函数，它以 box 结构体的引用作为形参，并显示每个成员的值。

b．请编写一个函数，它以 box 结构体的引用作为形参，并将 volume 成员设置为其他 3 条边的乘积。

习题解析：

C++中的引用变量主要应用于函数的形参。若以引用变量作为函数的形参，函数在执行过程中将使用实参的原始数据而不是其副本。本题可以这样定义引用参数。

```
void show_box(box& b)
{
        cout<<"The Box information:\n"<<"Maker: "<<b.maker<<endl;
        cout<<"Height: "<<b.height<<endl;
        cout<<"Width: "<<b.width <<endl;
        cout<<"Length: "<<b.length <<endl;
        cout<<"Volume: "<<b.volume <<endl;
}
void clac_volume(box& b)
{
        b.volume = b.height *b.width *b.length;
}
```

5．为了使函数 fill()和 show()以引用作为参数，需要对程序清单 7.15 做哪些修改?

习题解析：

程序清单 7.15 中的函数 fill()和 show()函数分别以指针与数组变量作为函数参数，为了将函数参数修改为引用，首先应当在 show()函数的参数列表中使用 const 关键字，此外，当指针和引用在函数内表示元素时，代码需要略做修改。完整代码如下。

```
#include <iostream>
#include <array>
#include <string>
//常量数据
const int Season = 4;
const std::array<std::string, Season> Sname =
        {"Spring","Summer","Fall","Winter"};

//修改数组对象的函数
void fill(std::array<double, Season>& da);

//使用数组对象的函数，不修改它
void show(const std::array<double, Season>& da);

int main()
{

    std::array<double, Season> expenses;
    fill(expenses);
    show(expenses);
    return 0;
}

void fill(std::array<double, Season> &da)
```

```
{
    using namespace std;
    for(int i = 0; i < Season; i++)
    {
        cout<<"Enter "<< Sname[i] <<" expenses: ";
        cin>>da[i];
    }
}
/*为了使fill()函数以引用作为参数，主要修改了数组元素的表示*/
void show(const std::array<double, Season> &da)
{
    using namespace std;
    double total = 0.0;
    cout<<"\nEXPENSES\n";

    for(int i = 0; i < Season; i++)
    {
        cout<< Sname[i] <<":$ "<<da[i]<<endl;
        total += da[i];
    }
    cout<<"Total Expenses:$ "<<total<<endl;
}
/*为了使show()函数以引用作为参数，不需要函数实现*/
```

6．指出下面每个功能是否可以使用默认参数或函数重载完成，或者这两种方法都无法完成，并提供合适的原型。

a．mass(density, volume)返回密度为 density、体积为 volume 的物体的质量，而 mass(denstity)返回密度为 density、体积为 $1.0m^3$ 的物体的质量。这些值的类型都为 double。

b．repeat(10, "I'm OK")将指定的字符串显示 10 次，而 repeat("But you're kind of stupid")将指定的字符串显示 5 次。

c．average(3, 6)返回两个 int 参数的平均值（int 类型），而 average(3.0, 6.0)返回两个 double 值的平均值（double 类型）。

d．mangle("I'm glad to meet you")根据是将值赋给 char 变量还是 char*变量，分别返回字符 I 和指向字符串“I'm mad to gleet you”的指针。

习题解析：

a．可以通过参数默认值实现这两个功能，也可以通过函数的重载来实现这个功能。

函数默认值方式如下。

```
double mass(double density, double volume = 1.0);
```

函数重载方式如下。

```
double mass(double density);

double mass(double density, double volume);
```

b．由于函数必须从右向左提供默认值，因此本题无法使用默认值，只能够使用函数的重载实现。

```
void repeat(int n, char* str);
void repeat(char* str);
```

c．题目中两个函数的参数类型不同，可以通过函数重载实现。

```
int average(int x, int y);
double average(double x, double y);
```

d．这样的函数的特征标相同，因此无法实现函数的重载。此外，对于函数 mangle("I'm glad to meet you")，返回的字符'I'和指向字符串"I'm glad to meet you"的指针都是 char*类型，即返回值也相同。

7．编写返回两个参数中较大值的函数模板。

习题解析：

函数模板是实现泛型编程的一个重要形式。函数模板的基本定义方式如下。

```
template<typenanem T> T Function_Name(T t);
```

其中，typename 也可以用旧标准下的 class 替代。题目要求实现返回较大值的函数，因此函数的定义如下。

```
template <typename  AnyType>
AnyType Bigger (const AnyType & x, const AnyType & y){
return x > y? x : y;
}
```

或者

```
template <class  T>
T Bigger (const T & x, const T & y){
return x > y? x : y;
}
```

8．给定复习题 7 的模板和复习题 4 的 box 结构体，实现一个模板的具体化，它接受两个 box 参数，并返回体积较大的一个。

习题解析：

box 结构体的定义及其他相关模板的定义如下。

```
struct box{
      char maker[40];
      float height;
      float width;
      float length;
      float volume;
};
```

当比较 box 对象的大小时，无法使用结构体对象直接进行比较，即无法使用 box1>box2 的形式比较。按照题意应当选择体积进行比较，因此应当进行模板的具体化。模板的具体化需要在原有模板函数的基础上，具体化特定类型的实现。下列是完整版本。

```
template<typename T> T& bigger(T&, T&);
//模板函数的原型声明
```

```
template<> box& bigger<box>(box&, box&);
//模板的具体化声明
template<typename T> T& bigger(T& x, T& y)
{
    return x>y?x:y;
}
//模板函数的实现
template<> box& bigger<box>(box& x, box& y){
    return x.volume > y.volume? x:y;
    //使用体积代替 box 对象从整体上比较大小
}
//模板函数具体化的实现
```

9．在下述代码（假定这些代码是一个完整程序的一部分）中，v1、v2、v3、v4 和 v5 分别是哪种类型？

```
int g(int x);
...
float m = 5.5f;
float& rm = m;
decltype(m) v1 = m;
decltype(rm) v2 = m;
decltype((m)) v3 = m;
decltype(g(100)) v4;
decltype(2.0*m) v5;
```

习题解析：

由于 C++语言使用了模板函数和模板类，因此声明中经常会发生需要使用中间变量但是因为模板中类型的不确定性而无法明确给出类型定义的情况。为了解决这个问题，C++中使用 decltype 关键字来动态地进行变量的声明。

本题的解答如下。

- 在 `decltype(m) v1 = m;` 中，变量 v1 的类型和 m 相同，为 float 类型。
- 在 `decltype(rm) v2 = m;` 中，变量 v2 的类型和 rm 相同，为 float&类型。
- 在 `decltype((m)) v3 = m;` 中，表达式 (m)是一个左值，因此变量 v3 的类型是指向 m 的引用，即 float&。
- 在 `decltype(g(100)) v4;` 中，表达式是一个函数调用，因此 v4 的类型与函数返回值相同，为 int 类型。
- 在 `decltype(2.0*m) v5;` 中，变量 m 是 float 类型，2.0 是 double 类型，运算中类型升级，表达式的类型为 double 类型，因此 v5 为 double 类型。

8.6 编程练习

1．编写通常接受一个参数（字符串的地址）并输出该字符串的函数。然而，如果提供了第 2 个参数（int 类型），且该参数不为 0，则该函数输出字符串的次数将为调用该函数的次数（注意，字符串的输出次数不等于第 2 个参数的值，而等于函数被调用的次数）。是的，这是一个非常可笑的函数，但它涉及本章介绍的一些方法。在一个简单的程序中使用该函数，以演示该函数是如何工作的。

编程分析：

本题中输出函数的定义需要使用默认参数，其中默认参数是第 2 个整型数据，默认值为 0。当输入参数非 0 时，函数应当输出字符串的次数为调用函数的次数，而函数的调用次数则需要使用函数内的静态变量进行记录。静态变量的定义使用 static 关键字。函数内的静态变量能够在函数返回后仍保留在程序存储区内，并在下一次调用函数时，再一次被函数继续使用。程序的完整代码如下。

```
/*第 8 章的编程练习 1*/
#include <iostream>
using namespace std;

void loop_print(const char* str, int n = 0);
/*函数声明，默认参数在声明处定义*/
int main()
{
     loop_print("Hello World!");
     loop_print("Hello World!");
     loop_print("Hello World!", 5);
     return 0;
}
void loop_print(const char* str, int n)
/*默认参数此处不表示*/
{
   static int func_count = 0;
   /*静态变量存储函数运行次数*/
   func_count++;
   if(n == 0)
   {
      cout<<"Arguments = 0 ;\n";
      cout<<str<<endl;
      /*若参数为 0，则输出一次*/
   }else{
       cout<<"Arguments != 0;\n";
      for(int i = 0;i < func_count; i++)
      {
         cout<<str<<endl;
      }
      /*若参数非 0 ，则使用静态变量循环输出*/
   }
}
```

2．CandyBar 结构体包含 3 个成员。第 1 个成员存储 candy bar 的品牌名称；第 2 个成员存储 candy bar 的重量（可能有小数）；第 3 个成员存储 candy bar 的热量（整数）。请编写一个程序，它使用一个这样的函数，即以 CandyBar 的引用、char 指针、double 值和 int 值作为参数，并用最后 3 个值设置相应的结构体成员。最后 3 个参数的默认值分别为 Millennium Munch、2.85 和 350。另外，该程序还包含一个以 CandyBar 的引用为参数并显示结构体内容的函数。请尽可能使用 const。

编程分析：

依照题义，定义 CandyBar。

```
struct CandyBar{
      string brand;
      float weight;
      int calorie;
};
```

程序需要使用两个函数，一个负责使用参数数据初始化 CandyBar 的结构体对象，另一个负责输出结构体对象的内容。依据函数功能，输出函数应当使用 const 关键字维护引用对象的数据安全。初始化函数需要设置默认参数值，因此两个函数的原型如下。

```
void create_candy(CandyBar& candy, string s = "Millennium Munch",float w = 2.85, int c = 350);
void show_candy(const CandyBar& candy);
```

完整代码如下。

```
/*第 8 章的编程练习 2*/
#include <iostream>
#include <string>
using namespace std;

struct CandyBar{
    string brand;
    float weight;
    int calorie;
};
void create_candy(CandyBar& candy, string s = "Millennium Munch",float w = 2.85, int c = 350);
void show_candy(const CandyBar& candy);
/*函数声明，默认参数在声明处指定*/
int main()
{
    CandyBar cb;
    create_candy(cb);
    /*使用默认参数，创建变量 cb*/
    show_candy(cb);
    create_candy(cb,"Nestle",1.2,200);
    /*使用非默认参数，创建变量 cb*/
    show_candy(cb);
    return 0;
}
void create_candy(CandyBar& candy, string s ,float w, int c)
{
    candy.brand = s;
```

```
    candy.weight = w;
    candy.calorie = c;
    /*因为使用了 string ，所以可以使用直接赋值形式，对于字符数组，需要调用函数*/
}
void show_candy(const CandyBar& candy)
{
    cout<<"The candybar is made by "<<candy.brand;
    cout<<" and its weight "<<candy.weight<<", ";
    cout<<candy.calorie <<" calorie"<<endl;
}
```

3．编写一个函数，它以一个指向 string 对象的引用作为参数，并将该 string 对象的内容转换为大写，为此可使用表 6.4 描述的函数 toupper()。然后编写一个程序，它通过使用一个循环实现用不同的输入来测试这个函数。该程序的运行情况如下。

```
Enter a string (q to quit): go away
Go AWAY
Next string (q to quit): good grief!
GOOD grief!
Next string (q to quit): q
Bye.
```

编程分析：

本题中函数的功能是将参数的 string 对象中的所有字符转换为大写字符。因为参数要使用引用类型，所以直接修改实参的字符串内容，不能使用 const 关键字修饰参数。通过以下语句声明函数的原型。

```
void uppercase(string& s);
```

完整代码如下。

```
/*第 8 章的编程练习 3*/
#include <iostream>
#include <string>
using namespace std;

void uppercase(string& s);
/*函数声明*/
int main()
{
    string st;
    cout<<"Enter a string (q to quit): ";
    getline(cin, st);
    while(st != "q")
    /*输入字符 q 退出循环*/
    {
        uppercase(st);
        cout<<st<<endl;
        cout<<"Next string (q to quit): ";
        getline(cin, st);
    }
```

```
    cout<<"Bye."<<endl;
    return 0;
}
void uppercase(string& s)
{
    for( int i = 0; i < s.size(); i++){
        s[i] = toupper(s[i]);
    }
    /*使用字符串的数组特性，逐一修改大小写*/
}
```

4．下面是一个程序框架。

```
#include <iostream>
using namespace std;
#include <cstring>              //为了使用 strlen()、 strcpy()
struct stringy {
    char *str;                  //指向字符串
    int ct;                     //字符串的长度（不包括'\0'）
};
// set()、 show()和 show() 的原型
int main()
{
    stringy beany;
    char testing[] = "Reality isn't what it used to be.";

    set(beany, testing);        //引用的第 1 个参数,
    //分配空间来保存 testing 的副本,
    //使 beany 的 str 成员指向新的块,
    //把 testing 复制到新的块，并设置 beany 的 ct 成员
    show(beany);                //指向成员字符串 1 次
    show(beany, 2);             //指向成员字符串 2 次
    testing[0] = 'D';
    testing[1] = 'u';
    show(testing);              //指向 testing 字符串 1 次
    show(testing, 3);           //指向 testing 字符串 3 次
    show("Done!");
    return 0;
}
```

请提供其中描述的函数和原型，从而完成该程序。注意，应有两个 show()函数，每个都使用默认参数。请尽可能使用 const 参数。set()使用 new 分配足够的空间来存储指定的字符串。这里使用的方法与设计和实现类时使用的方法相似。（可能必须修改头文件的名称，删除 using 编译指令，这取决于所用的编译器。）

编程分析：

题目给定的程序框架内包含了自定义函数 set()和 show()，从 set()函数的调用过程可知，其第 1 个参数 beany 是一个引用，为 beany 对象内成员分配空间来保存第 2 个参数 testing 的副本，功能是将第 2 个参数复制到第 1 个参数内。show()函数有 4 种调用形式，因此当第

1 个参数是 stringy 时，其第 2 个参数有默认值；第 1 个参数是字符串的引用时，其第 2 个参数也有默认值。但是由于字符串和 stringy 是两个不同类型，因此有函数重载。完整代码如下。

```
/*第 8 章的编程练习 4*/
#include <iostream>
using namespace std;
#include <cstring>
/*使用 C 风格字符串的处理函数，添加头文件 cstring*/
struct stringy {
    char *str;
    int ct;
};
void show(const string&, int n = 0);
void show(const stringy&,int n = 0);
void set(stringy&, char*);
/*函数原型声明，函数重载 */
int main()
{
    stringy beany;
    char testing[] = "Reality isn't what it used to be.";

    set(beany, testing);
    //第 1 个参数是一个引用，分配空间来保存 testing 副本，
    //使 beany 的 str 成员指向新的块，将 testing 复制到新的块，
    //并设置 beany 的 ct 成员
    show(beany);
    show(beany, 2);
    testing[0] = 'D';
    testing[1] = 'u';
    show(testing);
    show(testing, 3);
    show("Done!");
    /*beany 内的创建的动态存储未回收，可在程序结束前使用 delete 回收
     *例如：delete beany.str;*/
    return 0;
}
void show(const string& st, int n)
{
     if( n == 0) n++;
     for(int i = 0; i < n ; i++ )
     {
        cout<<st<<endl;
     }
}
/*输出 string 类型对象的信息*/
void show(const stringy& sty, int n)
{
     if( n == 0) n++;
     for(int i = 0; i < n; i++ )
```

```
        {
            cout<<sty.str<<endl;
        }
}
/*输出 stringy 类型对象的信息*/
void set(stringy& sty, char*st)
{
    sty.ct = strlen(st);
    sty.str = new char[sty.ct];
    /*通过 new 创建动态存储，此处不考虑回收*/
    strcpy(sty.str,st);
    /*复制字符串内容*/
}
```

5．编写模板函数 max5()，它以一个包含 5 个 T 类型元素的数组作为参数，并返回数组中最大的元素（由于长度固定，因此可以在循环中使用硬编码，而不必通过参数来传递）。在一个程序中使用该函数，将 T 替换为一个包含 5 个 int 值的数组和一个包含 5 个 double 值的数组，以测试该函数。

编程分析：

模板函数 max5()以 T 类型的数组作为元素，数组长度为固定长度；该函数的功能是查找数组内的最大元素，因此其返回值为类型 T。声明方式如下。

```
template<typename T> T max5(T[]);
```

其中函数并未指定数组的最大长度，如有必要，可以使用以下语句。

```
template<typename T> T max5(T[],int index = 5);
```

完整代码如下。

```
/*第 8 章的编程练习 5*/
#include <iostream>
using namespace std;

template<typename T> T max5(T[]);
/*模板函数的声明*/
int main()
{
    int arr[5] = {1,2,5,4,3};
    double arr_d[5] = {19.6,13,19.8,100.8,98.4};
    cout<<"The Max Element of int array: "<<max5(arr)<<endl;
    cout<<"The Max Element of double array: "<<max5(arr_d)<<endl;
    /*调用模板函数 统计数组中的最大值*/
return 0;
}
template<typename T> T max5(T st[])
{
    T max = st[0];
    for(int i = 0; i < 5; i++)
    {
        if(max < st[i]) max = st[i];
```

```
    }
    return max;
    /*通过循环，计算 5 个元素中的最大值，此处允许指定数组的长度，
     *否则需要通过参数传递数组长度*/
}
```

6. 编写模板函数 maxn()，它以由一个 T 类型元素组成的数组和一个表示数组元素数目的整数作为参数，并返回数组中最大的元素。在程序中对它进行测试，该程序使用一个包含 6 个 int 元素的数组和一个包含 4 个 double 元素的数组来调用该函数。程序还包含一个具体化，它以 char 指针数组和数组中的指针数量作为参数，并返回最长的字符串的地址。如果有多个这样的字符串，则返回其中第 1 个字符串的地址。使用由 5 个字符串指针组成的数组来测试该具体化。

编程分析：

题目在编程练习 5 的基础上，添加了模板函数中参数数组的长度值，因此可以使用如下声明。

```
template<typename T> T maxn(T[], int);
```

此外，模板函数还需要具体化 char 指针数组，因此具体化声明如下。

```
template<> char* maxn<char*>(char*[], int);
```

完整代码如下。

```
/*第 8 章的编程练习 6*/
#include <iostream>
#include <cstring>

using namespace std;

template<typename T> T maxn(T[], int);
template<> char* maxn<char*>(char*[], int);
/*模板参数的声明和模板的具体化声明*/
int main()
{
    int arr[5] = {1,2,5,4,3};
    double arr_d[5] = {19.6,13,19.8,100.8,98.4};
    string ss[] = {"Hello","Hello World!"};

    cout<<"The Max Element of int array: "<<maxn(arr,5)<<endl;
    cout<<"The Max Element of double array: "<<maxn(arr_d,5)<<endl;
    cout<<"The Max Element of string: "<<maxn(ss,2)<<endl;
    return 0;
}
template<typename T> T maxn(T st[], int n)
{
    T max = st[0];
    for(int i = 0; i < n; i++)
    {
        if(max < st[i]) max = st[i];
```

```
    }
    return max;
}
/*定义模板函数*/
template<> char*maxn<char*>(char*sst[], int n)
        {
    int pos = 0;
    for(int i = 0; i < n; i++)
    {
        if(strlen(sst[pos]) < strlen(sst[i]) ) pos = i;
    }
    return sst[pos];
}
/*定义模板函数的具体化*/
```

7．修改程序清单 8.14，使其使用两个名为 SumArray()的模板函数来返回数组元素的总和，而不是显示数组的内容。程序应显示 thing 的总和以及所有 debt 的总和。

编程分析：

程序清单 8.14 定义了两个 ShowArray()模板函数，用来显示数组信息，题目要求添加两个 SumArray()模板函数，以计算数组的元素总和并返回。结合原有的 ShowArray()模板函数，声明方式如下。

```
template<typename T> T SumArray(T arr[], int n);
template<typename T> T SumArray(T* arr[], int n);
```

相应代码如下。

```
/*第 8 章的编程练习 7*/
//tempover.cpp —— 重载模板
#include <iostream>

template <typename T>//模板 A
void ShowArray(T arr[], int n);

template <typename T>//模板 B
void ShowArray(T*arr[], int n);
/*原 ShowArray()的模板函数声明*/
template<typename T> T SumArray(T arr[], int n);
template<typename T> T SumArray(T* arr[], int n);
/*新添加的模板函数*/
struct debts
{
    char name[50];
    double amount;
};

int main()
{
    using namespace std;
    int things[6] = {13, 31, 103, 301, 310, 130};
```

```
    struct debts mr_E[3] = {
            {"Ima Wolfe", 2400.0},
            {"Ura Foxe", 1300.0},
            {"Iby Stout", 1800.0}
    };
    double *pd[3];
    //设置指针，指向结构体 mr_E 中的成员数
    for (int i = 0; i < 3; i++)
        pd[i] = &mr_E[i].amount;
    cout << "Listing Mr. E's counts of things:\n";
//things 是整型数组
    ShowArray(things, 6);  //使用模板 A
    cout << "Listing Mr. E's debts:\n";
//pd 是指针数组，指向双精度浮点数
    ShowArray(pd, 3);     //使用模板 B
    cout<<"The sum of things: "<<SumArray(things, 6)<<endl;
    cout<<"The sum of pd: "<<SumArray(pd, 3)<<endl;
/*调用模板函数，计算数组中元素的和，并输出元素的和*/
    return 0;
}
template <typename T>
void ShowArray(T arr[], int n)
{
    using namespace std;
    cout << "template A\n";
    for (int i = 0; i < n; i++)
        cout << arr[i] << ' ';
    cout << endl;
}
template <typename T>
void ShowArray(T *arr[], int n)
{
    using namespace std;
    cout << "template B\n";
    for (int i = 0; i < n; i++)
        cout << *arr[i] << ' ';
    cout << endl;
}
template<typename T>
T SumArray(T arr[], int n)
{
    using namespace std;
    T sum = 0;
    /*T 为模板参数输入类型，0 为 int 类型，通常可以进行类型转换，特定情况下可以使用
     *T sum = arr[0] - arr[0];
     *来标识此处未指定类型的初始化*/
    for (int i = 0; i < n; i++)
       sum += arr[i];
    return sum;
}
```

```
template<typename T>
T SumArray(T*arr[], int n)
{
    using namespace std;
    T sum = *arr[0] - *arr[0];
    for (int i = 0; i < n; i++)
        sum += *arr[i];
    return sum;

}
```

第 9 章　内存模型和名称空间

本章知识点总结

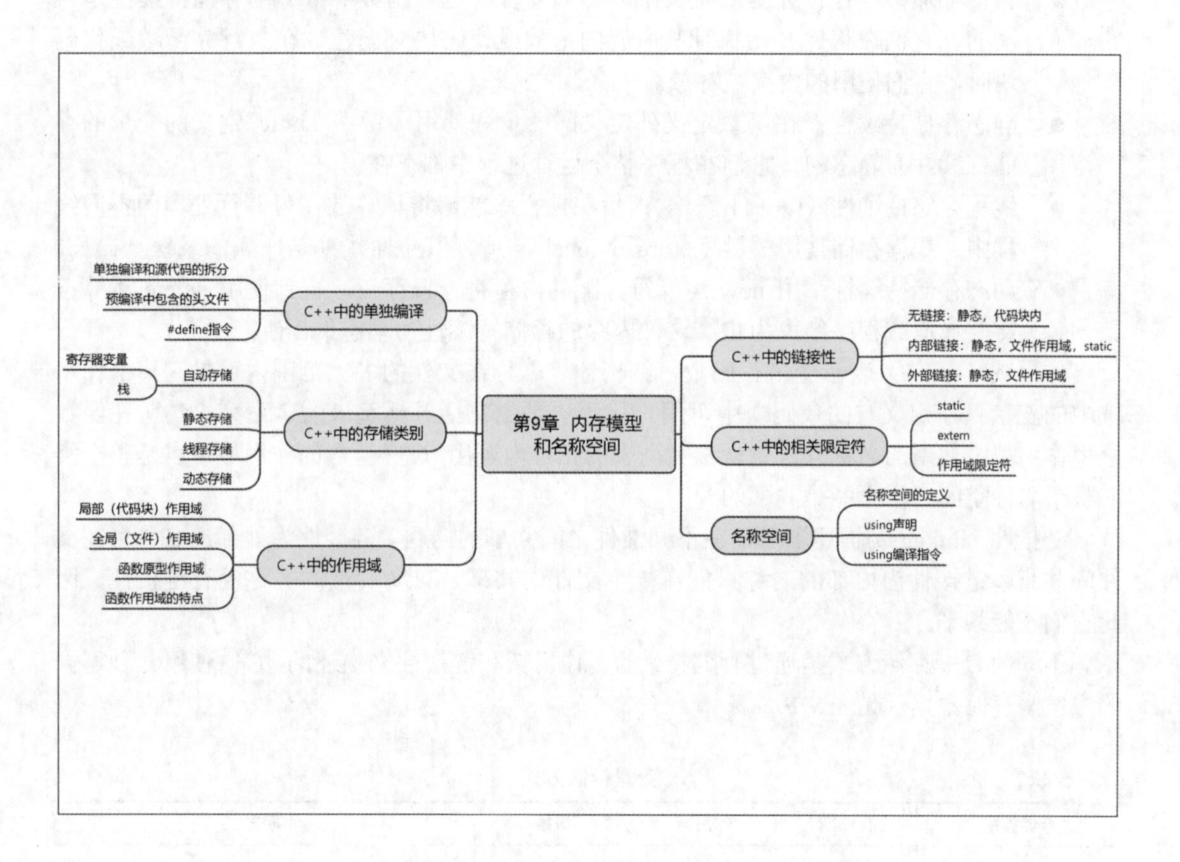

9.1　C++语言的多文件编译

在大规模软件的开发过程中，需要使用大量不同功能的函数模块，以及相关的常量、变量等。在一个文件内过于复杂的代码不利于程序员设计、更新和修改代码。因此 C++程序开发中建议将函数、变量依据其分属的不同功能模块，分别存储在不同的源代码文件中，在编译阶段编译器会对不同文件进行分别编译，形成中间文件，最后再由编译器统一链接成可执行文件。通常这些源代码文件分为 3 类：一类是以.h 结尾的头文件，通常包含定义的各种结构、类和函数原型；另一类文件是.cpp 结尾的文件，主要功能是实现.h 文件中定义的各种函数原型及相关代码；最后一类也是以.cpp 结尾的文件，即在前两种文件定义的函数和变量基础上实现程序的主要功能的文件，通常 main()函数单独定义在该文件内。在涉及多个文件的源代码中，同一个变量或者相关数据需要使用特殊的存储方式。

9.2 C++中的变量存储方式

C++语言使用多种方案来存储数据，这些方案的区别就在于数据保留在内存中的时间。

- 自动存储持续性：在函数定义中声明的变量（包括函数参数）的存储持续性是自动的。它们在程序开始执行其所属的函数或代码块时创建，在执行完函数或代码块时，它们使用的内存被释放。
- 静态存储持续性：在函数定义外定义的变量和使用关键字 static 定义的变量的存储持续性为静态的。它们在程序整个运行过程中都存在。
- 线程存储持续性（C++11 新增）：旨在使程序能够将计算放在可并行处理的不同线程中。线程存储使用关键字 thread_local 声明，其生命周期与所属的线程一样长。
- 动态存储持续性：用 new 运算符分配的内存将一直存在，直到使用 delete 运算符将其释放或程序结束为止。这种内存的存储持续性为动态的。

作用域（scope）表示变量、函数、结构体等名称在文件的多大范围内可见。局部作用域中的变量只在定义它的代码块中可用。代码块是由花括号括起来的一系列语句。在函数原型作用域中使用的名称只在包含参数列表的括号内可用。C++函数的作用域可以是整个类或整个名称空间，但不能是局部的。

链接性（linkage）描述了名称在不同文件之间共享的问题。链接性为外部的名称可在文件间共享，链接性为内部的名称只能在一个文件中共享。自动变量的名称没有链接性，因此它们不能共享。

不同的 C++存储方式是通过存储持续性、作用域和链接性来描述的。C++使用几种不同的方案来存储数据（见表 9.1）。

表 9.1 变量存储方式

存储类别	存储期	作用域	链接	声明方式
自动	自动	代码块	无	代码块内
寄存器	自动	代码块	无	代码块内，使用关键字 register
静态、无链接	静态	代码块	无	代码块内，使用关键字 static
静态、外部链接	静态	文件	外部	不在任何函数内
静态、内部链接	静态	文件	内部	不在任何函数内，使用关键字 static

9.3 C++中的名称空间

在应用 C++语言进行软件开发的过程中，随着开发规模的增大，开发过程中导入的第三方库、自定义的各种函数、变量、模板、类等都会急速增长。为了避免各种标识符的冲突，C++语言中采用名称空间来区分名称的作用域。通过不同的名称空间可以分离第三方库和用户自定义的名称，程序设计中只需要保证同一名称空间内的名称不冲突即可。名称空间的定义方式如下。

```
namespace Name_By_User{
int n;
};
```

这样定义在 Name_By_User 中的名称可以与其他名称空间内的名称相同,但是在使用这个名称空间内的数据时，需要通过预编译指令或者 using 语句显式地声明。其基本用法为

```
using namespace Name_By_User;//预编译指令
```

或者

```
using Name_By_User::n;
```

两者区别在于预编译指令能够一次导入该名称空间内所有名称，using 语句需要单独导入。此外，在名称冲突时，两者也会有部分差异，例如，名称空间和 using 声明的区域中存在相同的名称，如果在该区域使用 using 声明导入名称，则两个名称会发生冲突而出错。另外，如果使用 using 编译指令，则该区域的局部版本名称将会隐藏名称空间的版本。因此很多情况下认为 using 声明只导入需要的部分名称，它在使用上比 using 编译指令更安全。

9.4　复习题

1．对于下面的情况，应使用哪种存储方案?

a．homer 是函数的形参。

b．secret 变量由两个文件共享。

c．topsecret 变量由一个文件中的所有函数共享，但对于其他文件来说是隐藏的。

d．beencalled 记录包含它的函数被调用的次数。

习题解析：

a．函数的形参是自动变量的形式，因此 homer 变量的存储方案是自动存储。

b．若多个文件共享一个变量，应当在一个文件内将其定义为外部变量，其存储形式是静态外部链接。在其他需要使用 secret 变量的文件中使用 extern 关键字声明。

c．若 topsecret 变量在一个文件内共享且不能被其他文件访问，应当使用内部链接的静态变量，或者可以在未命名的名称空间中进行定义。

d．函数的函数体内定义的变量通常为自动变量，为了保存函数被调用次数，beencalled 需要使用本地静态变量来实现这个功能。

2．using 声明和 using 编译指令之间有何区别?

习题解析：

using 声明可以单独使用名称空间中某个特定的名称，其作用域与 using 所在的声明区

域相同。using 编译指令使名称空间中的所有名称可用。如果在全局中使用 using 编译指令，将使该名称空间中的名称在全局可用；如果在函数定义中使用 using 编译指令，将会在该函数中使该名称空间可用。此外，在名称冲突时，两者也会有部分差异，例如，名称空间和 using 声明的区域存在相同的名称，如果在该区域中使用 using 声明导入名称，则两个名称会发生冲突而出错。另外，如果使用 using 编译指令，则该区域的局部版本名称将会隐藏名称空间的版本。因此很多情况下认为 using 声明只导入需要的部分名称，它在使用上比 using 编译指令要更安全。

3．重新编写下面的代码，避免使用 using 声明和 using 编译指令。

```
#include <iostream>
using namespace std;
int main(){
    double x;
    cout<<"Enter value:";
    while(!(cin>>x)){
        cout<<"Bad input.Please enter a number:";
        cin.clear();
        while(cin.get()!='\n')
            continue;
    }
    cout<<"Value="<<x<<endl;
    return 0;
}
```

习题解析：

题目给定的程序代码中主要使用了 std 名称空间中的内容，如果不使用 using 声明和编译指令，那么可以直接在 std 内的名称前添加作用域运算符 std::。这里主要使用的名称包括 cin、cout、endl，因此修改后的代码如下。

```
#include <iostream>
int main(){
    double x;
    std::cout<<"Enter value:";
    while(!(std::cin>>x)){
        std::cout<<"Bad input.Please enter a number:";
        std::cin.clear();
        while(std::cin.get()!='\n')
            continue;
    }
    std::cout<<"Value="<<x<<std::endl;
    return 0;
}
```

4．重新编写下面的代码，使之使用 using 声明，而不是 using 编译指令。

```
#include<iostream>
using namespace std;
```

```
int main(){
      double x;
      cout<<"Enter value:";
      while(!(cin>>x)){
          cout<<"Bad input.Please enter an umber:";
          cin.clear();
          while(cin.get()!='\n')
              continue;
      }
      cout<<"Value="<<x<<endl;
}
```

习题解析：

为了使用 using 声明替代 using 编译指令，需要查询当前函数或者文件使用了哪些名称，并依次声明，本程序主要使用了 cin、cout、endl 这 3 个名字，因此要使用以下 using 声明。

```
using std::cout;
using std::cin;
using std::endl;
```

完整代码如下。

```
#include<iostream>
int main(){
     using std::cout;
     using std::cin;
     using std::endl;
     double x;
     cout<<"Enter value:";
     while(!(cin>>x)){
          cout<<"Bad input.Please enter a number:";
          cin.clear();
          while(cin.get()!='\n')
              continue;
          }
          cout<<"Value="<<x<<endl;
}
```

5．在一个文件中调用 average(3,6)函数时，它返回两个 int 型参数的 int 型平均值；在同一个程序的另一个文件中调用时，它返回两个 int 型参数的 double 型平均值。应如何实现?

习题解析：

在同一个程序的不同文件中使用不同函数，且由于两个 average()函数的参数相同（即特征标相同），因此不能重载函数。解决方案是定义不同的名称空间，调用时使用作用域解析；或者在每个文件中包含单独的静态函数定义，限制其是内部链接函数。

6．下面的程序由两个文件组成，该程序显示什么内容?

```
//file1.cpp

#include <iostream>
```

```
using namespace std;
void other();
void another();
int x=10;
int y;

int main(){
     cout<<x<<endl;
     {
         int x=4;
         cout<<x<<endl;
         cout<<y<<endl;
     }
     other();
     another();
     return 0;
}
void other(){
     int y=1;
     cout<<"Other:"<<x<<","<<y<<endl;
}

//file2.cpp
#include <iostream>
using namespace std;
extern int x;
namespace{
    int y=-4;
}
void another(){
    cout<<"another():"<<x<<","<<y<<endl;
}
```

习题解析：

为了分析该程序的输出，首先查看 main ()函数内的语句。

- `cout<<x<<endl;`语句输出本文件内外部链接的静态变量 x，所以程序会输出 10。

接下来的 3 条语句在一个单独的语句块内，该语句块中定义的 x 隐藏了外部链接的 x，因此赋值后输出 4，y 值未被隐藏，这时使用外部链接的静态变量 y，输出结果是初始化结果 0。

- `other();`语句调用文件内的函数 other()，该函数内部定义的局部变量 y 隐藏了外部变量 y，但 x 依然使用外部变量 x，输出结果为：`Other:10,1`。
- `another();`语句调用外部文件中的函数 another()，该函数在 file2.cpp 中，该文件使用了 file1.cpp 中的 x 变量，但是在文件的名称空间中定义了 `y=-4`，因此 other()函数的 x 为 file1.cpp 内外部链接的 x 变量，y 为 file2.cpp 内无链接的变量 y，输出结果为 `another():10-4`。

所以最终显示如下内容。

```
10
4
0
Other: 10，1
another():10-4
```

7．下面的代码将显示什么内容?

```
#include<iostream>
using namespace std;
void other();
namespace n1
{
    int x=1;
}
namespace n2
{
    int x=2;
}

int main(){
     using namespace n1;
     cout<<x<<endl;
     {
         int x=4;
         cout<<x<<","<<n1::x<<","<<n2::x<<endl;
     }
     using n2::x;
     cout<<x<<endl;
     other();
     return 0;
}
void other(){
     using namespace n2;
     cout<<x<<endl;
     {
         int x=4;
         cout<<x<<","<<n1::x<<","<<n2::x<<endl;
     }
     using n2::x;
     cout<<x<<endl;
}
```

习题解析：

为了分析该程序的输出，首先查看 main ()函数内的语句。

● `using namespace n1;`
`cout<<x<<endl;`

在 main()函数的前两条语句中，声明了名称空间 n1，因此第 2 条语句的输出 x 为 n1

内定义的 x，所以输出 1。

- {

```
int x=4;
cout<<x<<","<<n1::x<<","<<n2::x<<endl; }
```

以上语句首先定义了一个语句块，块内定义的变量 x 会隐藏其他名称空间中的同名变量，cout 输出语句中，通过作用域运算符指定的名称空间不会被块内的局部变量隐藏。因此输出 4,1,2。

- using n2::x;

```
cout<<x<<endl;
```

以上两条语句首先通过 using 声明使用 n2 内的名称，因此 cou 输出 2。

- other( );

以上语句调用函数 other()。函数内声明了名称空间 n2，因此首先输出 2。

在函数内的语句块中，再次造成名称的隐藏，因此输出局部变量 x 和作用域运算符修饰的 x，即输出 4,1,2。

函数最后再次使用 using n2::x，上一个语句块内局部变量的生命周期结束，不会造成名称的冲突或隐藏，因此输出 2。

最终程序输出以下结果。

```
1
4,1,2
2
2
4,1,2
2
```

9.5 编程练习

1．下面是一个头文件。

```
//golf.h—forpr9-1.cpp

const int Len = 40;
struct golf
{
    char fullname[Len];
    int handicap;
};
//非交互式版本:
//使用传递给函数的参数值，函数为提供的名称和 handicap 设置 golf 结构体
void setgolf(golf& g,const char* name,int hc);

//交互式版本:
//函数申请用户的名称和 handicap
//设置输入的 g 的个数
//如果输入名称，返回 1 ；如果名称是空字符串，返回 0
```

```
int setgolf(golf& g);

//函数重置 handicap 为新的值
void handicap(golf& g,int hc);

//函数显示 golf 结构体的信息
void showgolf(const golf& g);
```

注意，setgolf()被重载，可以这样使用其第 1 个版本：

```
golf ann;
setgolf(ann,"AnnBirdfree",24);
```

上述函数调用提供了存储在 ann 结构体中的信息。可以这样使用其第 2 个版本：

```
golf andy;
setgolf(andy);
```

上述函数将提示用户输入姓名和等级，并将它们存储在 andy 结构体中。这个函数可以（但是不一定必须）在内部使用第 1 个版本。

根据这个头文件创建一个多文件程序。其中的一个文件名为 golf.cpp，它提供了与头文件中的原型匹配的函数定义；另一个文件应包含 main()，并演示原型化函数的所有特性。例如，包含一个由用户输入的循环，并使用输入的数据来填充一个由 golf 结构体组成的数组，当数组被填满或用户将高尔夫选手的姓名设置为空字符串时，循环将结束。main()函数只使用头文件中原型化的函数来访问 golf 结构体。

编程分析：

题目提供了 golf.h 头文件，文件内定义了 golf 结构体和相关函数的原型，其中 setgolf()函数被重载：

```
setgolf(ann,"AnnBirdfree",24);
setgolf(andy);
```

第 1 个重载函数直接使用参数初始化 golf 结构体变量，第 2 个需要通过标准输入初始化结构体变量，两个函数都使用结构体引用作为参数。程序需要在 golf.cpp 文件中提供函数的定义，并创建一个包含 main()函数的.cpp 文件，与前两个文件进行多文件编译，并测试定义的函数。联合编译中.cpp 文件应当使用#include “golf.h” 头文件。

```
/*第 9 章的编程练习 1*/
//golf.cpp
#include <iostream>
#include <cstring>
#include "golf.h"
/*添加 golf.h 头文件*/
using namespace std;

int main(int argc, char *argv[])
{
    golf ann;
    setgolf(ann,"AnnBirdfree",24);

    golf andy;
    setgolf(andy);
```

```
    showgolf(ann);
    showgolf(andy);
}

void setgolf(golf& g,const char* name, int hc)
{
    strcpy(g.fullname, name);
    g.handicap = hc;
}
/*setgolf()函数的定义*/

int setgolf(golf& g)
{
    char name[Len];
    int hc;
    cout<<"Please enter the name: ";
    cin.getline(name,Len);

    cout<<"Please enter the handicap: ";
    while(!(cin >> hc))
    {
        cin.clear();
        while(cin.get() != '\n')
            continue;
        cout<<"Please enter the golf's handicap: ";
    }
    /*判断 hc 是否正确地输入整型数据*/
    if(name[0] != '\0')
    {
        setgolf(g,name,hc);
        return 1;
    }else{
        return 0;
    }
}
/*以交互方式创建 golf 对象*/

void handicap(golf& g,int hc)
{
    g.handicap = hc;
}

void showgolf(const golf& g)
{
    cout<<"Name : "<<g.fullname<<", Handicap is "<<g.handicap<<endl;
}
```

2．修改程序清单 9.9，要求用 string 对象代替字符数组。这样，该程序将不再需要判断输入的字符串是否过长，同时可以将输入字符串同字符串""进行比较，以判断输入内容是否为空行。

程序清单 9.9 如下。

```
//static.cpp —— 使用静态的局部变量
#include <iostream>
//常量
const int ArSize = 10;

//函数原型
void strcount(const char *str);

int main()
{
    using namespace std;
    char input[ArSize];
    char next;

    cout << "Enter a line:\n";
    cin.get(input, ArSize);
    while (cin)
    {
        cin.get(next);
        while (next != '\n')          //不合适的字符串
            cin.get(next);            //处理其余的
        strcount(input);
        cout << "Enter next line (empty line to quit):\n";
        cin.get(input, ArSize);
    }
    cout << "Bye\n";
    return 0;
}

void strcount(const char *str)
{
    using namespace std;
    static int total = 0;           //静态的局部变量
    int count = 0;                  //自动的局部变量
    cout << "\"" << str <<"\" contains ";
    while (*str++)                  //到达字符串末尾
        count++;
    total += count;
    cout << count << " characters\n";
    cout << total << " characters total\n";
}
```

编程分析：

程序清单 9.9 的 strcount() 函数通过静态变量 total 存储该函数调用后所统计的每一个字符串的长度和，并输出结果。程序中字符数组的长度为 10，因此只能读取和存储 9 个字符，修改为 string 类型后程序的适用范围更广。完整代码如下。

```
/*第 9 章的编程练习 2*/
//static.cpp ——使用静态的局部变量
#include <iostream>
#include <string>
/*添加相应头文件*/
//常量

//函数原型
void strcount(const std::string str);

int main()
{
    using namespace std;
    string input;
    /*替换原字符数组形式，使用 string 类型*/
    char next;

    cout << "Enter a line:\n";
    getline(cin,input);
    /*string 从 cin 读取输入，需要使用 getline()函数*/
    while (input != "")
    /*要判断字符串是否为空，可以直接使用比较运算符*/
    {
        strcount(input);
        cout << "Enter next line (empty line to quit):\n";
        getline(cin,input);
    }
    cout << "Bye\n";
    return 0;
}

void strcount(const std::string str)
{
    using namespace std;
    static int total = 0;           //静态局部变量
    int count = 0;                  //自动局部变量

    cout << "\"" << str <<"\" contains ";
    /*可以使用字符串末尾的空字符判断结束
     *while(str[count])             //到达字符串末尾
     *count++;
     *但是对于 string，要判断长度可以直接使用 string 的函数
     **/
    count  = str.length();
    total += count;
    cout << count << " characters\n";
    cout << total << " characters total\n";
}
```

3．下面是一个结构体的声明。

```
struct chaff{
       char dross[20];
       int slag;
};
```

编写一个程序，使用定位 new 运算符将一个包含两个这种结构体的数组放在一个缓冲区中。然后为结构体的成员赋值（对于 char 数组，使用函数 strcpy()），并使用一个循环来显示内容。一种方法是像程序清单 9.10 那样以一个静态数组作为缓冲区；另一种方法是使用常规 new 运算符来分配缓冲区。

编程分析：

C++语言中的 new 运算符实现了存储的动态分配，其本质是在存储堆中分配一个能够满足要求的内存块。定位 new 运算符则能够让程序员指定使用的内存的具体位置，即为 new 运算符提供具体的存储地址范围。定位 new 运算符直接使用传递给它的地址，它不负责判断哪些内存单元已被使用，也不查找未使用的内存块。使用方法如下。

```
char buffer[512];  //手动创建内存池
int* p2 = new (buffer) int[10];   //定位 new，分配的内存位于 buffer 中
```

在此基础上，程序的完整代码如下。

```
/*第 9 章的编程练习 3*/
#include <iostream>
using namespace std;
struct chaff{
    char dross[20];
    int slag;
};

int set_chaff(chaff&, char*, int);
void show_chaff(const chaff&);
/*函数声明*/

int main()
{
    char buffer[1024];
    /*创建缓冲区，使用定位 new
    *char*buffer =  new char[1024];
    *要使用动态存储创建缓冲区，需要在程序结束前使用 delete 回收存储空间
    **/
    char st[20];
    int slag, n = 0;
    chaff* pcf = new (buffer) chaff[2];
    /*使用定位 new 运算符，在 buffer 内分配存储单元*/
    cout<<"Enter String to set chaff: ";
    cin.getline(st,20);
    cout<<"Enter a number: ";
    cin>>slag;
    while(strlen(st) > 0)
    {
       cin.get();
```

```
        set_chaff(pcf[n++], st, slag);
        if(n >= 2) break;
        /*简易判断输入数组是否已满*/
        cout<<"Enter String to set chaff: ";
        cin.getline(st,20);
        cout<<"Enter a number: ";
        cin>>slag;
    }
    for(int i = 0; i < 2; i++ )
        show_chaff(pcf[i]);
    return 0;
}
int set_chaff(chaff& cf, char* str, int n)
{
    if(strlen(str) > 0)
    {
        strcpy(cf.dross, str);
        /*对于字符数组形式的字符串，可以直接复制*/
        cf.slag = n;
        return 1;
    }else{
        return 0;
    }
}
void show_chaff(const chaff& cf)
{
    cout<<"Chaff's Dross: " <<cf.dross<<endl;
    cout<<"Chaff's slag: "<<cf.slag<<endl;
}
```

4．请基于下面这个名称空间编写一个由 3 个文件组成的程序。

```
namespace SALES
{
    const int QUARTERS = 4;
    struct Sales
    {
        double sales[QUARTERS];
        double average;
        double max;
        double min;
    };
//把数组 ar 中的项（最多 4 项）复制到 s
//的 sales 成员中，计算 s，存储转入的项
//的平均值，最大值和最小值，
//如果 sales 有其他元素，把它们设置为 0
   void setSales(Sales& s, const double ar[], int n);
//收集 4 个季度的销量，在 s 的 sales 成员中存储它们，并计算均值、最大值和最小值
    void setSales(Sales& s);
//显示结构体中的所有信息
    void showSales(const Sales& s);
}
```

第 1 个文件是一个头文件，其中包含名称空间；第 2 个文件是一个源代码文件，它对

这个名称空间进行扩展，以提供这 3 个函数的定义；第 3 个文件声明两个 Sales 对象，并使用 setSales()的交互式版本为一个结构体提供值，然后使用 setSales()的非交互式版本为另一个结构体提供值。另外，第 3 个文件还使用 showSales()来显示这两个结构体的内容。

编程分析：

名称空间定了 Sales 结构体和 3 个函数，其中重载了以下两个函数。

```
void setSales(Sales& s, const double ar[], int n);
void setSales(Sales& s);
```

这里分别通过参数直接初始化 Sales 结构体对象和以交互形式初始化 Sales 结构体对象，并在 main( )函数内分别测试名称空间中的函数。完整代码如下。

```
/*第 9 章的编程练习 4*/
//sales.h
/*sales.h 头文件定义了 SALES 名称空间及相关的函数*/
namespace SALES
{
    const int QUARTERS = 4;
    struct Sales{
        double sales[QUARTERS];
        double average;
        double max;
        double min;
    };
    void setSales(Sales& s, const double ar[], int n);
    void setSales(Sales& s);
    void showSales(const Sales& s);
}


//sales.cpp
/*sales.cpp 定义了 sales.h 内的所有函数实现*/
#include <iostream>
#include "sales.h"

using namespace std;
using namespace SALES;
/*预编译指令包含 SALES 名称空间*/
void SALES::setSales(Sales& s, const double ar[], int n)
/*对于 SALES 名称空间内的函数实现，可以使用作用域运算符标识函数归属*/
{
    double sum = 0;
    if(n >= QUARTERS){
       for(int i = 0;i < QUARTERS; i++)
       {
          s.sales[i] = ar[i];
       }
    }else{
       for(int i = 0;i < n; i++)
```

```
        {
            s.sales[i] = ar[i];
        }
        for(int i = n;i < QUARTERS; i++)
        {
            s.sales[i] = 0;
        }
    }
    /*综合考虑在输入数据与 QUARTERS 不匹配的情况下，如何初始化数据
     *对于丢弃的数据、不足的数据，补 0*/
    s.max = s.min = s.sales[0];
    for(int i = 0;i < QUARTERS; i++)
    {
        sum += s.sales[i];
        if(s.min>s.sales[i]) s.min = s.sales[i];
        if(s.max<s.sales[i]) s.max = s.sales[i];
    }
    /*初始化最大值和最小值*/
    s.average = sum / QUARTERS;
}

void SALES::setSales(Sales& s)
{
    double ar[QUARTERS] = {};
    int i = 0;
    do{
        cout<<"Enter a number: ";
        if(!(cin>>ar[i]))
        {
            cin.clear();
            while(cin.get()!='\n') continue;
            cout<<"ERROE, Reenter a number: ";
            cin>>ar[i];
        }
        i++;
    }while(i < QUARTERS);
    /*交互输入 QUARTERS 个数据，并存储在数组内*/
    setSales(s,ar,4);
    /*通过重载函数初始化 Sales*/
}

void SALES::showSales(const Sales& s)
{
    cout<<"This Salse's quarter list info:"<<endl;
    for(int i = 0; i < QUARTERS; i++)
    {
        cout<<"No."<<i+1<<": sales: "<<s.sales[i]<<endl;
    }
    cout<<"AVERAGE: "<<s.average<<endl;
    cout<<"MAX: "<<s.max<<endl;
```

```
    cout<<"MIX: "<<s.min<<endl;
}

//main.cpp
/*包含 main()函数的主程序*/
#include <iostream>
#include "sales.h"

using namespace std;
using namespace SALES;
/*通过 using 预编译指令，添加 SALES 名称空间*/
int main()
{
    Sales s1, s2;
    double de[QUARTERS] = {12,23,45,67};
    setSales(s1,de,QUARTERS);
    showSales(s1);
    setSales(s2);
    showSales(s2);
    /*调用 SALES 内函数，初始化 s1 与 s2，并显示内容*/
    return 0;
}
```

第 10 章　对 象 和 类

本章知识点总结

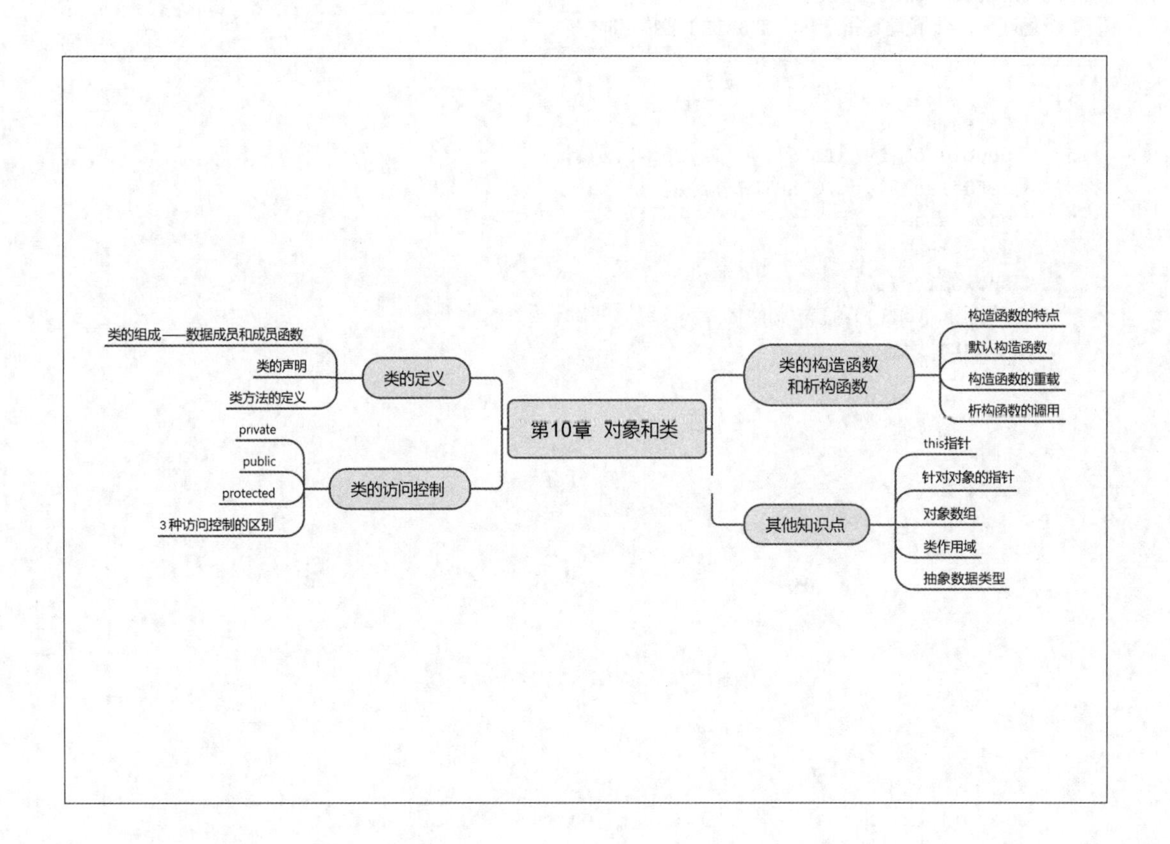

10.1　面向对象和类

在面向过程的程序设计中，开发人员的基本思路是分析程序的运行过程和每一步具体操作，然后再考虑如何细化并实现这些操作过程，以及细化过程中需要用到的数据和运算。而面向对象的程序设计思想则不同，它首先考虑程序中的数据，将数据作为程序的核心和关键，然后设计和表示数据，最后通过数据及数据的操作组合，实现相应的功能。面向对象的程序设计中的首要任务就是分析数据，并为每一个数据建立一个类型。这个类型不仅包含了数据的组成，而且包含了数据的操作或方法。这样由这个类型创建的每一个数据实体就具备了基本的组合能力，可以和其他数据进行信息的沟通和交流，从而组成一个复杂的系统。

C++中类的定义由 class 关键字表示，类内的数据对象称为数据成员，数据的操作函数称为成员函数，或者称为方法。类中的方法在定义中根据其对数据的操作类型分为两类：一类是读取数据成员的方法，通常会使用 get 表示；另一类是修改数据成员的方法，通常用 set 表示。定义完类之后，可以通过声明类变量来创建一个类的对象，或者通过指针和 new 创建类对象。所创建的每个新对象都有自己的存储空间，用于存储其内部变量和类成员。每一个对象都独立存储自己的数据成员，但同一个类的所有对象共享同一组类方法，即同一个类的每种方法只有一个副本。

10.2　C++中类的访问控制

在类中定义的名称的作用域为整个类，作用域为整个类的名称只在该类中已知，且类外是不可知的。类的作用域意味着不能从外部直接访问类的成员。也就是说，要访问或者使用类内的所有数据和方法，必须使用通过类生成的对象，以及成员运算符(.)、间接成员运算符(->)或者作用域解析运算符(::)。

除了类作用域能够控制和管理类内部的数据访问外，为了进一步实现类和对象的数据保护与内容的封装，C++在类的定义中使用关键字 private、public 和 protected 描述对类成员的访问控制。使用类对象的程序都可以直接访问公有部分，但只能通过公有成员函数来访问对象的私有成员。公有成员函数是程序和对象的私有成员之间的桥梁，提供了对象和程序之间的接口。也就是说，通常类内的数据成员需要设置称为私有成员，其基本访问和修改工作是由 get 与 set 类型的方法实现的，这样能够更加有效地维护类内的数据安全。

10.3　构造函数和析构函数

由于类内部的数据成员类型较复杂，可以包含数组、指针等复合类型，因此为了保证创建对象后数据成员的初始化，C++通过构造函数创建对象时，自动进行初始化工作。构造函数在构造新对象时将初始化的值赋给数据成员。同样，为了在对象销毁时能够安全地回收数据成员的存储单元，C++使用析构函数进行数据清理工作。构造函数具有特定格式，如构造函数名和类名相同，构造函数没有返回值（也不是 void）。如果用户未定义任何构造函数，系统会自动提供一个默认构造函数，默认构造函数是无参数的。但是如果用户定义了任何一个构造函数的重载实现，就必须重新定义并覆盖系统提供的默认构造函数；否则，编译器会发出警告。要表示析构函数的名称可以在类名前加～符号，析构函数也没有返回值，且没有参数。通常情况下析构函数的调用时间是由编译器决定的，而不能由用户显式地调用。

10.4 复习题

1．什么是类?

习题解析：

面向对象的思想把程序设计的关注点从程序运行的过程、步骤及其分解和组合，转向了分析程序中数据的属性和数据的操作。在这种思想的指导下 C++更加强调数据的抽象和设计。类就是一种用户自定义的关于数据的抽象的组合，它描述了程序中数据的基本构成（数据成员）以及对数据的访问和操作（成员函数或方法）。在类的基础上可以通过大量数据对象的组合和数据对象间的信息传递来组成完整的应用程序，实现最终的功能性目标。

2．类如何实现抽象、封装和数据隐藏?

习题解析：

类的定义本质上就是对数据的一种抽象过程，这种抽象一方面把复杂的数据结构分解为基本数据类型的组合，一方面定义和实现了对复杂数据类型的操作。因此，类的定义就是把程序中的数据还原成一种基本数据模型和相应操作方式的抽象化过程。C++语言中把类内的数据成员和成员函数通过类的作用域以及 3 种基本访问控制（public、private、protected）进行了数据封装，从类的外部无法访问类的基本数据，而只能通过类的对象实体和类内的公有成员才能够访问，这样类就实现了基本的数据隐藏和封装。

3．对象和类之间的关系是什么?

习题解析：

面向对象的程序设计中，类是一种数据及其操作的模型定义，也就是一种数据的抽象。通过类可以生成很多对象实体，同一个类生成的每一个对象实体都能够在存储空间内拥有自己的存储空间单元，实现数据的存储和访问操作。因此可以认为类是抽象模型，对象是由类这个数据抽象模型实例化之后创建的真实数据单元。

4. 除了类的函数成员是函数之外，类的函数成员与类的数据成员之间的区别还有什么?

习题解析：

从性质上看，类的数据成员用来描述类的基本数据构成，定义其存储空间和类型；而函数成员则是类的操作功能的定义。从存储上看，类所创建的每个新对象都有自己的存储空间，类的数据成员是每一个对象所独享的，但同一个类的所有对象共享同一组函数成员，即每种方法对于所有对象来说只有一个副本。

5. 定义一个类来表示银行账户。数据成员包括储户姓名、账号（两者使用字符串表示）和存款。成员函数执行如下操作。

- 创建一个对象并将其初始化。
- 显示储户姓名、账号和存款。
- 存入参数指定的存款。
- 取出参数指定的存款。

请提供类声明，而不用给出方法实现（编程练习 7 将要求编写实现）。

习题解析：

本题中的银行账号类包括 3 个数据成员，其中两个数据成员是字符串，存款应使用 double 类型数据。数据成员应设置为私有成员。成员函数包括构造函数、析构函数、输出账户信息的函数、存取款函数（对应存取款操作），因此可以设计如下类。

```
class BankAccount{
private:
    string fullname;
    string accid;
    double balance;
public:
    BankAccount();
    BankAccount(const string name, const string id, double bal);
    ~BankAccount();
    void init_account(const string name, const string id, double bal);
    void get_info() const;
    void deposit(double cash);
    void withdraw(double cash);
};
```

6. 类的构造函数在何时调用？类的析构函数呢？

习题解析：

类的构造函数是一个特殊的成员函数，在创建类对象时构造函数会被调用，用来初始化对象内的各数据成员。通常创建对象有两种情况。一种是直接创建类的变量，另外一种情况是通过 new 创建一个动态对象。这两种情况下都会自动调用构造函数。析构函数在类对象被回收、销毁时自动调用。对象被回收、销毁也有两种情况。一种是该变量在超过生命周期之后会被自动回收，另一种情况是动态对象通过 delete 回收时会自动调用析构函数。

7. 给出复习题 5 中银行账户类的构造函数的代码。

习题解析：

复习题 5 中银行账户类的构造函数原型如下。

```
BankAccount();
```

```
BankAccount(const string name, const string id, double bal);
```

其私有成员定义为 string 类型的字符串，必须包含头文件 string。string 的使用较简单，可以直接使用 string 的赋值功能。

```
BankAccount(const string name, const string id, double bal);
{
      fullname = name;
      accid = id;
      balance = bal;
}
BankAccount: :BankAccount ()
{
      fullname = "";
      accid = "";
      balance = 0;
}
```

如果使用了字符数组形式则需要使用 strncpy()，且包含头文件 cstring，例如，通过 strncpy(fullname, name)来实现数据成员的初始化。

8．什么是默认构造函数？拥有默认构造函数有何好处?

习题解析：

默认构造函数是指没有参数的构造函数，或者所有参数都有默认值的构造函数。默认构造函数能够创建类对象而不用初始化该对象，这种使用形式和基本数据类型类似，能够实现一种更加灵活和安全的对象创建方式。如果在定义类时没有定义构造函数，那么系统将会自动创建一个默认构造函数，但是如果用户定义了其他任何一个构造函数，就必须要手动定义一个默认构造函数。

9．修改 Stock 类的定义（stock20.h 中的版本)，使之包含返回各个数据成员值的成员函数。注意，返回公司名的成员函数不应为修改数组提供便利，也就是说，不能简单地返回 string 引用。

习题解析：

stock20.h 文件如下。

```
//stock20.h ——增强版本
#ifndef STOCK20_H_
#define STOCK20_H_
#include <string>

class Stock
{
private:
     std::string company;
     int shares;
     double share_val;
```

```
    double total_val;
    void set_tot() { total_val = shares *share_val; }
public:
    // Stock();        //默认构造函数
    Stock(const std::string & co, long n = 0, double pr = 0.0);
    ~Stock();          //不执行任何操作的析构函数
    void buy(long num, double price);
    void sell(long num, double price);
    void update(double price);
    void show()const;
    const Stock & topval(const Stock & s) const;
};

#endif
```

题目要求添加返回各个数据成员值的函数，因此应当包括以下函数。

```
int numshares() const ;
double shareval() const ;
double totalval() const ;
const string &co_name() const ;
```

这 4 个函数分别返回 4 个数据成员。完整代码如下。

```
#ifndef STOCK30_H_
#define STOCK30_H_
class Stock {
private:
    std::string company;
    long shares;
    double share_val;
    double total_val;

    void set_tot() {
       total_val = shares *share_val;
    }

public:
    Stock(); //默认构造函数
    Stock(const std::string &co, long n, double pr);
    ~Stock() {} //不执行任何操作的析构函数
    void buy(long num, double price);
    void sell(long num, double price);
    void update(double price);
    void show() const;
    const Stock &topval(const Stock &s) const;

    int numshares() const { return shares; }
    double shareval() const {
       return share_val;
    }
    double totalval() const {
```

```
        return total_val;
    }
    const string &co_name() const { return company; }
};
```

10．this 和*this 表示什么？

习题解析：

C++在类的定义中可以使用一个特殊的指针 this，该指针是指向由该类创建的对象本身的指针。请注意，该指针是指向类对象实体的指针，而不是针对类的。也就是说，this 本质上是每一个对象本身的地址，而*this 则是该对象本身。在类的定义过程中，可以通过 this 指针来调用成员函数的对象。this 虽然在类的定义中使用，但是它本质上是在类生成对象时关于具体对象的指针。

10.5 编程练习

1．为复习题 5 描述的类提供方法定义，并编写一个小程序来演示所有的特性。

编程分析：

复习题 5 定义了银行账号类，复习题 7 实现了该银行账号类的构造函数。本题需要在此基础上为剩余的函数提供所有的定义，主要包括存取款函数、输出账号信息的函数和初始化函数，并进行简单的功能测试。完整代码如下。

```
/*第 10 章的编程练习 1*/
#include <iostream>

using namespace std;
class BankAccount{
private:
    string fullname;
    string account;
    double deposit;
public:
    BankAccount();
    BankAccount(const string, const string, float);
    ~BankAccount();
    void init_account(const string, const string, float);
    void print_info() const;
    void save(float);
    void withdraw(float);
};
/*BankAccount 类的声明*/

int main()
{
```

```
    BankAccount ba("Nik","0001",1200);
    ba.print_info();
    ba.init_account("Nik Swit", "", 1500);
    ba.print_info();
    ba.save(223.5);
    ba.print_info();
    return 0;
}
/*在 main()函数内简单测试 BankAccount 类的功能*/

BankAccount::BankAccount()
{
    deposit = 0;
}
/*默认构造函数的定义*/
BankAccount::BankAccount(string name, string id, float f)
{
    fullname = name;
    account = id;
    deposit = f;
}
/*带参数构造函数的定义*/
BankAccount::~BankAccount()
{
    cout<<"All Done!"<<endl;
}
/*析构函数，仅表示对象的析构信息*/

void BankAccount::init_account(string name, string id, float f)
{
    cout<<"Initializing Account infomation..."<<endl;
    if(name != "") fullname = name;
    if(id != "") account = id;
    deposit = f;
}
/*初始化对象*/
void BankAccount::print_info() const
{
    cout<<"The Account info:"<<endl;
    cout<<"Full Name: "<<fullname<<endl;
    cout<<"Account ID: "<<account<<endl;
    cout<<"Deposit: "<<deposit<<endl<<endl;
}
/*输出账号信息 */
void BankAccount::save(float f)
{
    deposit += f;
}
/*存款函数， deposit 成员增加值 f */
void BankAccount::withdraw(float f)
```

```
{
    deposit -= f;
}
/*取款函数，deposit 成员减少值 f*/
```

2. 下面是一个非常简单的类定义。

```
class Person
{
private:
    static const int LIMIT = 25;
    string lname;                   //Person 的姓
    char fname[LIMIT];              //Person 的名
public:
    Person() {lname = ""; fname[0] = '\0';}                         //#1
    Person(const string & ln, const char *fn = "Heyyou");          //#2
    //以下方法显示 lname 与 fname
    void Show() const;          //先显示名，再显示姓
    void FormalShow() const;   //先显示姓，再显示名
};
```

它使用了一个 string 对象和一个字符数组，用户可以比较它们的用法。请提供未定义的方法的代码，以完成这个类。

再编写一个使用这个类的程序，它使用了 3 种可能的构造函数（没有参数、一个参数和两个参数）的调用以及两种显示方法。

下面是一个使用这些构造函数和方法的例子。

```
Person one;                     //使用 默认构造函数
Person two("Smythecraft");     //使用有一个默认参数的 #2
Person three("Dimwiddy", "Sam");   //使用 #2, 没有默认的 one.Show()
cout << endl;
one.FormalShow();
//etc. for two and three
```

编程分析：

题目提供了 Person 类的声明，以及该类的应用测试范例，因此需要实现 Person 类的所有函数，并在 main() 方法中使用上述函数调用形式进行测试。完整代码如下。

```
/*第 10 章的编程练习 2*/
#include <iostream>
#include <cstring>
using namespace std;
class Person
{
private:
    static const int LIMIT = 25;
    string lname;                       //Person 的姓
    char fname[LIMIT];                  //Person 的名
public:
    Person() {lname = ""; fname[0] = '\0';}                     //#1
    Person(const string & ln, const char *fn = "Heyyou");      //#2
```

```
    //以下方法显示 lname 与 fname
    void Show() const;          //先显示名，再显示姓
    void FormalShow() const;    //先显示姓，再显示名
};

int main()
{
    Person one;                              //使用默认构造函数
    Person two("Smythecraft");               //使用有一个默认参数的 #2
    Person three("Dimwiddy", "Sam");         //使用 #2，没有默认的 one.Show()
    cout << endl;
    one.FormalShow();
    //etc. for two and three
    two.FormalShow();
    three.FormalShow();
    return 0;
}

Person::Person(const string & ln, const char*fn)
/*在类外定义类内的成员函数需要使用作用域运算符*/
{
    lname = ln;
    strcpy(fname, fn);
    /*string 类型和字符数组类型需要复制数据的不同方法*/
}
void Person::Show() const
{
    if(lname == "" && fname[0] == '\0')
    {
        cout<<"No Name."<<endl;
    }else
        {
            cout<<"Person Name: "<<fname<<"."<<lname<<endl;
        }
    /*针对不同情况输出对象信息*/
}
void Person::FormalShow() const
{
    if(lname == "" && fname[0] == '\0')
    {
        cout<<"No Name."<<endl;
    }else{
        cout<<"Person Name: "<<lname<<"."<<fname<<endl;
        /*先输出姓，后输出名*/
    }
}
```

3．完成第 9 章的编程练习 1，但要用正确的 golf 类声明替换那里的代码。用带合适参数的构造函数替换 setgolf(golf&, const char*, int)，以提供初始值。保留 setgolf()的交互版本，

但要用构造函数来实现它。例如，setgolf()的代码应该获得数据，将数据传递给构造函数来创建一个临时对象，并将其赋给调用对象，即*this。

编程分析：

第 9 章的编程练习 1 定义了一个结构体 golf 及相应的结构体操作函数，基本信息如下。

```
const int Len = 40;
struct golf
{
    char fullname[Len];
    int handicap;
};
void setgolf(golf& g,const char*name,int hc);
int setgolf(golf& g);
void handicap(golf& g,int hc);
void showgolf(const golf& g);
```

如果要修改该结构体，将其定义为一个类，那么需要将结构体成员定义为私有数据成员，将设置 golf 结构体变量的函数修改和定义为构造函数，把 handicap()和 showgolf()修改为普通成员函数。完整代码如下。

```
/*第 10 章的编程练习 3*/
/*golf.h golf 类的声明文件*/
#include <iostream>

const int Len = 40;
class golf
{
private:
    char fullname[Len];
    int handicap;
public:
    golf();
    golf(const char*name,int hc);
    ~golf(){};
    /*可以使用默认析构函数，也可以定义空析构函数*/
    void sethandicap(int hc);
    void showgolf() const;
/*输出对象信息的函数，不修改数据成员，应添加 const 关键字*/
};



/*golf.cpp 包含类内成员函数的实现，
 *本例也包含了 main()函数*/
#include <iostream>
#include <cstring>
#include "golf.h"

using namespace std;
```

```
int main()
{
    golf ann("Ann Bird free",24);
    golf andy;

    ann.showgolf();
    andy.showgolf();
    return 0;
}

golf::golf(const char*name, int hc)
{
    strcpy(fullname, name);
    handicap = hc;
}/*对于构造函数的定义，应添加作用域运算符*/

golf::golf()
{
    char name[Len] = {'\0'};
    int hc;
    cout<<"Please enter the name: ";
    cin.getline(name,Len);

    cout<<"Please enter the handicap: ";
    while(!(cin>>hc))
    {
       cin.clear();
       while(cin.get() != '\n')
          continue;
       cout<<"Please enter the golf's handicap: ";
    }
    cout<<name<<"::"<<hc<<endl;
    strcpy(fullname, name);
    handicap = hc;
}
/*在默认构造函数中，通过交互方式，输入对象信息*/
void golf::sethandicap(int hc)
{
    handicap = hc;
}

void golf::showgolf( ) const
{
    cout<<"Name : "<<fullname<<", Handicap is "<<handicap<<endl;
}
/*输出对象信息的函数，不修改数据成员，应添加 const 关键字*/
```

4．完成第 9 章的编程练习 4，但将 Sales 结构体及相关的函数转换为一个类及其方法。用构造函数替换 setSales（sales&, double[], int）函数。用构造函数实现 setSales(Sales&)方法的交互版本。将类保留在名称空间 SALES 中。

编程分析：

第 9 章的编程练习 4 定义了 Sales 结构体和相应的操作函数，主要代码如下。

```
namespace SALES
{
    const int QUARTERS = 4;
    struct Sales
    {
        double sales[QUARTERS];
        double average;
        double max;
        double min;
    };
    void setSales(Sales& s, const double ar[], int n);
    void setSales(Sales& s);
    void showSales(const Sales& s);
}
```

题目要求将结构体修改为类的形式，因此可以取消 SALES 名称空间，将 4 个结构体成员设置成为私有函数，setSales()重载函数可以修改为类的构造函数。完整代码如下。

```
/*第 10 章的编程练习 4*/
/*sales.h 声明了 Sales 类
 **/
#include <iostream>

const int QUARTERS = 4;

class Sales{
private:
    double sales[QUARTERS];
    double average;
    double max;
    double min;
public:
    Sales();
    Sales(const double ar[], int n);
    ~Sales(){};
    /*可以定义空析构函数，也可以使用默认析构函数*/
    void showSales() const;
    /*输出函数对数据成员无修改操作，应当添加 const*/
};


/*sales.cpp 定义了 Sale 类的所有函数实现
 **/
#include <iostream>
#include "sales.h"
using namespace std;
```

```
Sales::Sales(const double ar[], int n)
{
    double sum = 0;
    if(n >= QUARTERS)
    {
        for(int i = 0;i < QUARTERS; i++)
        {
            sales[i] = ar[i];
        }
    }else{
        for(int i = 0;i < n; i++)
        {
            sales[i] = ar[i];
        }
        for(int i = n;i < QUARTERS; i++)
        {
            sales[i] = 0;
        }
    }
    max = min = sales[0];
    for(int i = 0;i < QUARTERS; i++)
    {
        sum += sales[i];
        if(min>sales[i]) min = sales[i];
        if(max<sales[i]) max = sales[i];
    }
    average = sum / QUARTERS;
}

Sales::Sales()
{
    int i = 0;
    double sum = 0;
    do{
       cout<<"Enter a number: ";
       if(!(cin>>sales[i]))
       {
           cin.clear();
           while(cin.get()!='\n') continue;
           cout<<"ERROE, Reenter a number: ";
           cin>>sales[i];
       }
       i++;
    }while(i < QUARTERS);
    max = min = sales[0];
    for(int i = 0;i < QUARTERS; i++)
    {
        sum += sales[i];
        if(min>sales[i]) min = sales[i];
        if(max<sales[i]) max = sales[i];
```

```
    }
    average = sum / QUARTERS;

}

void Sales::showSales() const
{
    cout<<"This Salse's quarter list info:"<<endl;
    for(int i = 0; i < QUARTERS; i++)
    {
        cout<<"No."<<i+1<<": sales: "<<sales[i]<<endl;
    }
    cout<<"AVERAGE: "<<average<<endl;
    cout<<"MAX: "<<max<<endl;
    cout<<"MIX: "<<min<<endl;
}

/*main.cpp 仅包含 main()函数的主程序
 **/
#include <iostream>
#include "sales.h"

using namespace std;

int main() {
    double de[QUARTERS] = {12,23,45,67};
    Sales s1(de,QUARTERS);
    Sales s2;
    s1.showSales();
    s2.showSales();
    /*简单使用类 Sale 创建对象，并进行基本功能测试*/
    return 0;
}
```

5．考虑下面的结构体声明。

```
struct customer{
    char fullname[35];
    double payment;
};
```

编写一个程序，它从栈中添加和删除 customer 结构体（栈用 Stack 类的声明表示）。每次 customer 结构体被删除时，其 payment 的值都将被添加到总数中，并报告总数。

注意，应该可以直接使用 Stack 类而不做修改；只需要修改 typedef 声明，使 Item 的类型为 customer，而不是 unsigned long 即可。

编程分析：

题目要求使用本章中 Stack 类的定义，将其中的元素定义为 custormer 对象。下面是原 Stack 类的声明。

```
//stack.h ——栈 ADT 的类定义
#ifndef STACK_H_
#define STACK_H_

typedef unsigned long Item;

class Stack
{
private:
    enum {MAX = 10};        //特定于类的常量
    Item items[MAX];        //保存栈中的元素
    int top;                //栈顶元素的索引
public:
    Stack();
    bool isempty() const;
    bool isfull() const;
    //如果栈已满，push() 返回 false ；否则，返回 true
    bool push(const Item & item);        //入栈
    //如果栈已空，pop() 返回 false ；否则，返回 true
    bool pop(Item & item);               //出栈
};
#endif
```

程序需要首先添加 custormer 结构体的定义，修改 typedef 声明，将元素定义为 custumer 结构体，并修改 payment 数据，添加一个存储总数的变量。其他部分可以直接使用 Stack 类的定义和实现。完整代码如下。

```
/*第 10 章的编程练习 5*/
/*在原有代码的基础上添加题目要求的部分内容*/
//stack.h
#ifndef STACK_H_
#define STACK_H_

struct customer{
    char fullname[35];
    double payment;
};
/*添加 customer 结构体的声明*/
typedef customer Item;
/*修改原 Stack 内的元素 unsigned long 为 customer*/

class Stack
{
private:
    enum {MAX = 10};        //特定于类的常量
    Item items[MAX];        //保存栈中的元素
    int top;                //栈顶元素的索引
    /*也可以在栈类内定义 Stack 使用过的所有元素的 payment 数据，
     *但是这样会失去 Stack 类的通用性。另外，需要修改构造函数和 pop() 函数
     *double sum_payment;
```

```
    *因此推荐在主程序中添加该变量计算出栈数据的payment*/
public:
    Stack();
    bool isempty() const;
    bool isfull() const;
    //如果栈已满，push() 返回 false ; 否则，返回 true
    bool push(const Item & item);        //入栈
    //pop() returns false ; 否则，返回 true
    bool pop(Item & item);               //出栈
};
#endif

/*修改原有的stack.cpp，添加all_payment数据的处理*/
//stack.cpp — Stack 类的成员函数
#include "stack.h"
Stack::Stack()    //新建空栈
{
    top = 0;
}

bool Stack::isempty() const
{
    return top == 0;
}

bool Stack::isfull() const
{
    return top == MAX;
}

bool Stack::push(const Item & item)
{
    if (top < MAX)
    {
       items[top++] = item;
       return true;
    }
    else
       return false;
}

bool Stack::pop(Item & item)
{
    if (top > 0)
    {
       item = items[--top];
       return true;
    }
    else
       return false;
```

```
}

/*包含 main()函数的主程序，测试定义的 Stack 类的用法*/
#include <iostream>
#include "stack.h"

using namespace std;

int main()
{
    Stack st; //create an empty stack
    customer cust;
    double sum_payment = 0;
    char select;
    /*定义变量，对出栈元素的 payment 求和*/
    cout<<"Select a (add), p (pop) or q(quit) :";
    while(cin.get(select) && select != 'q')
    {
        while(cin.get() != '\n') continue;
        if(select == 'a')
        {
            cout << "Enter a customer's name : ";
            cin.getline(cust.fullname, 35);
            cout << "Enter a customer's payment : ";
            cin >> cust.payment;
            while(cin.get() != '\n') continue;
            st.push(cust);
            cout<<"Item pushed.\n";
        }
        if(select == 'p')
        {
            st.pop(cust);
            sum_payment += cust.payment;
            cout<<"Pop Item's info:\nName : "<<cust.fullname<<endl;
            cout<<"Payment: "<<cust.payment<<endl;
            cout<<"Now, sum of payments: "<<sum_payment<<endl;
        }
        cout<<"Select a (add), p (pop) or q(quit) :";
    }
/*简单模拟出栈和入栈 */
    cout<<"Bye\n";
    return 0;
}
```

6. 下面是一个类声明。

```
class Move{
private:
    double x;
```

```
    double y;
public:
    Move(double a = 0, double b = 0);       //sets x, y to a, b
    void showmove() const;                  //shows current x,y values
    Move add(const Move & m) ;
    //把m、x与x相加，得到新的x,
    //把m、y与y相加，得到新的y,
    //创建新的Move对象，初始化为新的x、y值，并返回它
    void reset(double a = 0, double b = 0);            //重置 x、y 为 a 与 b
};
```

请提供成员函数的定义和测试这个类的程序。

编程分析：

题目给出了类 Move 的声明文件，Move 类的数据成员有 double 类型的 x、y 两个变量。成员函数有构造函数、显示函数、重置函数以及 add()函数，其中关键在于 add()函数的设计，需要返回一个新的 Move 对象。完整代码如下。

```
/*第10章的编程练习 6*/
#include <iostream>

using namespace std;
class Move{
private:
    double x;
    double y;
public:
    Move(double a = 0, double b = 0);
    void showmove() const;
    Move add(const Move & m) ;
    void reset(double a = 0, double b = 0);
};

int main(int argc, char *argv[])
{
    Move a, b(12.5,19);
    double x, y;
    a.showmove();
    b.showmove();
    /*a 使用构造函数的默认值输出对象的信息，b 使用初始化参数输出对象的信息*/
    cout<<"Enter X and Y: ";
    cin>>x>>y;
    cout<<"Reste Object A:"<<endl;
    a.reset(x,y);
    a.showmove();
    b.showmove();
    /*a 调用 reset()函数，设置用户输入的数据*/
    cout<<"Object A add B:"<<endl;
    a = a.add(b);
    a.showmove();
```

```
    b.showmove();
    /*a 调用 add()函数，与 b 相加，把返回值赋值给 a*/
    return 0;
}
Move::Move(double a, double b)
{
    x = a;
    y = b;
}
void Move::showmove() const
{
    cout<<"Current x = "<<x<<", y = "<<y<<endl;
}
Move Move::add(const Move& m)
{
    Move temp;
    temp.x = x + m.x;
    temp.y = y + m.y;
    return temp;
}
void Move::reset(double a, double b)
{
    x = a;
    y = b;
}
```

7. Betelgeusean plorg 在数据上有以下特征。

- plorg 的名称不超过 19 个字符。
- plorg 的满意指数（CI）是一个整数。

在操作上有以下特征。

- 新的 plorg 将有名称，其 CI 值为 50。
- plorg 的 CI 可以修改。
- plorg 可以报告其名称和 CI。

请编写 Plorg 类的声明（包括数据成员和成员函数原型）来表示 plorg，并编写成员函数的函数定义。然后编写一个小程序，以演示 Plorg 类的所有特性。

编程分析：

plorg 类包含两个数据成员，一个是描述名称的字符串，另一个是描述 CI 的整型数据。相关的成员函数包括构造函数、修改 CI 的函数和输出信息的函数。程序需要实现所有的函数并进行简单测试。完整代码如下。

```
/*第 10 章的编程练习 7*/
#include <iostream>
using namespace std;

const int SIZE = 19;
class plorg{
```

```
private:
    char name[SIZE];
    int CI;
    /*两个数据成员，name 使用字符数组*/
public:
    plorg(const char st[] = "Plorga", int ci = 50);
    /*构造函数带参数默认值，默认 CI 为 50*/
    void reset_ci(int n);
    void print_info() const;
    /*输出数据信息函数应添加 const 关键字*/
};

int main()
{
    plorg pl;
    pl.print_info();
    pl.reset_ci(98);
    pl.print_info();
    plorg pm("Stenom",87);
    pm.print_info();
    return 0;
}
plorg::plorg(const char st[], int ci)
{
    strcpy(name,st);
    CI = ci;
}
void plorg::reset_ci(int n)
{
    CI = n;
}
void plorg::print_info() const
{
    cout<<"plorg name: "<<name<<", CI = "<<CI<<endl;
}
```

8. 可以将简单列表描述如下。

- 可存储 0 或多个某种类型的列表。
- 可创建空列表。
- 可在列表中添加数据项。
- 可确定列表是否为空。
- 可确定列表是否已满。
- 可访问列表中的每一个数据项，并对它执行某种操作。

可以看到，这个列表确实很简单，例如它不允许插入或删除数据项。

请设计一个 List 类来表示这种抽象类型。应提供头文件 list.h 和实现文件 list.cpp，前者包含类定义，后者包含类方法的实现。还应该创建一个简短的程序来使用这个类。

该列表的规范很简单，旨在简化这个编程练习。可以选择使用数组或链表来实现该列表，但公有接口不应依赖于所做的选择。也就是说，公有接口不应有数组索引、节点指针等。

应使用通用概念来表达创建列表、在列表中添加数据项等操作。对于访问数据项以及执行操作，通常应使用以函数指针作为参数的函数来处理。

```
void visit(void (*pf)(Item &));
```

其中，pf 指向一个以 Item 引用作为参数的函数（而不是成员函数），Item 是列表中数据项的类型。visit()函数将该函数用于列表中的每个数据项。

编程分析：

列表是程序设计中一种非常重要的数据结构（数据的组织形式）。列表内主要是存储相同类型数据的元素，可以使用数组的形式或者指针的链状结构形式实现。本题要求实现的主要功能为创建列表、添加数据项、判满、判空和遍历。由于列表的具体数据组织方式可以选择数组或者链表，因此公有接口必须保证通用性，隐藏具体的设计细节。其中设计的难点在于遍历函数，由于 Item 数据项的不确定性，因此为了实现其通用性，应当使用函数指针的形式，每当用户创建一个特定类型的列表时，都需要提供一个元素的访问函数，并将该函数作为参数代入列表的遍历函数内。程序的完整代码如下。

```
/*第 10 章的编程练习 8*/
/*list.h ——列表的声明*/
typedef unsigned long Item;
/*此处声明 unsigned long 为 Item 元素的类型*/

void visit_Item(Item&);
/*针对每一个类型需要定义对应的数据访问函数，此处是输出功能*/
class List
{
private:
    enum {MAX = 10};
    Item items[MAX];        //holds list items
    /*此处使用数组形式维护列表*/
    int top;                //栈顶列表项的索引
public:
    List();
    bool isempty() const;
    bool isfull() const;
    bool add(const Item & item);   //向列表中添加元素
    void visit(void (*pf)(Item &));
};

/*list.cpp 列表类的具体定义*/
#include "list.h"

List::List(){
    top = 0;
}

bool List::isempty() const{
    return top == 0;
```

```
}

bool List::isfull() const{
    return top == MAX;
}

bool List::add(const Item & item) {
    if (top < MAX){
        items[top++] = item;
        return true;
    }
    else
        return false;
}

void List::visit(void (*pf)(Item &)){
    for (int i = 0; i < top; i++)
        pf(items[i]);
}

void visit_Item(Item& item){
    cout<<"The Item's info: "<<item<<endl;
}
/*visit_Item()函数是全局函数，并非 List 类内的函数*/


#include <iostream>
#include "list.h"
using namespace std;

int main()
{
    List list;
    Item item = 0;
    cout<<"Enter the unsigned long number: ";
    cin>>item;
    while(item != 0)
    {
        cin.get();
        list.add(item);
        cout<<"Enter the unsigned long number: ";
        cin>>item;
    }
    cout<<"Now end of add element, start to visit:"<<endl;
    list.visit(visit_Item);
    return 0;
}
```

第 11 章　使　用　类

本章知识点总结

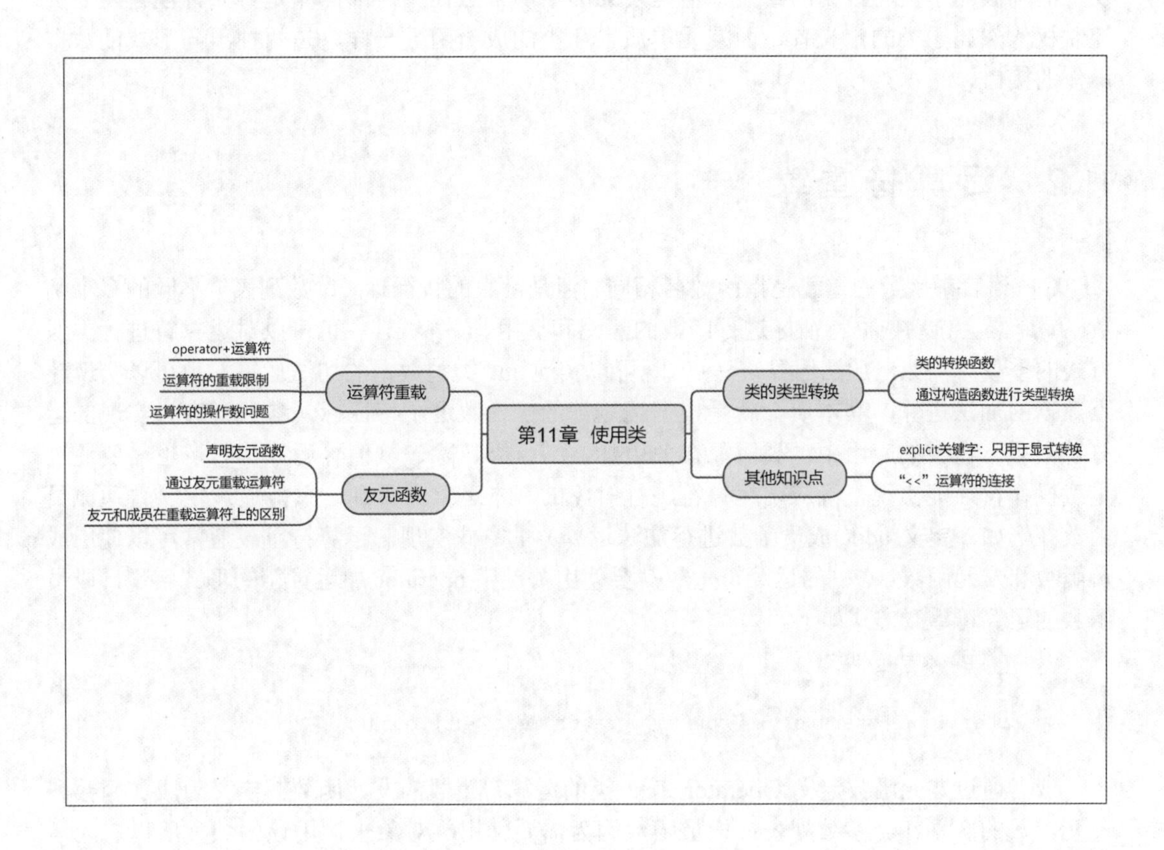

11.1　类的友元函数

C++语言中，类是一种实现数据隐藏和封装的重要方式。类内部通过对私有成员的访问控制以及通过成员函数间接访问私有成员的形式，隐藏了数据成员的细节描述。但是在某些特定情况下，类内部过于严格的访问控制方式也限制了程序设计的灵活性。因此为了更加灵活地实现不同对象之间数据的访问，C++又提供了友元的方式，实现了一种从类外部访问类内部数据的形式。友元有 3 种类型，分别是友元函数、友元类和友元成员函数。创建一个类的友元函数，需要首先在类的定义内声明该友元函数的原型，且通过原型添加关键字 friend，以表明友元关系，例如：

```
class Friend_Test{
...
friend void f();
}
```

类定义内的 friend 关键字就表明函数 f()是 Friend_Test 类的友元函数。除了在类内的声明之外，还需要在类外定义并实现该函数，但是函数定义和实现中不使用类限定符和 friend 关键字（因为该类并不是类内数据，且友元关系应当在类内声明）。在友元函数 f()的内部可以直接访问 Friend_Test 类内的数据成员和方法，其访问权限和成员函数相同。从表面上看，友元的确破坏了类对数据的封装，但是友元的声明依然在类内部，即是否允许创建友元的控制权依然属于类的定义者，而友元本质上和类的成员函数一样，是两种不同形式的数据操作的接口。

11.2 运算符重载

C++语言中函数的重载是指函数名相同、但是函数的特征标（参数列表）不同的多个函数，编译器会自动依据特征标进行函数的匹配和调用。C++语言中也可以对运算符进行类似方式的重载。例如，加法运算符（+）既可以实现 int 类型数据的加法运算，也可以实现浮点数据的加法运算，其本质上就是根据运算符的操作数这个特征标（参数列表）实现的运算符重载。在类的应用中，类对象在使用加、减、乘、除、赋值等运算符时，也需要通过重载操作符才能实现正确应用运算符进行对象的基本运算。否则，只能通过定义普通成员函数（例如，定义 add()成员函数进行加法运算）的方式实现。运算符重载通常可以使用成员函数和友元函数。C++的运算符重载需要使用关键字 operator 加运算符的形式实现。成员函数的运算符重载方式如下。

```
class Operator_Test{
...
Operator_Test  operator + (const Operator_Test& ot) const;
}
```

或者通过友元函数定义 Operator_Test 类的运算符重载的友元函数。定义为成员函数和友元函数的区别在于参数列表：成员函数因为需要使用类对象来调用该函数，所以二元运算符" + "只需要一个参数即可，另一个操作数是调用运算符的对象本身；对于二元运算符，友元函数则需要两个参数来实现。C++中关于运算符重载的规定如下。

- 重载后的运算符至少必须有一个操作数是用户定义的类型。
- 在使用运算符时不能违反运算符原来的句法规则。
- 不能创建新运算符。
- 不能重载 sizeof、成员运算符、作用域运算符等特定的运算符。
- 赋值运算符、下标运算符、函数运算符等运算符只能通过成员函数进行重载。

此外，“<<”运算符可以通过成员函数和友元函数实现重载。但是，如果通过成员函数重载该运算符，那么运算符调用中需要将类对象放置在运算符左侧，即以 class_object.operator<<()

的类似形式进行调用，无法使用 cout<<class_object 的形式，因此通常将其通过友元函数进行重载，且友元函数的第 1 个参数需要定义为 ostream 类型，这样才能保证使用 cout 对象在左侧进行运算符调用。最后，为了实现“<<”运算符的连续输出的组合，可以在运算符的重载函数中返回基本输入/输出对象的引用，即：

```
ostream& operator<<(ostream& os, const Class_Test& t){
...
        return os;
}
```

成员函数和友元函数在运算符重载中的另一个区别就是，当以成员函数重载运算符时，第 1 个操作数都应该是类对象，因此在二元运算符重载中还需要考虑运算符调用中两个操作数的具体类型和函数原型中两个参数的位置。这一点在编程设计和调用重载运算符时应当特别关注。

11.3 类的类型转换

在 C++程序设计中，用户通常可以使用自动类型转换或者强制类型转换的方式对类对象进行数据类型转换。为了实现这样的数据类型转换功能，在类的定义中需要定义 C++的一个类型转换函数。类型转换函数可以理解为定义或者描述当前类如何转换为其他指定数据类型的函数，功能是使得当前类的对象能够从当前类型转换为另一种相关的数据类型。定义类型转换函数的方式如下。

```
operator TypeName();
```

其中，TypeName 可以替换成需要的数据类型名，且转换函数必须定义为成员函数，转换函数不能指定返回类型，也不能有参数。

类的类型转换函数是用户定义的强制类型转换函数。通过这个类型转换函数程序可以使用强制类型转换的方式进行数据类型转换。应用中可以使用强制类型转换的形式将类对象转换为其他类型，也可以利用隐式类型转换的形式实现。

类型转换函数的功能是将类对象转换为目标数据类型，实际应用中还可以将其他数据类型转换为当前的类对象，这种转换是通过只有一个参数的类的构造函数实现的。例如：

```
class Class_Test{
...
Class_Test(double& n)
}
```

通过参数为 double 类型的构造函数名，就可以间接实现 double 类型的数据向 Class_Test 类型的转换。在构造函数声明中使用 explicit 可防止隐式转换，而只允许显式转换。

11.4 复习题

1．使用成员函数为 Stonewt 类重载乘法运算符，该运算符将数据成员与 double 类型的值相乘。注意，当用英石[①]和磅[②]表示时，需要进位。也就是说，将 10 英石 8 磅乘以 2 等于 21 英石 2 磅。

习题解析：

程序清单 11.16 定义了 Stonewt.h，程序清单 11.17 定义了 Stonewt.cpp，题目要求以成员函数的形式重载乘法运算符，利用数据成员和参数计算英石和磅。因此需要在头文件与源代码文件分别添加声明和定义。乘法运算的进位需要通过枚举数据 Lbs_per_stn 实现，需要重新设置其全部数据成员，计算比较麻烦。其中比较简便的方式是通过构造函数，直接输入对象的 pounds 值，并由构造函数完成每个成员的计算和赋值。重载运算符的声明如下。

```
Stonewt operator*(double multi);
```

注意，这种方法使用构造函数建立新对象，再通过返回值赋值。如果用户自己计算所有函数，可以参考 Stonewt(double lbs)函数的具体实现，但是返回值可以为 void，并直接使用数据成员计算。

```
class Stonewt{
private:
      enum {Lbs_per_stn = 14};              //1 英石=14 磅
      int stone;                            //整数形式的英石
      double pds_left;                      //小数形式的磅
      double pounds;                        //以磅为单位的整体重量
public:
      Stonewt(double lbs);                  //针对 double 类型磅值的构造函数
      Stonewt(int stn, double lbs);         //针对 int 类型英石值和 double 类型磅值的构造函数
      Stonewt();                            //默认构造函数
      ~Stonewt();
      void show_lbs() const;                //以磅为单位显示重量
      void show_stn() const;                //以英磅为单位显示重量
      Stonewt operator*(double multi);      //乘法运算符重载
};

//在 Stonewt.cpp 文件下添加如下重载函数的实现

Stonewt Stonewt:: operator*(double multi){
      return Stonewt(multi* pounds);
}
```

2．友元函数与成员函数之间的区别是什么？

习题解析：

从其性质上看，成员函数是类定义的一部分，由类生成的所有对象均会共享类的一组成员函数。从功能上看，成员函数能够访问类内部的所有数据成员。而友元函数并不是类定义的一部分，而是一个具备特定的类访问权限的函数，友元函数从功能上说也能够直接

① 1 英石=14 磅。——编者注

② 1 磅=0.45 千克。——编者注

访问所有类的数据成员，但是并不能隐式访问，而必须通过成员运算符用于参数传递的对象。

3．非成员函数必须是友元才能访问类成员吗?

习题解析：

首先从访问控制上说，使用类对象的程序能够直接通过类对象访问所有的公有数据接口，但无法访问类的私有数据接口。通过类的友元函数能够直接访问类内的所有数据成员和函数，包括具有私有访问权限的数据成员和成员函数。因此，片面地认为非成员函数必须是友元函数才能访问类成员是不正确的。

4．使用友元函数为 Stonewt 类重载乘法运算符，该运算符将 double 值与 Stone 值相乘。

习题解析：

和复习题 1 不同的是，本题要求使用友元函数进行乘法操作符的重载，因此需要首先在 Stonewt.h 内类定义中添加友元函数的声明，并且在 Stonewt.cpp 文件中添加友元函数的实现。相对于以友元函数重载乘法运算符，在使用成员函数进行类似运算（double 值乘以 Stone 值）时，由于成员函数导致其第 1 个操作数只能是 Stone 类型，因此只能实现 Stone 值乘以 double 值。若通过友元函数实现，函数的参数应当有 Stonewt 类型和 double 类型，返回值为 Stonewt。具体实现如下。

```
class Stonewt{
private:
     enum {Lbs_per_stn = 14};        //1 英石=14 磅
     int stone;                      //整数形式的英石
     double pds_left;                //小数形式的磅
     double pounds;                  //以磅为单位的整体重量
public:
     Stonewt(double lbs);            //针对 double 类型磅值的构造函数
     Stonewt(int stn, double lbs);//针对 int 类型英石值和 double 类型磅值的构造函数
     Stonewt();                      //默认构造函数
     ~Stonewt();
     void show_lbs() const;          //以磅为单位显示重量
     void show_stn() const;          //以英石为单位显示重量

     //添加友元函数的声明，函数的参数为 Stonewt 类型和 double 类型
     friend Stonewt operator*(double multi, const Stonewt& s);
};

//把友元函数的定义添加至 Stonewtc.cpp 文件中
Stonewt operator*(double multi, const Stonewt& s){
     return Stonewt(multi*s.pounds);
}
```

5．哪些运算符不能重载?

习题解析：

C++语言中的运算符重载有一定的限制，包括重载运算符至少有一个参数是用户自定义的类型，使用的运算符不能违反原运算符的语法规则，不能创建新的运算符，部分运算符不能重载。不能重载的运算符如下。

- sizeof——sizeof 运算符。
- .——成员运算符。
- .*——成员指针运算符。
- ::——作用域解析运算符。
- ?:——条件运算符。
- typeid——一个 RTTI 运算符。
- const_cast——强制类型转换运算符。
- dynamic_cast——强制类型转换运算符。
- reintepret_cast——强制类型转换运算符。
- static_cast——强制类型转换运算符。

6．在重载运算符=、()、[ ]和->时，有什么限制?

习题解析：

C++语言中运算符重载的基本限制包含复习题 5 列出的 4 条限制，即重载运算符至少有一个参数是用户自定义的类型，使用的运算符不能违反原运算符的语法规则，不能创建新的运算符，部分运算符不能重载。此外，大多数运算符可以通过成员或非成员函数进行重载，但=、()、[ ]和->运算符只能通过成员函数进行重载，不能通过友元函数进行重载。

7．为 Vector 类定义一个转换函数，将 Vector 类转换为一个 double 类型的值，后者表示矢量的长度。

习题解析：

类的转换函数是将用户自定义的类转换为基本数据类型或者其他自定义类型的转换方法的函数，其主要功能说明如何进行数据类型转换。转换函数的定义主要有几个限制，即转换函数必须是类方法，转换函数不能有参数，不能指定转换函数的返回类型。其函数原型可以表示为 operator Type_Name()。

Vector 类的声明和定义在程序清单 11.13 和程序清单 11.14 中，需要添加的转换函数如下。

```
class Vector{
//其他相关代码
…;
operator double(){ return mag;};
//为了将 Vector 转换成 double 类型，只需要返回其表示长度的数据成员 mag 即可
}
```

11.5　编程练习

1．修改程序清单 11.15，使之将一系列连续的随机漫步者位置写入文件中。对于每个位置，用步号进行标识。另外，使该程序将初始条件（目标距离和步长）以及结果小结写入该文件中。该文件的内容如下。

```
Target Distance: 100, Step Size: 20
0: (x,y) = (0, 0)
1: (x,y) = (-11.4715, 16.383)
2: (x,y) = (-8.68807, -3.42232)
...
26: (x,y) = (42.2919, -78.2594)
27: (x,y) = (58.6749, -89.7309)
After 27 steps, the subject has the following location:
(x,y) = (58.6749, -89.7309)
or
(m,a) = (107.212, -56.8194)
Average outward distance per step = 3.97081
```

编程分析：

题目要求修改程序清单 11.15，因此程序还需要使用程序清单 11.13 的 vect.h 和程序清单 11.14 的 vect.cpp 两个文件。程序清单 11.15 当前已经能够完成所有的随机漫步者指定距离的运行工作，因此题目要求添加文件读写功能，将随机漫步者行走的编号、位置按照指定格式输出到文件中。因此，修改工作主要集中在添加 fstream 头文件的预编译指令、将标准输出修改为文件输出和最后的输出格式化。完整代码如下。

```
/*第 11 章的编程练习 1*/
//vect.h —— 具有<<并采用不同坐标类的 Vector 类
#ifndef VECTOR_H_
#define VECTOR_H_
#include <iostream>
namespace VECTOR
{
    class Vector
    {
    public:
        enum Mode {RECT, POL};
        //RECT 表示直角坐标，POL 表示极坐标
    private:
        double x;          //横坐标
        double y;          //纵坐标
        double mag;        //向量的长度
        double ang;        //向量的方向，以度表示
        Mode mode;         //RECT 或 POL
        //用于设置值的私有方法
        void set_mag();
```

```
        void set_ang();
        void set_x();
        void set_y();
    public:
        Vector();
        Vector(double n1, double n2, Mode form = RECT);
        void reset(double n1, double n2, Mode form = RECT);
        ~Vector();
        double xval() const {return x;}          //报告 x 值
        double yval() const {return y;}          //报告 y 值
        double magval() const {return mag;}      //报告幅值
        double angval() const {return ang;}      //报告角度
        void polar_mode();                       //使用 POL 坐标
        void rect_mode();                        //使用 RECT 坐标
        //运算符重载
        Vector operator+(const Vector & b) const;
        Vector operator-(const Vector & b) const;
        Vector operator-() const;
        Vector operator*(double n) const;
        //友元
        friend Vector operator*(double n, const Vector & a);
        friend std::ostream & operator<<(std::ostream & os, const Vector & v);
    };

}   //名称空间 VECTOR 结束
#endif


//vect.cpp —— Vector 类的方法
#include <cmath>
#include "vect.h"   //包括 <iostream>
using std::sqrt;
using std::sin;
using std::cos;
using std::atan;
using std::atan2;
using std::cout;

namespace VECTOR
{
    //以弧度为单位，计算角度
    const double Rad_to_deg = 45.0 / atan(1.0);
    //应约为 57.2957795130823

    //私有方法
    //根据 x 与 y 计算幅值
    void Vector::set_mag()
    {
        mag = sqrt(x *x + y *y);
    }
```

```
void Vector::set_ang()
{
    if (x == 0.0 && y == 0.0)
        ang = 0.0;
    else
        ang = atan2(y, x);
}

//根据极坐标设置 x
void Vector::set_x()
{
     x = mag *cos(ang);
}

//根据极坐标设置 y
void Vector::set_y()
{
     y = mag *sin(ang);
}

//私有方法
Vector::Vector()              //默认构造函数
{
    x = y = mag = ang = 0.0;
    mode = RECT;
}

//如果 form 等于 RECT，根据直角坐标构造向量;
//如果 form 等于 ROL，根据极坐标构造向量
Vector::Vector(double n1, double n2, Mode form)
{
     mode = form;
     if (form == RECT)
     {
         x = n1;
         y = n2;
         set_mag();
         set_ang();
     }
     else if (form == POL)
     {
         mag = n1;
         ang = n2 / Rad_to_deg;
         set_x();
         set_y();
     }
     else
     {
         cout << "Incorrect 3rd argument to Vector() -- ";
```

```
            cout << "vector set to 0\n";
            x = y = mag = ang = 0.0;
            mode = RECT;
        }
    }

//如果 form 等于 RECT，根据直角坐标构造向量；
//如果 form 等于 ROL，根据极坐标构造向量
void Vector:: reset(double n1, double n2, Mode form)
{
    mode = form;
    if (form == RECT)
    {
        x = n1;
        y = n2;
        set_mag();
        set_ang();
    }
    else if (form == POL)
    {
        mag = n1;
        ang = n2 / Rad_to_deg;
        set_x();
        set_y();
    }
    else
    {
        cout << "Incorrect 3rd argument to Vector() -- ";
        cout << "vector set to 0\n";
        x = y = mag = ang = 0.0;
        mode = RECT;
    }
}

Vector::~Vector()              //析构函数
{
}

void Vector::polar_mode()      //使用极坐标
{
    mode = POL;
}

void Vector::rect_mode()       //使用直角坐标
{
    mode = RECT;
}

//重载运算符
//两个向量相加
```

```
    Vector Vector::operator+(const Vector & b) const
    {
        return Vector(x + b.x, y + b.y);
    }

    //向量 a 减向量 b
    Vector Vector::operator-(const Vector & b) const
    {
        return Vector(x - b.x, y - b.y);
    }

    //对向量的符号取反
    Vector Vector::operator-() const
    {
        return Vector(-x, -y);
    }

    //向量乘以 n
    Vector Vector::operator*(double n) const
    {
        return Vector(n *x, n *y);
    }

    //友元模式
    //向量 a 乘以 n
    Vector operator*(double n, const Vector & a)
    {
        return a *n;
    }

    //如果 mode 等于 RECT，显示直角坐标，
    //如果 mode 等于 ROL，显示极坐标
    std::ostream & operator<<(std::ostream & os, const Vector & v)
    {
        if (v.mode == Vector::RECT)
            os << "(x,y) = (" << v.x << ", " << v.y << ")";
        else if (v.mode == Vector::POL)
        {
            os << "(m,a) = (" << v.mag << ", "
               << v.ang *Rad_to_deg << ")";
        }
        else
            os << "Vector object mode is invalid";
        return os;
    }

} //名称空间 VECTOR 结束


//randwalk.cpp —— 使用 Vector 类
```

```
//通过 vect.cpp 文件进行编译
#include <iostream>
#include <cstdlib>      //rand()、 srand() 的原型
#include <ctime>        //time() 的原型
#include <fstream>

#include "vect.h"

int main()
{
    using namespace std;
    using VECTOR::Vector;

    //创建输出文件对象 fout
    ofstream fout;
    //创建磁盘文件 randwalk.txt，保存输出的数据
    fout.open("randwalk.txt");

    srand(time(0));    //随机种子值生成器
    double direction;
    Vector step;
    Vector result(0.0, 0.0);
    unsigned long steps = 0;
    double target;
    double dstep;
    cout << "Enter target distance (q to quit): ";
    while (cin >> target)
    {
        cout << "Enter step length: ";
        if (!(cin >> dstep))
            break;

        //使用 fout 对象和重定向操作符将字符串输出到文件中，使用方法和 cout 类似
        fout << "Target Distance: " << target << ", Step Size: " << dstep << endl;

        while (result.magval() < target)
        {
            //需要记录漫步者的每一次步进号和 result 数值
            //使用 fout 对象和重定向操作符将行走的编号和坐标输出到文件中
            fout << steps<<" : " << result << endl;
            //vect 类重载了<<操作符，因此可以直接使用 fout 输出到文件中

            direction = rand() % 360;
            step.reset(dstep, direction, Vector::POL);
            result = result + step;
            steps++;
        }
        cout << "After " << steps << " steps, the subject "
                                "has the following location:\n";
        cout << result << endl;
```

```
        //使用 fout 对象和重定向操作符将字符串输出到文件中
        fout << "After " << steps << " steps, the subject "
                            "has the following location:\n";
        fout << result << endl;

        result.polar_mode();
        cout << " or\n" << result << endl;
        cout << "Average outward distance per step = "
            << result.magval()/steps << endl;

        //使用 fout 对象和重定向操作符将字符串输出到文件中
        fout << " or\n" << result << endl;
        fout << "Average outward distance per step = "
           << result.magval()/steps << endl;

        steps = 0;
        result.reset(0.0, 0.0);
        cout << "Enter target distance (q to quit): ";
    }
    cout << "Bye!\n";
    cin.clear();
    while(cin.get() != '\n')
        continue;
    return 0;
}
```

2．对 Vector 类的头文件（程序清单 11.13）和实现文件（程序清单 11.14）进行修改，使其不再存储向量的长度和角度，而在调用 magval()和 angval()时计算它们。应保留公有接口不变（公有方法及其参数不变），但对私有部分（包括一些私有方法）和方法进行修改。然后，使用程序清单 11.15 对修改后的版本进行测试，结果应该与以前相同，因为 Vector 类的公有接口与原来相同。

编程分析：

题目要求删除 Vector 类的私有数据成员 mag 和 ang，删除这两个数据成员后，同时还需要删除 set_mag()和 set_ang()函数。为了保持类型原有的长度和角度计算功能，需要修改相关的公有成员函数 magval()和 angval()。函数有两种修改方法，可以继续以内联函数的形式在 vect.h 文件中修改，或者作为普通成员函数在 vect.cpp 内新建其定义。完整代码如下。

```
/*第 11 章的编程练习 2*/
//vect.h -- Vector class with <<, mode state
#ifndef VECTOR_H_
#define VECTOR_H_
#include <iostream>
namespace VECTOR
{
    class Vector
    {
```

```
public:
    enum Mode {RECT, POL};
    //RECT 表示直角坐标, POL 表示极坐标
private:
    double x;           //横坐标
    double y;           //纵坐标
    Mode mode;          //RECT 或 POL
    /*按照题目要求删除 mag 和 ang 两个数据成员
    double mag;         //length of vector
    double ang;         //direction of vector in degrees
   */
    //private methods for setting values
    /*删除 mag 和 ang 数据成员后，相应的两个私有成员函数也将不再使用
    void set_mag();
    void set_ang();
   */
    void set_x(double mag, double ang);       //通过用户输入的 mag 和 ang 求 x
    void set_y(double mag, double ang);       //通过用户输入的 mag 和 ang 求 y
public:
    Vector();
    Vector(double n1, double n2, Mode form = RECT);
    void reset(double n1, double n2, Mode form = RECT);
    ~Vector();
    double xval() const {return x;}       //报告 x 值
    double yval() const {return y;}       //报告 y 值
    /*成员函数 magval()和 angval()可以使用两种方式实现
    可以修改原内联函数，直接计算数值并返回
    double magval() const ( return sqrt(x *x + y *y);)
    double angval() const {
        if (x == 0.0 && y == 0.0)
            return 0.0;
        else
            return atan2(y, x);
        }
    也可以修改为非内联函数，并在 vect.cpp 中定义
   */
    double magval() const;                //报告幅值
    double angval() const;                //报告角度
    void polar_mode();                    //使用极坐标
    void rect_mode();                     //使用直角坐标
    /运算符重载
    Vector operator+(const Vector & b) const;
    Vector operator-(const Vector & b) const;
    Vector operator-() const;
    Vector operator*(double n) const;
    //友元
    friend Vector operator*(double n, const Vector & a);
    friend std::ostream & operator<<(std::ostream & os, const Vector & v);
};
```

```
}   //名称空间 VECTOR 结束
#endif
```

```
//vect.cpp —— Vector 类的方法
#include <cmath>
#include "vect.h"   //包括 <iostream>
using std::sqrt;
using std::sin;
using std::cos;
using std::atan;
using std::atan2;
using std::cout;

namespace VECTOR
{
    //以弧度为单位，计算角度
    const double Rad_to_deg = 45.0 / atan(1.0);
    //应约为 57.2957795130823

    //私有方法
    //根据 x 与 y 计算幅值
    /*删除 set_mag()和 set_ang()函数
    void Vector::set_mag()
    {
        mag = sqrt(x*x + y*y);
    }

    void Vector::set_ang()
    {
        if (x == 0.0 && y == 0.0)
            ang = 0.0;
       else
            ang = atan2(y, x);
    }
    如果使用非内联函数，则此处应当添加 magval()和 angval()函数的实现
   */
    double Vector::magval() const
    {
        return sqrt(x*x + y*y);
    }

    double Vector::angval() const
    {
        if (x == 0.0 && y == 0.0)
            return 0.0;
        else
            return atan2(y, x);
```

```
}

//根据极坐标设置 x
void Vector::set_x(double mag, double ang)
{
    x = mag *cos(ang);
}

//根据极坐标设置 y
void Vector::set_y(double mag, double ang)
{
    y = mag *sin(ang);
}

//公有方法
Vector::Vector()              //默认构造函数
{
    x = y = 0.0;
    mode = RECT;
}

//如果 form 等于 RECT，根据直角坐标构造向量；
//如果 form 等于 POL，根据极坐标构造向量
Vector::Vector(double n1, double n2, Mode form)
{
    mode = form;
    if (form == RECT)
    {
        x = n1;
        y = n2;
    }
    else if (form == POL)
    {
        set_x(n1, n2/Rad_to_deg);
        set_y(n1, n2/Rad_to_deg);
    }
    else
    {
        cout << "Incorrect 3rd argument to Vector() -- ";
        cout << "vector set to 0\n";
        x = y = 0.0;
        mode = RECT;
    }
}

//如果 form 等于 RECT（默认值），根据直角坐标重置向量；
//如果 form 等于 POL，根据极坐标重置向量
void Vector:: reset(double n1, double n2, Mode form)
{
    mode = form;
```

```
    if (form == RECT)
    {
        x = n1;
        y = n2;
    }
    else if (form == POL)
    {
        set_x(n1, n2/Rad_to_deg);
        set_y(n1, n2/Rad_to_deg);
    }
    else
    {
        cout << "Incorrect 3rd argument to Vector() --";
        cout << "vector set to 0\n";
        x = y = 0.0;
        mode = RECT;
    }
}

Vector::~Vector()              //析构函数
{
}

void Vector::polar_mode()      //使用极坐标
{
    mode = POL;
}

void Vector::rect_mode()       //使用直角坐标
{
    mode = RECT;
}

//运算符重载
//两个向量相加
Vector Vector::operator+(const Vector & b) const
{
    return Vector(x + b.x, y + b.y);
}

//向量 a 减去向量 b
Vector Vector::operator-(const Vector & b) const
{
    return Vector(x - b.x, y - b.y);
}

//对向量的符号取反
Vector Vector::operator-() const
{
    return Vector(-x, -y);
```

```
    }

    //向量乘以 n
    Vector Vector::operator*(double n) const
    {
        return Vector(n*x, n*y);
    }

    //友元方法
    //向量 a 乘以 n
    Vector operator*(double n, const Vector & a)
    {
        return a*n;
    }

    //如果 mode 等于 RECT，显示直角坐标；
    //如果 mode 等于 POL，显示极坐标
    std::ostream & operator<<(std::ostream & os, const Vector & v)
    {
        if (v.mode == Vector::RECT)
            os << "(x,y) = (" << v.x << ", " << v.y << ")";
        else if (v.mode == Vector::POL)
        {
            os << "(m,a) = (" << v.magval() << ", "
               << v.angval() *Rad_to_deg << ")";
        }
        else
              os << "Vector object mode is invalid";
        return os;
    }

} //名称空间 VECTOR 结束

//randwalk.cpp —— 使用 Vector 类
//通过 vect.cpp 文件进行编译
#include <iostream>
#include <cstdlib>      //rand()、 srand() 的原型
#include <ctime>        //time() 的原型
#include "vect.h"
int main()
{
    using namespace std;
    using VECTOR::Vector;
    srand(time(0));     //随机种子值生成器
    double direction;
    Vector step;
    Vector result(0.0, 0.0);
    unsigned long steps = 0;
    double target;
    double dstep;
```

```
    cout << "Enter target distance (q to quit): ";
    while (cin >> target)
    {
        cout << "Enter step length: ";
        if (!(cin >> dstep))
            break;

        while (result.magval() < target)
        {
            direction = rand() % 360;
            step.reset(dstep, direction, Vector::POL);
            result = result + step;
            steps++;
        }
        cout << "After " << steps << " steps, the subject "
                            "has the following location:\n";
        cout << result << endl;
        result.polar_mode();
        cout << " or\n" << result << endl;
        cout << "Average outward distance per step = "
            << result.magval()/steps << endl;
        steps = 0;
        result.reset(0.0, 0.0);
        cout << "Enter target distance (q to quit): ";
    }
    cout << "Bye!\n";
    return 0;
}
```

3．修改程序清单 11.15，使之报告 N 次测试中的最高、最低和平均步数（其中 N 是用户输入的整数），而不是报告每次测试的结果。

编程分析：

题目要求在程序清单 11.15 的基础上，统计和报告 N 次测试中的最高、最低和平均步数。修改只涉及对 N 次随机漫步的步长的统计和分析，代码中需要添加相应的统计分析变量，且在 N 次测试结束后输出最后的统计结果，由于 vecct.h 和 vect.cpp 不需要修改，因此下面只列出 randwalk.cpp 的完整代码。

```
/*第 11 章的编程练习 3*/
//randwalk.cpp —— 使用 Vector 类
//通过 vect.cpp 文件进行编译
#include <iostream>
#include <cstdlib>      //rand()、 srand() 的原型
#include <ctime>        //time() 的原型
#include "vect.h"
int main()
{
    using namespace std;
    using VECTOR::Vector;
```

```
    srand(time(0));     //随机种子值生成器
    double direction;
    Vector step;
    Vector result(0.0, 0.0);
    unsigned long steps = 0;
    double target;
    double dstep;
    /*
    定义变量，记录最高、最低和平均步数
  */
    unsigned long Max = 0;
    unsigned long Min = 0;
    unsigned long Sum = 0;
    unsigned int count = 0;

    cout << "Enter target distance (q to quit): ";
    while (cin >> target)
    {
        cout << "Enter step length: ";
        if (!(cin >> dstep))
            break;

        while (result.magval() < target)
        {
            direction = rand() % 360;
            step.reset(dstep, direction, Vector::POL);
            result = result + step;
            steps++;
        }
        cout << "After " << steps << " steps, the subject "
                                    "has the following location:\n";

       cout << result << endl;
       result.polar_mode();
       cout << " or\n" << result << endl;
       cout << "Average outward distance per step = "
            << result.magval()/steps << endl;
       /*每次计算完成后，统计最高、最低和平均步数。steps 置 0
        *
        **/
       if (Max < steps) Max = steps;
       if (Min == 0) Min = Max;
       if (Min > steps) Min = steps;
       Sum += steps;
       count++;
       steps = 0;
       result.reset(0.0, 0.0);
       cout << "Enter target distance (q to quit): ";
    }
    cout << "Your input "<<count<<" times, and statistics info :"<<endl;
```

```
    cout << "Max Step = " << Max << endl;
    cout << "Mix Step = " << Min << endl;
    cout << "Average Step =  " << Sum/count << endl;

    cout << "Bye!\n";
    cin.clear();
    while(cin.get() != '\n')
        continue;
    return 0;
}
```

4．重新编写最后的 Time 类示例（程序清单 11.10、程序清单 11.11 和程序清单 11.12），使用友元函数来实现所有的重载运算符。

编程分析：

C++语言中可以使用类的成员函数和友元函数两种形式进行运算符重载，但是其中 =、( )、[ ]和->运算符只能通过成员函数进行重载，不能通过友元函数进行重载。其他可重载运算符中，友元函数和成员函数的最主要区别在于两者的函数参数，以及运算符在调用中的操作数。作为成员函数，第 1 个操作数必须是类对象，且同一个运算符的函数参数比友元函数少一个。此外，输入/输出重定向运算符的调用方式是 cout<<user_class_object，因此为了维护该习惯，通常将其定义为友元函数的形式。若使用成员函数，则用户自定义的类对象将在运算符<<左侧。此外，为了保证乘法运算符重载中操作数的顺序，重载函数 `Time operator*(double n) const;`应当替换为友元函数的形式：

```
friend Time operator*(const Time & s, double n);
```

完整代码如下。

```
/*第 11 章的编程练习 4*/
//mytime3.h —— 具有友元的 Time 类
#ifndef MYTIME3_H_
#define MYTIME3_H_
#include <iostream>

class Time
{
private:
    int hours;
    int minutes;
public:
    Time();
    Time(int h, int m = 0);
    void AddMin(int m);
    void AddHr(int h);
    void Reset(int h = 0, int m = 0);

    /*修改原有的成员函数的操作符重载方式，在使用友元函数时，注意修改函数的参数，
     新添加一个 Time 类型的参数。为了保持和原有成员函数类似，实现下面的友元函数
```

```
    并修改返回值。友元函数不需要使用 const 关键字描述函数的属性
  */
    friend Time operator+(const Time & s, const Time & t);
    friend Time operator-(const Time & s, const Time & t);
    friend Time operator*(const Time & s, double n);

    /*以下为原有的友元函数*/
    friend Time operator*(double m, const Time & t)
    { return t *m; }   //inline definition
    friend std::ostream & operator<<(std::ostream & os, const Time & t);

};
#endif
//mytime3.cpp —— 实现 Time 类的方法
#include "mytime3.h"

Time::Time()
{
    hours = minutes = 0;
}

Time::Time(int h, int m )
{
    hours = h;
    minutes = m;
}

void Time::AddMin(int m)
{
    minutes += m;
    hours += minutes / 60;
    minutes %= 60;
}

void Time::AddHr(int h)
{
    hours += h;
}

void Time::Reset(int h, int m)
{
    hours = h;
    minutes = m;
}
/*以下以友元函数形式实现的操作符重载函数较简单，只需要在原有函数的基础上
将成员函数使用的私有数据成员替换为新添加参数的对象成员即可，由于使用友元函数的形式，
因此该函数可以访问所有的数据成员和成员方法
*/
Time operator+(const Time & s, const Time & t)
{
```

```
    Time sum;
    sum.minutes = s.minutes + t.minutes;
    sum.hours = s.hours + t.hours + sum.minutes / 60;
    sum.minutes %= 60;
    return sum;
}

Time operator-(const Time & s, const Time & t)
{
    Time diff;
    int tot1, tot2;
    tot1 = t.minutes + 60 *t.hours;
    tot2 = s.minutes + 60 *s.hours;
    diff.minutes = (tot2 - tot1) % 60;
    diff.hours = (tot2 - tot1) / 60;
    return diff;
}

Time operator*(const Time & s, double mult)
{
    Time result;
    long totalminutes = s.hours *mult *60 + s.minutes *mult;
    result.hours = totalminutes / 60;
    result.minutes = totalminutes % 60;
    return result;
}

std::ostream & operator<<(std::ostream & os, const Time & t)
{
    os << t.hours << " hours, " << t.minutes << " minutes";
    return os;
}
//usetime3.cpp 使用 Time 类的第 4 个版本
//同时编译 usetime3.cpp 与 mytime3.cpp
#include <iostream>
#include "mytime3.h"

int main()
{
    using std::cout;
    using std::endl;
    Time aida(3, 35);
    Time tosca(2, 48);
    Time temp;

    cout << "Aida and Tosca:\n";
    cout << aida<<"; " << tosca << endl;
    temp = aida + tosca;     //operator+()
    cout << "Aida + Tosca: " << temp << endl;
    temp = aida*1.17;  //member operator*()
```

```
    cout << "Aida *1.17: " << temp << endl;
    cout << "10.0 *Tosca: " << 10.0 *tosca << endl;
    return 0;
}
```

5．重新编写 Stonewt 类（程序清单 11.16 和程序清单 11.17），使它有一个状态成员，由该成员控制对象应转换为英石格式、整数磅格式还是浮点磅格式。重载<<运算符，使用它来替换 show_stn()和 show_lbs()方法。重载加法、减法和乘法运算符，以便可以对 Stonewt 值进行加、减、乘运算。编写一个使用所有类方法和友元的小程序，来测试这个类。

编程分析：

题目要求在程序清单 11.16 和程序清单 11.17 的基础上完成以下几个主要功能调整。首先，添加一个描述对象转换状态的数据成员，此处可以参考程序清单 11.13，使用枚举类型添加相应的数据成员。然后，重载输出重定向运算符，替换 show_stn()和 show_lbs()两个负责输出的方法。最后，重载加、减、乘运算符，实现 Stonewt 对象的加、减、乘运算。如果要为类重载运算符，并以非类的项作为其第 1 个操作数，则可以用友元函数来反转操作数的顺序

```
/*第 11 章的编程练习 5*/
//stonewt.h —— Stonewt 类的定义
#ifndef STONEWT_H_
#define STONEWT_H_
class Stonewt
{
public:
    enum Style{STONE, POUNDS, FLOATPOUNDS};
private:
    enum {Lbs_per_stn = 14};        /1 英石等于 14 磅
    int stone;                      //整型的英石值
    double pds_left;                //double 类型的磅值
    double pounds;                  //以磅为单位的整体重量
    Style style;

public:
    Stonewt(double lbs);            //针对 double 类型磅值的构造函数
    Stonewt(int stn, double lbs);   //针对 int 类型的英石值和 double 类型的磅值的构造函数
    Stonewt();                      //默认构造函数
    ~Stonewt();

    /*删除 show_lbs()和 show_stn()函数，使用操作符 << 重载实现相应功能
    void show_lbs() const;          //以磅为单位显示重量
    void show_stn() const;          //以英石为单位显示重量
   */
    void Set_Style(Style m);
    Stonewt operator+(const Stonewt & s) const;
    Stonewt operator-(const Stonewt & s) const;
    Stonewt operator*(double n) const;
```

```
    /*当前乘法运算符重载在使用成员函数时会限定乘法中两个操作数的顺序，double 值必须在右侧
     *如果需要把 double 值放在左侧，还需要重新定义一个友元函数来重载乘法运算符
     **/
    friend std::ostream & operator<<(std::ostream & os, const Stonewt & s);
};
#endif
//stonewt.cpp —— Stonewt 方法
#include <iostream>
#include "stonewt.h"
//根据 double 值构造 Stonewt 对象
Stonewt::Stonewt(double lbs)
{
    stone = int (lbs) / Lbs_per_stn;    //整数除法
    pds_left = int (lbs) % Lbs_per_stn + lbs - int(lbs);
    pounds = lbs;
    style = POUNDS;
}

//根据 int 型英石值和 double 型磅值构造 Stonewt 对象
Stonewt::Stonewt(int stn, double lbs)
{
    stone = stn;
    pds_left = lbs;
    pounds =  stn *Lbs_per_stn +lbs;
    style = FLOATPOUNDS;
}

Stonewt::Stonewt()          //默认构造函数, wt = 0
{
    stone = pounds = pds_left = 0;
    style = STONE;
}

Stonewt::~Stonewt()         //析构函数
{
}

void Stonewt::Set_Style(Style m)
{
    style = m;

}

/*以下为运算符重载函数的实现
 **/
Stonewt Stonewt::operator+(const Stonewt & s)const
{
    Stonewt temp;
    temp.pounds = pounds + s.pounds;
    temp.stone = temp.pounds / Lbs_per_stn;
```

```
    temp.pds_left=int(temp.pounds)%Lbs_per_stn + temp.pounds - int(temp.pounds);
    temp.style = this->style;
    return temp;
}

Stonewt Stonewt::operator-(const Stonewt & s)const
{
    Stonewt temp;
    temp.pounds = pounds - s.pounds;
    temp.stone = temp.pounds / Lbs_per_stn;
    temp.pds_left=int(temp.pounds)%Lbs_per_stn + temp.pounds - int(temp.pounds);
    temp.style = this->style;
    return temp;
}

Stonewt Stonewt::operator*(double n) const
{
    Stonewt temp;
    temp.pounds = pounds *n;
    temp.stone = temp.pounds / Lbs_per_stn;
    temp.pds_left=int(temp.pounds)%Lbs_per_stn + temp.pounds - int(temp.pounds);
    temp.style = this->style;
    return temp;
}


std::ostream & operator<<(std::ostream & os, const Stonewt & s)
{
    if (s.style == Stonewt::STONE)
    {
        double st = s.stone + s.pds_left / Stonewt::Lbs_per_stn;
        os << st << " stone\n";
    }
    if (s.style == Stonewt::POUNDS)
        os << s.pounds << " pounds\n";

    if (s.style == Stonewt::FLOATPOUNDS)
        os << s.stone << " stone, " << s.pds_left << " pounds\n";
    return os;
}

#include <iostream>
#include "stonewt.h"

using namespace std;

int main()
{
    Stonewt incognito = 275; //使用构造函数初始化
    cout<<"incognito: "<<incognito<<endl;
```

```
    Stonewt wolfe(285.7);    //与 Stonewt wolfe = 285.7 相同
    cout<<"wolfe: "<<wolfe<<endl;
    Stonewt taft(21, 8);
    cout<<"taft: "<<taft<<endl;

    incognito = 276.8;      //使用构造函数转换
    cout<<"incognito: "<<incognito<<endl;

    cout<<"wolfe: "<<wolfe*2.3<<endl;
    taft = incognito + wolfe + 200;
    cout<<"taft: "<<taft<<endl;
    wolfe.Set_Style(Stonewt::FLOATPOUNDS);
    wolfe = wolfe*2.3;
    cout<<"wolfe: "<<wolfe*2.3<<endl;
    return 0;
}
```

6．重新编写 Stonewt 类（程序清单 11.16 和程序清单 11.17），重载 6 个关系运算符。运算符对 pounds 成员进行比较，并返回一个布尔值。编写一个程序，它声明一个包含 6 个 Stonewt 对象的数组，并在数组声明中初始化前 3 个对象。然后使用循环来读取用于设置剩余 3 个数组元素的值。接着报告最小的元素、最大的元素以及大于或等于 11 英石的元素的数量（最简单的方法是创建一个 Stonewt 对象，并将其初始化为 11 英石，然后将该对象同其他对象进行比较）。

编程分析：

题目要求先在程序清单 11.16 和程序清单 11.17 的基础上，重载 Stonewt 类的 6 个关系运算符，即>、>= 、<、<=、!=、==，实现 Stonewt 类的对象之间的关系比较。关系运算符的返回值是布尔值，可以使用成员函数的形式进行定义。Stonewt 类中的关系比较只需要比较 pounds 数据成员即可。最后创建 Stonewt 的对象数组，并利用关系运算符比较对象的大小。如果需要 Stonewt 对象直接和整型数据进行关系运算，那么还需要继续重载关系运算符，其中参数使用 int 类型数据。完整代码如下。

```
/*第 11 章的编程练习 6*/
//stonewt.h —— Stonewt 类的定义
#ifndef STONEWT_H_
#define STONEWT_H_
class Stonewt
{
private:
    enum {Lbs_per_stn = 14};
    int stone;
    double pds_left;
    double pounds;
public:
    Stonewt(double lbs);
    Stonewt(int stn, double lbs);
```

```
    Stonewt();          //默认构造函数
    ~Stonewt();

    /*重载 6 个关系运算符
     **/
    bool operator<(const Stonewt & s) const;
    bool operator<=(const Stonewt & s) const;
    bool operator>(const Stonewt & s) const;
    bool operator>=(const Stonewt & s) const;
    bool operator==(const Stonewt & s) const;
    bool operator!=(const Stonewt & s) const;

    void show_lbs() const;
    void show_stn() const;
};
#endif

//stonewt.cpp —— Stonewt 类的方法
#include <iostream>
using std::cout;
#include "stonewt.h"

//根据 double 值构造 Stonewt 对象
Stonewt::Stonewt(double lbs)
{
    stone = int (lbs) / Lbs_per_stn;  //整数除法
    pds_left = int (lbs) % Lbs_per_stn + lbs - int(lbs);
    pounds = lbs;
}

//根据 int 型英石值和 double 型磅值构造 Stonewt 对象
Stonewt::Stonewt(int stn, double lbs)
{
    stone = stn;
    pds_left = lbs;
    pounds = stn *Lbs_per_stn +lbs;
}

Stonewt::Stonewt()     //默认构造函数, wt = 0
{
    stone = pounds = pds_left = 0;
}

Stonewt::~Stonewt()     //析构函数
{
}

/*重载运算符的实现
 **/
bool Stonewt::operator<(const Stonewt & s) const
```

```
{
    return pounds < s.pounds;
}

bool Stonewt::operator<=(const Stonewt & s) const
{
    return pounds <= s.pounds;
}

bool Stonewt::operator>(const Stonewt & s) const
{
    return pounds > s.pounds;
}

bool Stonewt::operator>=(const Stonewt & s) const
{
    return pounds <= s.pounds;
}

bool Stonewt::operator==(const Stonewt & s) const
{
    return pounds == s.pounds;
}

bool Stonewt::operator!=(const Stonewt & s) const
{
    return pounds != s.pounds;
}

//以英石为单位显示重量
void Stonewt::show_stn() const
{
    cout << stone << " stone, " << pds_left << " pounds\n";
}

//以磅为单位显示重量
void Stonewt::show_lbs() const
{
    cout << pounds << " pounds\n";
}


#include <iostream>
#include "stonewt.h"

const int SIZE = 6;
using namespace std;

int main()
{
```

```
    Stonewt stone_arr[SIZE] = {253.6, Stonewt(8, 0.35), Stonewt(23, 0)};
    double input;
    Stonewt eleven = Stonewt(11, 0.0);
    Stonewt max = stone_arr[0];
    Stonewt min = stone_arr[0];
    int num = 0;

    for (int i = 3; i < SIZE; i++)
    {
        cout << "enter the No." << i+1 << "'s element info(in pounds): " ;
        cin >> input;
        stone_arr[i] = Stonewt(input);
        while(cin.get() != '\n')
            continue;
    }

    for (int i = 0; i < SIZE; i++)
    {
        if (max < stone_arr[i])  max = stone_arr[i];
        if (min > stone_arr[i])  min = stone_arr[i];
        if (stone_arr[i] > eleven)
            num++;
    }

    cout << "The weight max : ";
    max.show_stn();

    cout << "\nThe weight min: ";
    min.show_stn();

    cout << "\nHeavy than eleven : " << num << endl;
    return 0;
}
```

7. 复数由两个部分组成，分别是实部和虚部。复数的一种书写方式是(3.0，4.0)，其中，3.0 是实部，4.0 是虚部。假设 $a = (A, B\mathrm{i})$，$c = (C, D\mathrm{i})$，则下面是一些复数运算。

加法：$a+c=(A+C,(B+D)\mathrm{i})$

减法：$a-c=(A-C,(B-D)\mathrm{i})$

乘法：$a*c=(A*C-B*D,(A*D+B*C)\mathrm{i})$

数乘：$x*c=(x*C, x*D\mathrm{i})$

共轭：$\sim a=(A, -B\mathrm{i})$

请定义一个复数类，以便下面的程序可以使用它来获得正确的结果。

```
#include <iostream>
using namespace std;
#include "complex0.h"      //避免与 complex.h 混淆
int main()
{
   complex a(3.0, 4.0);   //初始化为 (3,4i)
```

```
    complex c;
    cout << "Enter a complex number (q to quit):\n";
    while (cin >> c)
    {
        cout << "c is " << c << '\n';
        cout << "complex conjugate is " << ~c << '\n';
        cout << "a is " << a << '\n';
        cout << "a + c is " << a + c << '\n';
        cout << "a - c is " << a - c << '\n';
        cout << "a *c is " << a *c << '\n';
        cout << "2 *c is " << 2 *c << '\n';
        cout << "Enter a complex number (q to quit):\n";
    }
    cout << "Done!\n";
    return 0;
}
```

注意，必须重载运算符“<<”和“>>”。标准 C++头文件 complex 提供了比这个示例更广泛的复数支持，因此应将自定义的头文件命名为 complex0.h，以免发生冲突。应尽可能使用 const。下面是该程序的运行情况。

```
Enter a complex number (q to quit):
real: 10
imaginary: 12
c is (10,12i)
complex conjugate is (10,-12i)
a is (3,4i)
a + c is (13,16i)
a - c is (-7,-8i)
a *c is (-18,76i)
2 *c is (20,24i)
Enter a complex number (q to quit):
real: q
Done!
```

请注意，经过重载后，cin>>c 将提示用户输入实部和虚部。

编程分析：

运算符重载中，依据题目要求使用 const 关键字，并且通过返回值获得运算结果。因此需要在运算符重载中新建临时对象，并返回该临时对象，尤其是使用一元操作符～共轭重载成员函数时，应使用无参数的重载函数，且共轭结果也需要通过临时对象返回。实数乘以复数的形式 2 *c is (20,24i)表示需要通过友元函数实现该运算符重载。

```
/*第 11 章的编程练习 7*/
//complex0.h
#ifndef COMPLEX0_H_
#define COMPLEX0_H_

class complex
```

```
{
private:
    double real;
    double imaginary;
public:
    complex(double real = 0.0, double imaginary = 0.0);
    ~complex();

    complex operator+(const complex & c) const;
    complex operator-(const complex & c) const;
    complex operator*(const complex & c) const;
    complex operator~() const;

    friend complex operator*(double x, const complex & c);
    friend std::istream & operator>>(std::istream & is, complex & c);
    friend std::ostream & operator<<(std::ostream & os, const complex & c);
};
#endif

//complex.cpp
#include <iostream>
#include "complex0.h"
complex::complex(double realnum, double imagnum)
{
    real = realnum;
    imaginary = imagnum;
}

complex::~complex()
{
}

complex complex::operator+(const complex & c) const
{
    return complex(real + c.real,imaginary + c.imaginary);
}

complex complex::operator-(const complex & c) const
{
    return complex(real - c.real,imaginary - c.imaginary);
}

complex complex::operator*(const complex & c) const
{
    complex temp;
    temp.real = real *c.real - imaginary *c.imaginary;
    temp.imaginary = real *c.imaginary + imaginary *c.real;
    return temp;
}
```

```
complex operator*(double x, const complex & c)
{
    return complex(x *c.real,x *c.imaginary);
}
complex complex::operator~() const
{
    return complex(real,-imaginary);
}

std::istream & operator>>(std::istream & is, complex & c)
{
    std::cout << "real: ";
    if (!(is >> c.real))
        return is;
    std::cout << "imaginary: ";
    is >> c.imaginary;
    return is;
}

std::ostream & operator<<(std::ostream & os, const complex & c)
{
    os << "(" << c.real << ", " << c.imaginary << "i)";
    return os;
}

#include <iostream>
using namespace std;
#include "complex0.h"  //避免与 complex.h 混淆

int main()
{
    complex a(3.0, 4.0);    //initialize to (3,4i)
    complex c;
    cout << "Enter a complex number (q to quit):\n";
    while (cin >> c)
    {
        cout << "c is " << c << '\n';
        cout << "complex conjugate is " << ~c << '\n';
        cout << "a is " << a << '\n';
        cout << "a + c is " << a + c << '\n';
        cout << "a - c is " << a - c << '\n';
        cout << "a *c is " << a *c << '\n';
        cout << "2 *c is " << 2 *c << '\n';
        cout << "Enter a complex number (q to quit):\n";
    }
    cout << "Done!\n";
    return 0;
}
```

第 12 章　类和动态内存分配

本章知识点总结

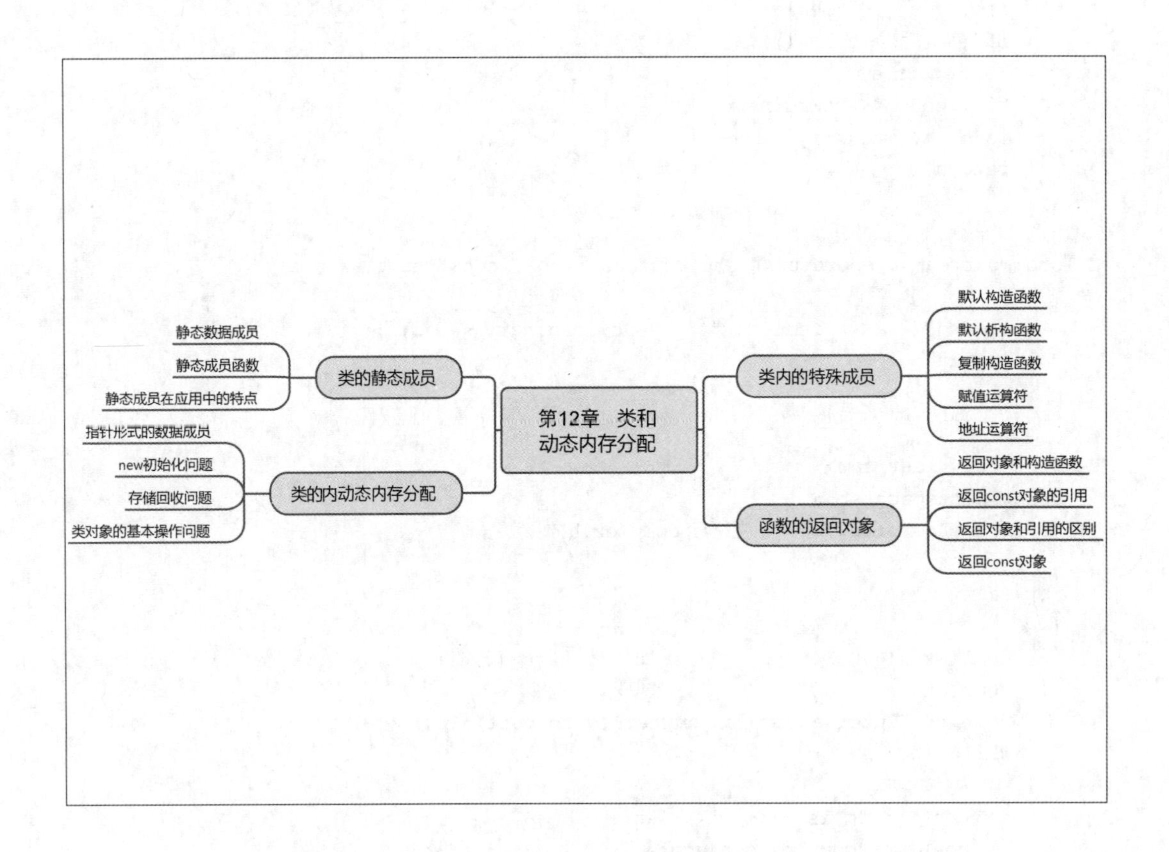

12.1　类中的静态数据成员和函数

类的数据成员也可以通过 static 关键字定义成静态数据类型。类中的静态成员并不是类对象的一部分，而是单独存储的。即类的所有对象共享一个类的静态数据成员。对于类的静态数据成员，在实际使用中，不能够在类的声明中初始化静态成员变量，而需要在类声明之外，单独使用赋值语句初始化静态数据成员，初始化过程中需要使用域运算符且不使用 static 关键字。但是如果静态成员是 const 整数类型或者枚举类型，则可以在类声明中初始化。

类的声明中也可以将成员函数声明为静态类型的成员函数，静态类型的成员函数和静态数据成员类似，是所有类对象所共有的，因此在使用中不能使用类对象调用静态成员函数。如果静态函数是公有成员，那么应当使用类作用域的形式直接调用静态函数。由于静态成员函数的存储类别关系，在静态成员函数中也不能够调用类对象的普通数据成员，而只能在函数体内使用静态数据成员。

12.2　类中的动态存储形式

类的数据成员也可以声明为指针形式，并在构造函数内使用 new 进行动态存储分配。对于动态存储，还必须在析构函数中使用 delete 释放内存，这样才不会出现内存泄露（内存泄露是指被使用的存储单元无法回收和再利用，从而导致整个系统的存储空间越来越少的现象）。对于动态存储类型的数据变量，由于其数据存储单元的管理形式，除在构造函数和析构函数的相关操作中需要注意外，还需要在一系列的类成员函数中进行动态存储管理，其中主要包括复制构造函数、赋值运算符重载和地址运算符重载等。

C++语言在类的定义中会自动提供构造函数、析构函数、复制构造函数、赋值运算符和地址运算符的默认定义形式。其中默认的函数和运算符定义在处理类内动态数据成员时都无法自动实现动态存储形式的管理。例如，复制构造函数用于将一个对象复制到新创建的类对象当中，也就是说，它用于对象的初始化过程中，而不是一个常规的复制过程。类的复制构造函数的原型如下。

```
Class_Name(const Class_Name&);
```

当用户未定义复制构造函数时，系统将创建一个默认的构造函数并逐个复制非静态成员。这样默认的构造函数在复制动态数据存储的情况下则会发生错误，因为复制过程会将地址直接复制到新对象的数据成员中，这样两个对象将会使用同一个动态存储单元，这种复制对象的过程一般也称为浅拷贝。为了在复制构造函数中正确实现对象的动态存储，需要在复制构造函数内使用 new 针对新对象创建动态存储，并进行数据的复制，这种复制称为深拷贝。此外，对于类的赋值运算符，也存在类似的深拷贝问题，而这些需要实现深拷贝的操作都需要用户通过自定义函数实现。简单地表述就是类内动态存储的数据成员在进行对象的复制、构建和析构过程都需要用户自定义相关函数，这些函数主要包括构造函数、析构函数、复制构造函数等。

12.3　类中成员函数的返回对象问题

类内的成员函数的返回值可以指定返回对象或者返回对象的引用两种形式。如果使用对象作为返回值，那么函数在调用过程中将会调用类的复制构造函数实现返回值的复制操作，因此以引用作为返回值能够提高程序运行效率。但是当返回对象是函数内部的局部变量时，则不应该使用引用的方式返回，因为局部变量在函数结束后会被系统自动调用的析构函数回收，该引用对象将不复存在。总之，如果方法或者函数要返回局部对象，则应返回对象，而不是指向对象的引用；如果函数要返回一个没有可用的复制构造函数的类的对象，则必须返回一个指向这个对象的引用。

在返回对象引用时，通常添加 const 关键字提高程序的运行效率，但是在重载赋值运算符和"<<"运算符使用返回对象的引用时通常不添加 const 关键字。此外，特定情况下使用 const 限定返回值为对象类型也能够限制某些临时对象的不合理操作，例如：

```
force1 + force2 = net;
```

如果重载加法运算符且返回值为 const 类型，则该赋值语句不合法；否则，语句合法，但是该加法运算符生成的临时对象在赋值后被销毁，因此该语句并无实际意义。

12.4　复习题

1．假设 String 类有如下私有成员。

```
class String
{
      private:
      char* str;
      int len;
//…
}
```

a．下述默认构造函数有什么问题?

```
String::String(){}
```

b．下述构造函数有什么问题?

```
String::String(cosnt char*s)
{
      str = s;
      len = strlen(s);
}
```

c．下述构造函数有什么问题?

```
String::String(const char*s)
{
      strcpy(str,s);
      len = strlen(s);
}
```

习题解析：

类 String 的定义中，私有数据成员包括一个字符指针 str，因此在创建对象时必须通过构造函数对字符指针进行动态存储的初始化工作。除构造函数外，析构函数、复制构造函数、赋值运算符等多个类方法中的动态存储都需要用户自己处理。在这个基础上我们可以分析题目。

a 问的默认构造函数可以将 str 初始化为空，或者以 new 进行动态内存分配。但是默认构造函数通常将指针初始化为空（C 格式为 NULL，C++11 格式为 nullptr)，因此可以修改为：

```
String::String(){str = nullptr}
```

b 问的构造函数直接使用数据成员赋值为形参 s，String 类对象并没有创建新的字符串，只是将 str 指针指向了参数输入的字符串地址，这样对象的使用过程中以及后来的析构函数中都会引起多个指针指向同一个字符串，造成系统的运行错误，因此应当使用 new 新建字符串，并通过 strcpy()函数进行复制，修改如下。

```
String::String(cosnt char*s)
{
        len = std::strlen(s);
        str = new char[len+1];
        std::strcpy(str, s);
}
```

c 问的构造函数使用 strcpy()函数进行数据的复制，但是应当在复制前对 str 进行动态存储申请操作，否则 str 指针并未指向确定的存储单元，直接进行数据复制会导致程序的运行错误，正确的方式如下。

```
String::String(cosnt char*s)
{
        len = std::strlen(s);
        str = new char[len+1];
        std::strcpy(str, s);
}
```

2．如果你定义了一个类，其指针成员是使用 new 初始化的，请指出可能出现的 3 个问题以及如何纠正这些问题。

习题解析：

类内数据成员的动态存储管理中可能出现的 3 个问题即对象销毁、赋值和复制构造的问题。首先，使用 new 进行动态存储分配，当对象在生命周期外被销毁后，对象的成员指针指向的数据单元可能保留在内存中，这将占用空间，造成内存泄露。因此，类定义中应当使用 delete 删除构造函数中通过 new 分配的内存。其次，当程序中有通过 new 分配的动态存储对象时，如果通过赋值形式进行对象的赋值或者通过现有对象进行复制构造，也会产生默认赋值函数只是浅拷贝（仅复制指针的地址数据，而不是指向的存储单元的内容）。因此，含有动态存储的对象需要重新定义复制构造函数，并且需要重载赋值运算符实现深拷贝。

3．如果没有显式提供类方法，编译器将自动生成哪些类方法？请描述这些隐式生成的函数的行为。

习题解析：

类的定义中如果没有显示提供类方法，那么系统将会提供一些默认方法，主要包括默认构造函数、默认析构函数、复制构造函数、赋值运算符和地址运算符。默认构造函数不完成任何工作，但允许声明数组和未初始化的对象。默认复制构造函数和默认赋值运算符使用成员赋值。默认析构函数也不完成任何工作。隐式地址运算符返回调用对象的地址（即 this 指针的值）。

4．找出并改正下述类声明中的错误。

```
class nifty
{
    //数据
    char personality[];
    int talents;
    //方法
    nifty();
    nifty (char *s);
    ostream& operator<<(ostream& os , nifty & n);
}

nifty:: nifty()
{
    personality = NULL;
    talents = 0;
}
nifty:: nifty(char*s)
{
    personality = new char[strlen(s)];
    personality = s;
    talents = 0;
}
ostream& nifty::operator<<(ostream& os , nifty & n)
{
    os<<n;
}
```

习题解析：

题目定义了类 nifty，数据成员为整型数据 talents 和字符数组 personality，并重载了运算符<<进行输出。首先，在类的权限控制上，未标注的均默认为 private，因此会影响成员函数的相关功能。其次，类的构造函数混用了字符指针和字符数组的相关运算。修改如下，修改部分使用注释标注。

```
class nifty
{
    //数据
    /*默认权限为 private，为了使代码清晰可读，建议添加*/
private:
    char* personality;
    /*修改为字符指针，也可以修改为定长的字符数组，
    为了使用字符指针，必须添加析构函数、复制构造函数，重载赋值运算符
    题目仅添加了析构函数。其余函数可以自行添加
    */
    int talents;
//方法
/*添加 public 关键字，相关方法应当是公有的*/
public:
```

```
    nifty();
    /*对于构造函数的参数，应当添加 const 关键字，用于放置函数内的误操作*/
    nifty(const char *s);
    /*为了使用字符指针作为数据成员，需要添加析构函数*/
    ~nifty();
    /*<<重载运算符应当使用友元函数*/
    ostream& operator<<(ostream& os,const nifty& n);
};
/*添加分号*/
nifty:: nifty()
{
    personality = NULL;
    /*若使用字符指针，则此处不用修改；若使用字符数组形式，则需要修改*/
    talents = 0;
}
nifty:: nifty(const char*s )
{
    personality = new char[std::strlen(s) + 1];
    std::strcpy(personality,s);
    talents = 0;
    /*对于字符指针需要修改字符串的复制方式 ，使用 new 动态分配内存，并
    使用 strcpy()函数进行数据复制*/
}
nifty:: ~nifty(){
    if(personality != NULL) delete[] personality;
}
ostream& operator<<(ostream& os , nifty & n)
{
    /*运算符重载中需要重新实现输出方式，不能直接使用 nifty 对象输出*/
    os<<n.personality<<endl;
    os<<n.,talents<<endl;
    return os;
    /*必须返回 os 对象，实现运算符插入操作*/
}
```

5. 对于下面的类声明，回答下面的问题。

```
class Golfer
{
private:
    char *fullname; //指向包含高尔夫的名称的字符串
    int games;      //保存玩的高尔夫游戏的个数
    int *scores;    //指向高尔夫分数数组的第一个元素
public:
    Golfer();
    Golfer(const char* name, int g = 0);
    Golfer(const Golfer & g);
    ~Golfer();
};
```

a. 下列各条语句将调用哪些类方法?

```
1  Golfer nancy;//#1
2  Golfer lulu("Little Lulu"); //#2
3  Golfer roy("Roy Hobbs", 12); //#3
4  Golfer *par = new Golfer;
5  Golfer next = lulu;
6  Golfer hazzard = "Weed Thwacker";
7  *par = nancy;
8  nancy = "Nancy Putter";
```

b．很明显，类需要有另外几个方法才能更有用，但是类需要哪些方法才能防止数据被损坏呢?

习题解析：

- 第 1 条语句创建 Golfer 对象变量，未使用参数，将使用默认构造函数。
- 第 2 条语句创建对象，使用字符串作为参数，调用 `Golfer(const char* name, int g = 0)` 构造函数，并使用了默认值 `g = 0`。
- 第 3 条语句创建对象，使用字符串和整型数据作为参数，调用 `Golfer(const char* name, int g = 0)` 构造函数。
- 第 4 条语句使用 new 创建对象，未使用参数，因此将调用默认构造函数。
- 第 5 条语句使用赋值运算符创建对象，因此调用复制构造函数 `Golfer(const Golfer & g)`。
- 第 6 条语句通过字符串创建对象，调用 `Golfer(const char* name, int g = 0)` 构造函数，并使用了默认值 `g = 0`。
- 在第 7 条语句中，*par 对象已经存在，因此此处赋值运算符调用默认的赋值运算符进行赋值。
- 在第 8 条语句中，nancy 对象已经存在，因此此处使用默认赋值运算符进行赋值，但是在赋值之前会调用 `Golfer(const char* name, int g = 0)` 构造函数创建一个临时对象。

由于某些代码会调用默认的赋值运算符，会造成指针变量仅复制指向地址值，无法复制存储单元中的全部数据，因此为了使代码能够正确执行，还需要重载赋值运算符。

12.5 编程练习

1．对于下面的类声明，为这个类提供实现，并编写一个使用所有成员函数的小程序。

```
class Cow {
     char name[20];
     char *hobby;
     double weight;
public:
     Cow();
     Cow(const char *nm, const char *ho, double wt);
     Cow(const Cow c&);
```

```
    ~Cow();
    Cow & operator=(const Cow & c);
    void ShowCow() const;              //显示所有关于奶牛的数据
};
```

编程分析：

题目给定的 Cow 类的声明中，数据成员 name 使用字符数组表示，因此可以不需要考虑其动态存储的分配问题，只需要在赋值、复制中使用对应字符串操作函数即可。数据成员 hobby 是一个字符指针，因此需要在构造函数、析构函数、赋值、复制中考虑动态存储问题。完整代码如下。

```
/*第 12 章的编程练习 1*/
//cow.h
#ifndef COW_H_
#define COW_H_

class Cow {
    char name[20];
    char *hobby;
    double weight;
public:
    Cow();
    Cow(const char *nm, const char *ho, double wt);
    Cow(const Cow & c);
    ~Cow();
    Cow & operator=(const Cow & c);
    void ShowCow() const;              //显示奶牛的所有数据
};
#endif


//Cow.cpp
#include "cow.h"
using namespace std;

Cow::Cow()
{
    name[0] = '\0';
    hobby = nullptr;
    /*hobby 也初始化为一个空字符串，即
     *hobby = new char[1];
     *hobby[0] = '/0';
     *这样在部分函数内可以省略对空指针的判断
     **/
    weight = 0.0;
}
```

```
Cow::Cow(const char *nm, const char *ho, double wt)
{
    strncpy(name, nm, 20);
    if(strlen(nm) >= 20) name[19] = '\0';
    else name[strlen(nm)] = '\0';
    /*使用 strncpy()函数，通过第 3 个参数，限制输入字符的长度，
     *并设置字符串末尾为空字符*/
    hobby = new char[strlen(ho)+1];
    strcpy(hobby, ho);
    /*使用 new 创建动态存储，可以直接进行数据复制*/
    weight = wt;
}

Cow::Cow(const Cow & c)
{
    strcpy(name, c.name);
    hobby = new char[strlen(c.hobby)+1];
    strcpy(hobby, c.hobby);
    weight = c.weight;
}

Cow::~Cow()
{
    delete[] hobby;
}

Cow & Cow::operator=(const Cow & c)
{
    if (this == &c)
       return *this;
    if(hobby != nullptr) delete[] hobby;
    hobby = new char[strlen(c.hobby)+1];
    /*当 hobby 使用 new 进行动态存储分配时，必须确保原指针为空；否则，会产生
     *内存泄露*/
    strcpy(name, c.name);
    strcpy(hobby, c.hobby);
    weight = c.weight;
    return *this;
}

void Cow::ShowCow() const
{
    if(hobby == nullptr){
        /*如果为空指针，则
         *cout<<hobby;
         *无法通过该指针寻址到字符串，会出现运行时错误*/
        cout<<"This Cow's info is Empty!"<<endl;
        return;
    }else{
```

```
        cout << "This is Information of COW."<<endl;
        cout << "Name: " << name << endl;
        cout << "Hobby: " << hobby << endl;
        cout << "Weight: " << weight << endl;
    }
}

//usecow.cpp
#include <iostream>
#include <cstring>
#include "cow.h"

using namespace std;
int main()
{
    cout<<"Initialize, and show No.1:"<<endl;
    Cow cow_list[2] = {{"Tom", "Sleeppy", 200}};
    cow_list[0].ShowCow();

    cout<<"No.2 's default values: "<<endl;
    cow_list[1].ShowCow();

    cout<<"Now using operator =(). to init No.2"<<endl;
    cow_list[1] = cow_list[0];
    cow_list[1].ShowCow();

    cout<<"Now, using copy constructor to init No.3."<<endl;
    Cow copy(cow_list[0]);
    copy.ShowCow();

    return 0;
}
```

2．通过完成下面的工作来改进 String 类的声明（即将 String1.h 升级为 String2.h）。

a．对+运算符进行重载，使之可将两个字符串合并成一个。

b．提供一个 Stringlow()成员函数，将字符串中所有的字母字符转换为小写（注意 cctype 系列的字符函数）。

c．提供 String()成员函数，将字符串中所有字母字符转换成大写。

d．提供一个这样的成员函数，它接受一个 char 参数，返回该字符在字符串中出现的次数。使用下面的程序来测试你的工作。

```
//pe12_2.cpp
#include <iostream>
using namespace std;
#include "string2.h"
int main()
{
    String s1(" and I am a C++ student.");
    String s2 = "Please enter your name: ";
```

```
    String s3;
    cout << s2;                      //重载的 << 运算符
    cin >> s3;                       //重载的 >> 运算符
    s2 = "My name is " + s3;         //重载的 =、 + 运算符
    cout << s2 << ".\n";
    s2 = s2 + s1;
    s2.stringup();                   //把字符串转换为大写
    cout << "The string\n" << s2 << "\ncontains " << s2.has('A')
         << " 'A' characters in it.\n";
    s1 = "red";     //String(const char *), 然后 String & operator=(const String&)
    String rgb[3] = { String(s1), String("green"), String("blue")};
    cout << "Enter the name of a primary color for mixing light: ";
    String ans;
    bool success = false;
    while (cin >> ans)
    {
        ans.stringlow();             //把字符串转换为小写
        for (int i = 0; i < 3; i++)
        {
            if (ans == rgb[i])       //重载的== 运算符
            {
                cout << "That's right!\n";
                success = true;
                break;
            }
        }
        if (success)
            break;
        else
            cout << "Try again!\n";
    }
    cout << "Bye\n";
    return 0;
}
```

输出应与下面相似。

```
Please enter your name: Fretta Farbo
My name is Fretta Farbo.
The string
MY NAME IS FRETTA FARBO AND I AM A C++ STUDENT.
contains 6 'A' characters in it.
Enter the name of a primary color for mixing light: yellow
Try again!
BLUE
That's right!
Bye
```

编程分析：

题目要求在程序清单 12.4 和程序清单 12.5 的基础上，定义 String 类（为避免和系统预

定义 string 类冲突，首字母需要大写），定义的 String 类不仅需要实现基本的构造函数、析构函数和题目要求的+运算符重载，还要实现 stringlow()函数（用于将字符串转换成小写）、stringup()成员函数（用于将字符串转换成大写），以及一个字符匹配函数。由于程序清单 12.4 中 String 的数据成员包括一个动态存储的字符指针，因此在程序设计中应当注意存储管理。完整代码如下。

```
/*第 12 章的编程练习 2*/
#ifndef STRING2_H_
#define STRING2_H_
#include <iostream>
using std::ostream;
using std::istream;

class String
{
private:
    char *str;
    int len;
    static int num_strings;
    static const int CINLIM = 80;
public:
//构造函数和其他方法
    String(const char *s);      //构造函数
    String();                   //默认构造函数
    String(const String &);     //用于复制的构造函数
    ~String();                  //析构函数
    int length () const { return len; }
//重载运算符的方法
    String & operator=(const String &);
    String & operator=(const char *);
    char & operator[](int i);
    const char & operator[](int i) const;
    /*添加新的运算符重载，+ 使用友元函数的方式实现，
     *依据题目要求添加 stringlow()和 stringup()函数，
     *函数无返回值，无参数，直接修改私有成员 str*/
    //String operator+(const String &s) const;
    //可以使用成员函数重载两个 String 的加法，也可以用友元函数
    friend String operator+(const char *s, const String &st);
    friend String operator+(const String &s, const String &st);
    /*题目要求实现字符指针和 String 的加法，以及两个 String 对象的相加，因此需要两个运算符重载
     *函数，两者在函数的参数上有所不同，可以使用成员函数或者友元函数*/
    void stringlow();
    void stringup();
    int has(char c) const;
    /**/

    friend bool operator<(const String &st, const String &st2);
    friend bool operator>(const String &st1, const String &st2);
    friend bool operator==(const String &st, const String &st2);
    friend ostream & operator<<(ostream & os, const String & st);
```

```
    friend istream & operator>>(istream & is, String & st);

    static int HowMany();
};
#endif
```

```
//string2.cpp
//string2.cpp —— String 类的方法
#include <cstring>
#include "string2.h"                //包括 <iostream>
using std::cin;
using std::cout;

int String::num_strings = 0;

int String::HowMany()
{
    return num_strings;
}

String::String(const char *s)
{
    len = std::strlen(s);
    str = new char[len + 1];
    std::strcpy(str, s);
    num_strings++;
}

String::String()                    //默认构造函数
{
    len = 4;
    str = new char[1];
    str[0] = '\0';
    num_strings++;
}

String::String(const String & st)
{
    num_strings++;
    len = st.len;
    str = new char [len + 1];
    std::strcpy(str, st.str);
}

String::~String()
{
    --num_strings;
```

```
    delete [] str;
}

String & String::operator=(const String & st)
{
    if (this == &st)
        return *this;
    delete [] str;
    len = st.len;
    str = new char[len + 1];
    std::strcpy(str, st.str);
    return *this;
}

String & String::operator=(const char *s)
{
    delete [] str;
    len = std::strlen(s);
    str = new char[len + 1];
    std::strcpy(str, s);
    return *this;
}

char & String::operator[](int i)
{
    return str[i];
}

const char & String::operator[](int i) const
{
    return str[i];
}
/*在头文件内注释该函数，因此，在 cpp 文件内也需要注释
String String::operator+(const String &s) const
{
    String result;
    result.len = len + s.len;
    result.str = new char[result.len + 1];
    strcpy(result.str, str);
    strcat(result.str, s.str);
    return result;
}*/
String operator+(const String &s, const String &st)
{
    String result;
    result.len = s.len + st.len;
    result.str = new char[result.len + 1];
    strcpy(result.str, s.str);
    strcat(result.str, st.str);
    return result;
```

```
}
String operator+(const char *st, const String &s)
{
    String result;
    result.len = s.len + strlen(st);
    result.str = new char[result.len + 1];
    strcpy(result.str, st);
    strcat(result.str, s.str);
    return result;
}

void String::stringlow()
{
    for (int i = 0; i < len; i++)
        str[i] = tolower(str[i]);
}

void String::stringup()
{
    for (int i = 0; i < len; i++)
        str[i] = toupper(str[i]);
}
int String::has(char c) const{
    int result = 0;
    for (int i = 0; i < len; i++)
    {
        if (str[i] == c)
            result++;
    }
    return result;
}

bool operator<(const String &st1, const String &st2)
{
    return (std::strcmp(st1.str, st2.str) < 0);
}

bool operator>(const String &st1, const String &st2)
{
    return st2 < st1;
}

bool operator==(const String &st1, const String &st2)
{
    return (std::strcmp(st1.str, st2.str) == 0);
}

ostream & operator<<(ostream & os, const String & st)
{
```

```
    os << st.str;
    return os;
}

istream & operator>>(istream & is, String & st)
{
    char temp[String::CINLIM];
    is.get(temp, String::CINLIM);
    if (is)
       st = temp;
    while (is && is.get() != '\n')
       continue;
    return is;
}

//pe12_2.cpp
#include <iostream>
using namespace std;
#include "string2.h"
int main()
{
    String s1(" and I am a C++ student.");
    String s2 = "Please enter your name: ";
    String s3;
    cout << s2;
    cin >> s3;
    s2 = "My name is " + s3;
    cout << s2 << ".\n";
    s2 = s2 + s1;
    s2.stringup();
    cout << "The string\n" << s2 << "\ncontains " << s2.has('A')
        << " 'A' characters in it.\n";
    s1 = "red";
    String rgb[3] = { String(s1), String("green"), String("blue")};
    cout << "Enter the name of a primary color for mixing light: ";
    String ans;
    bool success = false;
    while (cin >> ans)
    {
        ans.stringlow();
        for (int i = 0; i < 3; i++)
        {
            if (ans == rgb[i])
            {
                cout << "That's right!\n";
                success = true;
                break;
            }
        }
        if (success)
```

```
            break;
        else
            cout << "Try again!\n";
    }
    cout << "Bye\n";
    return 0;
}
```

3．重写程序清单 10.7 和程序清单 10.8 描述的 Stock 类，使之使用动态分配的内存，而不是 string 类对象来存储股票名称。另外，使用重载的 operator<<()定义代替 show()成员函数。再使用程序清单 10.9 测试新的程序。

编程分析：

程序清单 10.7 和程序清单 10.8 中定义的 Stock 类使用 string 类存储股票名称，为了修改该数据成员为 char*company，需要修改构造函数、析构函数，sell()、buy()等函数可以不用修改。把成员函数 show() 替换为运算符重载 operator<<() 。完整代码如下。

```
/*第 12 章的编程练习 3*/
//stock20.h —— 增强版
#ifndef STOCK20_H_
#define STOCK20_H_
#include <string>

class Stock
{
private:
    char*company;
    /*使用字符指针，实现动态存储的分配*/
    int shares;
    double share_val;
    double total_val;
    void set_tot() { total_val = shares *share_val; }
public:
    Stock();        //默认构造函数
    Stock(const char*co, long n = 0, double pr = 0.0);
    ~Stock();       //不执行任何操作的析构函数
    void buy(long num, double price);
    void sell(long num, double price);
    void update(double price);
    const Stock & topval(const Stock & s) const;
    friend std::ostream& operator<<(std::ostream& os, const Stock &stock);
    /*重载运算符<<,实现原 show()函数的功能*/
};

#endif

//stock20.cpp —— 增强版
#include <iostream>
```

```
#include "stock20.h"
using namespace std;
//constructors
Stock::Stock()          //默认构造函数
{
    company = new char[8];
    strcpy(company, "no name");
    /*默认构造函数使用字符串"no name"初始化 company,
     *因此先使用 new 动态存储分配 8 个字符*/
    shares = 0;
    share_val = 0.0;
    total_val = 0.0;
}

Stock::Stock(const char*co, long n, double pr)
{
    company = new char[strlen(co)+1];
    strcpy(company, co);
    /*依据参数 co 动态分配 company 的内存*/
    if (n < 0)
    {
        std::cout << "Number of shares can't be negative; "
                  << company << " shares set to 0.\n";
        shares = 0;
    }
    else
        shares = n;
    share_val = pr;
    set_tot();
}

//类析构函数
Stock::~Stock()         //quiet 空的类析构函数
{
    if(company != nullptr) delete[] company;
    /*析构函数需要回收 company 占用的内存*/
}

//其他方法
void Stock::buy(long num, double price)
{
    if (num < 0)
    {
        std::cout << "Number of shares purchased can't be negative. "
                << "Transaction is aborted.\n";
    }
    else
    {
        shares += num;
        share_val = price;
```

```
        set_tot();
    }
}

void Stock::sell(long num, double price)
{
    using std::cout;
    if (num < 0)
    {
        cout << "Number of shares sold can't be negative. "
            << "Transaction is aborted.\n";
    }
    else if (num > shares)
    {
        cout << "You can't sell more than you have! "
             << "Transaction is aborted.\n";
    }
    else
    {
        shares -= num;
        share_val = price;
        set_tot();
    }
}

void Stock::update(double price)
{
    share_val = price;
    set_tot();
}

ostream& operator<<(ostream& os, const Stock &stock)
{
    /*为了使用原 show()函数的代码，需要修改其输出语句 std::cout 为参数的 os 对象*/
    using std::ios_base;
    //set format to #.###
    ios_base::fmtflags orig =
            os.setf(ios_base::fixed, ios_base::floatfield);
    std::streamsize prec = os.precision(3);

    os << "Company: " << stock.company
        << "  Shares: " << stock.shares << '\n';
    os << "  Share Price: $" << stock.share_val;
    //设置格式为 #.##
    os.precision(2);
    os << "  Total Worth: $" << stock.total_val << '\n';

    //还原原始格式
    os.setf(orig, ios_base::floatfield);
    os.precision(prec);
```

```
    return os;
}

const Stock & Stock::topval(const Stock & s) const
{
    if (s.total_val > total_val)
        return s;
    else
        return *this;
}

#include <iostream>
#include "stock20.h"

const int STKS = 4;
int main()
{

        Stock stocks[STKS] = {
                Stock("NanoSmart", 12, 20.0),
                Stock("Boffo Objects", 200, 2.0),
                Stock("Monolithic Obelisks", 130, 3.25),
                Stock("Fleep Enterprises", 60, 6.5)
        };

        std::cout << "Stock holdings:\n";
        int st;
        for (st = 0; st < STKS; st++)
            std::cout<<stocks[st];
        const Stock *top = &stocks[0];
        for (st = 1; st < STKS; st++)
            top = &top->topval(stocks[st]);
        std::cout << "\nMost valuable holding:\n";
        std::cout<< *top;
    return 0;
}
```

4. 请看程序清单 10.10 定义的 Stack 类的变量。

```
typedef unsigned long Item;

class Stack
{
private:
    enum {MAX = 10};
    Item *pitems;
    int size;
    int top;
public:
    Stack(int n = MAX);
    Stack(const Stack & st);
```

```
    ~Stack();

    bool isempty() const;
    bool isfull() const;
    bool push(const Item & item);
    bool pop(Item & item);

    Stack & operator=(const Stack & st);
};
```

正如私有成员表明的，这个类使用动态分配的数组来保存栈中的项。请重新编写方法，以适应这种新的表示法，并编写一个程序来演示所有的方法，包括复制构造函数和赋值运算符。

编程分析：

程序清单 10.10 内定义的 Stack 类使用数组形式实现栈，题目要求修改为动态存储形式来重新定义该类，因此需要修改相关函数，主要包括构造函数、析构函数、入栈和出栈的操作。完整代码如下。

```
/*第 12 章的编程练习 4*/
/*使用题目给定的头文件 stack.h*/
//stack.cpp
#include "stack.h"
using namespace std;
Stack::Stack(int n)
{
    pitems  = new Item[n];
    size = n;
    top = 0;
}
Stack::Stack(const Stack & st)
{
    pitems = new Item[st.size];
    /*复制构造函数通过参数 st 复制对象,
     *数组内数据通过循环复制*/
    for(int i = 0; i < st.top; i++)
    {
        pitems[i] = st.pitems[i];
    }
    size = st.size;
    top = st.top;
}
Stack::~Stack()
{
    if(pitems != nullptr)
       delete[] pitems;
}
bool Stack::isempty() const
{
    if(top == 0) return true;
```

```
    else return false;
}
bool Stack::isfull() const
{
    if(top == size) return true;
    else return false;
}
bool Stack::push(const Item &item)
{
    if(!isfull())
    {
       pitems[top++] = item;
       return true;
    }else
       {
          return false;
       }
}
bool Stack::pop(Item &item)
{
    if(!isempty())
    {
       item = pitems[--top];
       return true;
    }else
       {
          return false;
       }
}
Stack& Stack::operator=(const Stack &st)
{
    if (this == &st)
       return *this;
    pitems = new Item[st.size];
    for (int i = 0; i < size; i++)
       pitems[i] = st.pitems[i];
    size = st.size;
    top = st.top;
    return *this;
}

#include <iostream>
#include "stack.h"
using namespace std;

const int MAX = 5;
int main() {
    Stack st(MAX);
    Item item;
    for(int i = 0 ; i < MAX ; i++)
```

```
    {
        cout << "Enter a unsigned long number : ";
        cin >> item;
        while(cin.get() != '\n') continue;
        st.push(item);
        cout<<"Item pushed.\n";
    }
    Stack st_new(st);
    for(int i = 0 ; i < MAX ; i++)
    {
        st_new.pop(item);
        cout<<"Item poped: "<<item<<endl;;
    }
    /*简单模拟出栈和入栈*/
    cout << "Bye\n";
    return 0;
}
```

5．Heather 银行进行的研究表明，ATM 客户希望排队时间不超过 1min。使用程序清单 12.10 中的模拟，求出要使平均等候时间为 1min，每小时到达的客户数应为多少（试验时间不短于 100h）？

编程分析：

程序清单 12.10 是一个使用队列模拟银行 ATM 排队的基础模型。ATM 在服务客户时，如果客户较多，则需要客户排队。队伍的形式是先来先得。如果没有人，则可以直接服务；如果人数过多，客户可能无法等候，会先离开。每位客户的服务时间不确定，需要由随机数确定。本题中需要确定平均等候时间为 1min，即所有客户的等待时间除以总人数不超过 60s。模拟的基本算法需要测试并确定每小时到达的客户数为何值才能满足这个标准，因此需要使用一个循环，以测试每小时到达人数改变的条件下的平均等候时间。最终确定一个最接近值。bank.cpp 的完整代码如下，其余 queue.h 和 queue.cpp 文件未做修改，可直接使用。

```
/*第 12 章的编程练习 5*/
/*使用程序清单 12.10 和程序清单 12.11 中的 queue.h 头文件与 queue.cpp 定义文件*/
//bank.cpp
#include <iostream>
#include <cstdlib>
#include <ctime>
#include "queue.h"
const int MIN_PER_HR = 60;
const int MIN_SIM_HOURS = 150;

bool newcustomer(double x);

int main()
{
    using std::cin;
```

```
    using std::cout;
    using std::endl;
    using std::ios_base;

    std::srand(std::time(0));

    cout << "Case Study: Bank of Heather Automatic Teller\n";
    cout << "Enter maximum size of queue: ";
    int qs;
    cin >> qs;
    Queue line(qs);
    cout << "The number of simulation hours >= 100. "<<endl;
    /*固定最短时间为 MIN_SIM_HOURS，当前常量 MIN_SIM_HOURS 为 150*/
    int hours = MIN_SIM_HOURS;
long cyclelimit = MIN_PER_HR *hours;

    //cout << "Enter the average number of customers per hour: ";
    double perhour = 0;
    Item temp;
double average_wait = 0;
/*为了检测该临界值，从 perhour = 1 开始循环，进行模拟和计算，当排队时间
 *大于或等于 1min 时停止循环，此时输出的最后一组数据就是当前 ATM 的排队状态*/
//进行模拟
    while(average_wait < 1)
    {
        double min_per_cust;
        long turnaways = 0;
        long customers = 0;
        long served = 0;
        long sum_line = 0;
        int wait_time = 0;
        long line_wait = 0;
        perhour++;
        if(!line.isempty()) line.dequeue(temp);
         /*每次进入循环，首先清空 ATM 的排队队列*/
        for (int cycle = 0; cycle < cyclelimit; cycle++)
        {
            min_per_cust = MIN_PER_HR / perhour;
            if (newcustomer(min_per_cust))
            {
                if (line.isfull())
                    turnaways++;
                else
                {
                    customers++;
                    temp.set(cycle);
                    line.enqueue(temp);
                }
            }
            if (wait_time <= 0 && !line.isempty())
```

```
            {
                line.dequeue (temp);
                wait_time = temp.ptime();
                line_wait += cycle - temp.when();
                served++;
            }
            if (wait_time > 0)
                wait_time--;
            sum_line += line.queuecount();
        }
        average_wait = (double) line_wait / served;
        if(average_wait < 1){
            if (customers > 0)
            {
               cout << "customers accepted: " << customers << endl;
               cout << "  customers served: " << served << endl;
               cout << "         turnaways: " << turnaways << endl;
               cout << "average queue size: ";
               cout.precision(2);
               cout.setf(ios_base::fixed, ios_base::floatfield);
               cout << (double) sum_line / cyclelimit << endl;
               cout << " average wait time: "
                     << (double) line_wait / served << " minutes\n";
            }
            else
               cout << "No customers!\n";
            cout<<"The average "<<perhour<<" of arrival per hour, and average wait time is "
                <<average_wait<<endl;
        }
    }

    cout << "Done!\n";
    return 0;
}

bool newcustomer(double x)
{
    return (std::rand() *x / RAND_MAX < 1);
}
```

6. Heather 银行想要了解，如果再开设一台 ATM，情况将如何。首先，请对模拟情况进行修改，以包含两个队列。假设当第 1 台 ATM 前的排队人数少于第 2 台 ATM 时，客户将排在第 1 队；否则，将排在第 2 队。然后，再求出要使平均等候时间为 1 分钟，每小时到达的客户数应该为多少（注意，这是一个非线性问题，即将 ATM 数量加倍，不能保证每小时处理的客户数量也翻倍，并确保客户等候的时间少于 1min）？

编程分析：

在编程练习 5 的基础上，题目要求添加一台 ATM，进行排队模拟。首先，当在两台 ATM

前排队时，客户排队的基本策略是选择人数较少的 ATM 并排队，自动保持两队人数均衡。其次，在计算等候时间时需要考虑两个队伍的所有人数，因此需要针对两台 ATM 分别设置一些变量。bank.cpp 的完整代码如下，queue.h 和 queue.cpp 等文件未做修改，可直接使用。

```
/*第 12 章的编程练习 6*/
/*使用程序清单 12.10 和程序清单 12.11 的 queue.h 头文件与 queue.cpp 定义文件*/
//bank.cpp
//通过 queue.cpp 进行编译
#include <iostream>
#include <cstdlib>
#include <ctime>
#include "queue.h"
const int MIN_PER_HR = 60;
const int MIN_SIM_HOURS = 1500;

bool newcustomer(double x);

int main()
{
    using std::cin;
    using std::cout;
    using std::endl;
    using std::ios_base;
    std::srand(std::time(0));

    cout << "Case Study: Bank of Heather Automatic Teller\n";
    cout << "Enter maximum size of queue: ";
    int qs;
    cin >> qs;
    Queue line_one(qs);
    Queue line_two(qs);

    cout << "The number of simulation hours >= 100. "<<endl;
    /*固定最短时间为 1h*/
    int hours = MIN_SIM_HOURS;

    long cyclelimit = MIN_PER_HR *hours;

    //cout << "Enter the average number of customers per hour: ";
    double perhour = 0;
    double min_per_cust;
    Item temp;
    long turnaways;
    long customers, customers_one, customers_two;
    long served;
    long sum_line_one, sum_line_two;
    int wait_time_one, wait_time_two;
    long line_wait;
    double average_wait = 0;
/*分别定义两个队列需要的变量，部分变量可共用*/
```

```
while(average_wait < 1.0 ){
    turnaways = 0;
    customers = customers_one = customers_two = 0;
    served = 0;
    sum_line_one = sum_line_two = 0;
    wait_time_one = wait_time_two = 0;
    line_wait = 0;
    perhour++;
    min_per_cust = MIN_PER_HR / perhour;
    while(!line_one.isempty()) line_one.dequeue(temp);
    while(!line_two.isempty()) line_two.dequeue(temp);
    /*每次进入循环，首先清空 ATM 排队队列*/
   for (int cycle = 0; cycle < cyclelimit; cycle++)
   {

       if (newcustomer(min_per_cust))
       {
           if (line_one.isfull() && line_two.isfull())
               turnaways++;
           else
           {
               customers++;
               temp.set(cycle);
               if(line_one.queuecount() < line_two.queuecount()) {
                   line_one.enqueue(temp);
                   customers_one++;
               }else {
                   line_two.enqueue(temp);
                   customers_two++;
               }
              /*平均处理两个 ATM 排队的人数*/
           }
       }
       if (wait_time_one <= 0 && !line_one.isempty())
       {
            line_one.dequeue (temp);
            wait_time_one = temp.ptime();
            line_wait += cycle - temp.when();
            served++;
       }
       if (wait_time_two <= 0 && !line_two.isempty())
       {
            line_two.dequeue (temp);
            wait_time_two = temp.ptime();
            line_wait += cycle - temp.when();
            served++;
       }
       if (wait_time_one > 0) wait_time_one--;
```

```
        if (wait_time_two > 0) wait_time_two--;
        sum_line_one += line_one.queuecount();
        sum_line_two += line_two.queuecount();
    }
    average_wait = (double) (line_wait)/ (served);
    if(average_wait < 1){
        if (customers > 0)
        {
            cout << "customers accepted: " << customers << endl;
            cout << "  customers served: " << served << endl;
            cout << "         turnaways: " << turnaways << endl;
            cout << "average queue size: ";
            cout.precision(2);
            cout.setf(ios_base::fixed, ios_base::floatfield);
            cout << (double) (sum_line_one + sum_line_two)/ cyclelimit << endl;
            cout << " average wait time: "
                 << average_wait << " minutes\n";
        }
        else
            cout << "No customers!\n";
        cout<<"The average "<<perhour<<" of arrival per hour, and average wait time is "
            <<average_wait<<endl;
    }
  }

  cout << "Done!\n";
  return 0;
}


bool newcustomer(double x)
{
    return (std::rand() *x / RAND_MAX < 1);
}
```

第 13 章 类 继 承

本章知识点总结

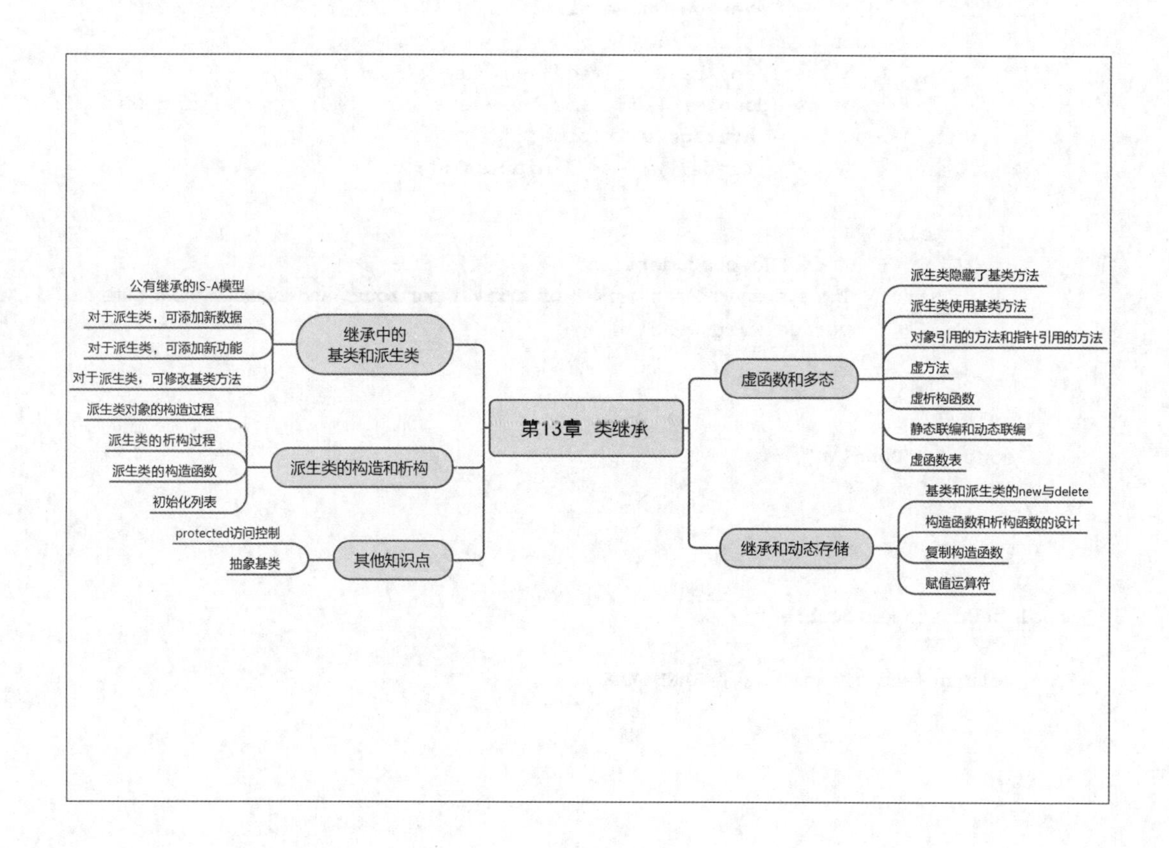

13.1 C++中的继承

C++中的继承是实现程序设计中代码重用的一个重要途径。继承是指以现有的某个类为基础派生出新的类，派生类能够继承原有类的所有特征（通常是指数据成员和成员函数）。原有的类称作基类或者父类，新类称作派生类或者子类。除继承基类的所有特征外，还可以为派生类添加新的数据成员和成员函数，或者修改基类原有的成员函数。继承的基本语法如下。

```
class base{}
```

```
class derive : public base{}
```

public 表示使用公有继承的方式继承基类的属性。基类的访问控制在派生过程中也会产生作用，简单说就是派生类的对象不能够直接访问基类中私有属性的数据和方法，但是可以通过基类的公有方法间接访问基类的私有数据。

当创建一个派生类的对象时，必须首先创建基类的对象。也就是说，派生类的构造过程需要分两步。首先调用基类的构造函数（或者使用默认基类的构造函数）创建基类对象，然后通过派生类的构造函数创建派生类的对象。在派生类的构造函数中，应当通过成员初始化列表将基类的信息传递给基类的构造函数，派生类的构造函数应当初始化派生类新增的数据成员。因为基类的构造函数早于派生类的构造函数运行，所以不能够在派生类的构造函数内部调用基类的构造函数。C++语言中派生类的初始化列表的基本结构如下。

```
derived::derived(type1 x, type2 y):base(x,y){
}
```

对于析构函数则相反。程序首先调用派生类的析构函数，然后调用基类的析构函数。

派生类在使用 public 关键字继承基类时实现的是一种公有继承关系。这种继承关系是一种 IS-A 关系，即可以认为派生类的对象是一种特殊的基类对象，用户可以像使用基类对象那样使用派生类对象。这种公有继承的 IS-A 关系和之后章节的私有继承和包含关系有区别，具体区别和用法会在第 14 章详细论述。

13.2　继承中的多态性和虚函数

类的派生过程中，基类的指针可以在不进行显式类型转换的情况下指向派生类的对象，基类的引用可以在不进行显式类型转换的情况下引用派生类的对象。但是基类指针或者引用只能调用基类方法，不能调用派生类的方法。类的派生过程中，为使派生类的对象和基类的对象的同名方法能够实现不同的操作行为，即称为继承中的多态性，即同一个方法的行为会依据上下文内容确定。为了实现多态性，首先可以在派生类中重新定义基类的方法（通常称为函数的隐藏），或者使用虚方法。实际应用中，派生类覆盖基类的成员函数，会引起在派生类对象中隐藏了基类的同名函数（即派生类对象无法直接调用基类的同名函数），同时可以在派生类的函数中通过基类的作用域运算符调用基类的隐藏函数。但是如果使用指针或者引用调用对象，那么程序将根据引用或者指针的类型，而不是对象的类型选择成员函数（基类指针可以指向基类对象或者派生类对象）。为了实现以对象为标准选择成员函数，需要使用虚方法。虚方法的基本形式如下。

```
class base{
...
        virtual void f(){}
}
```

通过虚函数，程序将根据指针或者引用所指向对象的类型来选择合适的成员函数。如果析构函数不是虚函数，那么将只调用指针类型对应的析构函数，这样将不会根据当前对象的具体类型调用相应的析构函数，将无法保证对象被正确销毁。因此程序设计中建议将析构函数定义为虚函数。

纯虚函数是指未提供实现并且以“= 0”标识的虚函数，例如：

```
virtual Type_Name Fun_Name(Para List……)  = 0;
```

含有纯虚函数的类称为抽象类。抽象类是没有办法创建对象的，只有通过继承在派生类中实现所有的纯虚函数，才能够创建对象。实践中经常可以为抽象类的纯虚函数提供方法的定义（实现），这种情况通常出现在纯虚析构函数中，因为派生类的对象在析构过程中最终还会利用到基类的析构函数，所以必须实现纯虚的析构函数。程序设计中抽象类的理念更像是一种必须要实现的接口的模型，迫使派生类实现纯虚函数的接口的一种标准格式。

13.3 静态联编和动态联编

程序在运行时具体使用哪一个可用函数和哪一段代码是由编译器确定的，编译器在编译过程中将会根据具体情况选择和确定函数代码，这样的操作称为联编（很多教材中称为绑定）。由于 C++中为函数重载的功能编译器会在编译过程中依据代码选择匹配的函数，因此这种在编译阶段实现和确定的函数代码的匹配称为静态联编（静态绑定）。然而，虚函数的引入会导致在编译阶段编译器无法确定最终调用者是何种对象，因此编译阶段将无法完成联编的工作。这项工作将会被延迟到程序运行时，再根据运行时的具体情况选择合适的虚函数并进行调用，因此这种效果称为动态联编（动态绑定）。根据两者的运行阶段和基本原理可以看出，动态联编的运行效率远远低于静态联编，因此程序设计中，为了提高程序的运行效率，应当尽量使用非虚函数，以避免动态联编。

编译器在处理虚函数时，会给每个对象都添加一个虚函数列表（Virtual Function Table，VTBL）。该虚函数列表中存储了该类对象声明的虚函数的地址。该虚函数列表会在继承过程中逐步补充派生类的虚函数。在对象调用虚函数时，系统会查找对象的虚函数列表，并在找到匹配的虚函数后执行，这种虚函数的查找和运行方式会增加程序占用的存储空间，并提高函数调用和执行的效率。

类的构造函数不能是虚函数，因为构造函数的调用原理决定了派生类的对象生成之前必须要先调用基类的构造函数，所以在继承过程中构造函数并不存在动态联编问题。而析构函数则应当定义成虚函数，这样才能正确地根据指针的指向类型选择合适的析构函数。此外，友元函数不能是虚函数，因为友元函数不是类的成员函数。如果派生类没有重新定义基类函数，那么派生对象将直接使用基类的函数；如果派生类位于派生链中，那么将使用最新的虚函数版本。

需要注意的是，派生过程中重新定义基类的方法不是一种重载过程，而是一种隐藏关系。如果重新定义了派生类中的函数，无论参数列表是否相同都会产生隐藏关系，因此，如果要重新定义基类的方法，应确保函数原型一致，以避免调用中的二义性引起编译器的警告。但是派生类函数的返回值可以从基类指针修改为派生类指针，这种情况称作返回类型的协变。如果基类的某个函数是重载函数，那么必须在派生类中重新定义所有同名的重载函数；否则，隐藏特性可能会在函数调用中引发编译器的错误。

13.4　继承中的其他知识点

除 public 和 private 外，类内部的权限控制还有 protected 访问权限。protected 权限与 private 类似，在类外无法直接访问，只能通过 public 的函数间接访问。但是区别在于继承过程中，派生类可以直接访问 protected 成员，而不能访问 private 成员。也就是说，对于类外来说，protected 和 private 类似；对于继承来说，protected 和 public 类似。

当基类中存在动态内存分配时，即数据成员有用户自定义的构造函数、析构函数和复制构造函数的情况下，如果派生类没有新增动态内存的数据成员，那么继承过程中，派生类会自动调用基类的相关自定义方法对动态内存进行处理，派生类可以直接使用派生类的默认构造函数、析构函数和复制构造函数等。但是当派生类在基类的基础上增加了新的动态内存分配类型时，就必须为派生类显式地定义构造函数、析构函数、复制构造函数和赋值运算符。对于析构函数，只需要处理新增数据，构造函数需要通过初始化成员列表进行基类构造函数的初始化，赋值运算符则必须显式地调用基类的赋值运算符进行处理。

《C++ Primer Plus（第 6 版）中文版》在第 13 章总结了以公有方式派生的类的对象可以通过多种方式来使用基类的方法的列表，这些要点需要在应用中熟练掌握。

- 派生类对象自动使用继承而来的基类方法（如果派生类没有重新定义相关方法）。
- 派生类的构造函数自动调用基类的构造函数。
- 派生类的构造函数自动调用基类的默认构造函数（如果没有在成员初始化列表中指定其他构造函数）。
- 派生类构造函数显式地调用成员初始化列表中指定的基类构造函数。
- 派生类方法可以使用作用域解析运算符来调用公有的和受保护的基类方法。
- 派生类的友元函数可以通过强制类型转换，将派生类的引用或指针转换为基类的引用或指针，然后使用该引用或指针来调用基类的友元函数。

13.5　复习题

1．派生类从基类那里继承了什么？

习题解析：

在类的继承和派生中，C++中的派生类能够继承基类的所有数据成员和大部分成员函数。但是基类中不同访问控制权限的成员在派生类中的访问权限也不相同。公有成员直接成为派生类的公有成员，派生类的对象可以直接访问；基类的保护成员成为派生类的保护成员；基类的私有成员被派生类继承，但派生类不能直接访问，只能通过基类的公有方法间接访问。

2．派生类不能从基类那里继承什么？

习题解析：

C++的继承过程中，派生类能够从基类继承所有数据成员和大部分成员函数，但是基类的构造函数、析构函数、赋值运算符、友元函数和友元类不能继承。基类的构造函数能够在派生类的构造过程中默认调用，也可以通过派生类的初始化列表显式调用。在派生类对象的析构过程中，系统先调用派生类的析构函数，后调用基类的析构函数。

3．假设 baseDMA ::operator=()函数的返回类型为 void，而不是 baseDMA &，将会产生什么后果？如果返回类型为 baseDMA，而不是 baseDMA &，又将有什么后果?

习题解析：

如果赋值运算符在重载中返回值的类型为 void，仍可以使用单个赋值，但不能使用连锁赋值，即可以使用 `baseDMA a = b`，但不能使用 `baseDMA a = b = c`。

因为 `b = c` 的结果为 void。如果赋值运算符返回一个对象，而不是引用，则该方法在执行过程中会进行函数的返回值向被赋值对象的赋值过程，因此运行效率将会降低。

4．在创建和删除派生类对象时，构造函数和析构函数调用的顺序是怎样的?

习题解析：

类的继承过程中，派生类不会继承基类的构造函数和析构函数，但是派生类的对象在创建过程中会先调用基类的构造函数，创建基类对象，然后以此类推，最后调用派生类的构造函数。析构过程正好相反，先调用派生类的析构函数，再依次调用基类的析构函数，最后完成对象的销毁。

5．如果没有为派生类添加任何数据成员，它是否需要构造函数?

习题解析：

C++语言中每一个类都必须有自己的构造函数。即使没有数据成员和自定义的构造函数，系统也会自动生成一个默认构造函数，并在创建对象时调用该构造函数。

6．如果基类和派生类定义了同名的方法，当派生类对象调用该方法时，被调用的将是哪个方法?

习题解析：

如果派生类和基类都定义了同名的函数，将会导致派生类的同名方法隐藏基类的方法。通常情况下派生类的对象只会调用派生类中新定义的方法，不会调用基类中的同名方法。只有当派生类没有重新定义同名方法或使用基类的作用域运算符时，派生类才会调用基类方法。调用形式为 d.Base::f()。

7．在什么情况下，派生类应定义赋值运算符?

习题解析：

如果派生类中存在动态存储，即使用 new 或 new[]运算符来初始化类的某些指针类型的数据成员，那么该类必须定义一个赋值运算符，这样该类对象在赋值过程中才能针对动态存储分配的数据进行所有数据单元的内容复制（深拷贝），而不是仅仅进行指针变量所存储地址值的复制（浅拷贝）。简单来看就是如果默认赋值不能正确进行，就必须重新定义赋值运算符。

8．可以将派生类对象的地址赋给基类的指针吗？可以将基类对象的地址赋给派生类的指针吗?

习题解析：

类的继承过程中，可以将派生类对象的地址赋给基类的指针，这样在使用基类的指针操作派生类对象时，对于非虚函数的同名函数，将根据指针的类型调用基类的同名函数，而不是派生类的同名函数；如果使用虚函数，则会根据对象的类型调用派生类的同名方法，而不是简单根据指针类型来调用。基类对象的地址可以通过强制类型转换，赋值给派生类的指针（向下转换），但是这样使用指针很不安全，不推荐使用。

9．可以将派生类对象赋给基类对象吗？可以将基类对象赋给派生类对象吗?

习题解析：

C++中可以将派生类对象赋给基类对象，因为可以认为派生类对象中存在一个基类的对象，通常也说这是一种 IS-A 关系，即表示派生类对象也是一种基类对象的关系。因此被包含的基类对象可以通过基类的赋值运算符给基类对象赋值，且派生类中新增的数据成员都不会传递给基类对象。通常情况下，基类的对象是不能够给派生类对象赋值的，但是当派生类定义了转换运算符（即类型转换，将包含以基类引用作为唯一参数的构造函数）或使用基类作为参数的赋值运算符时（重新定义赋值运算符），这种赋值才是可行的。

10．假设定义了一个函数，它以基类对象的引用作为参数。为什么该函数也可以使用派生类对象作为参数?

习题解析：

C++语言中允许将派生类对象当作基类的对象使用，因为 C++语言中的继承关系可以看作 IS-A 关系，派生类对象是可以当作一个基类的对象使用的，所以 C++允许基类引用指向从该基类派生而来的任何类型。

11．假设定义了一个函数，它以基类对象作为参数（即函数按值传递基类对象）。为什么该函数也可以使用派生类对象作为参数?

习题解析：

C++的公有继承关系通常也是一种 IS-A 关系，即派生类对象是可以作为一种基类对象使用的。函数中作为参数对象是按值传递的，实参向形参的复制过程是调用复制构造函数的过程。因此派生类对象将调用基类的复制构造函数，最终将生成一个新的基类对象，作为函数的临时对象参与运算，其生命周期是局部生命周期。

12．为什么通常按引用传递对象比按值传递对象的效率更高?

习题解析：

C++中按值传递对象的过程是一个对象的复制过程，其优点是保护对象的源不被修改。若按引用传递对象，则传递的对象本身，并没有对象的复制过程，这样可以节省存储空间和节约时间，尤其对于大型对象能够大幅提高运行效率。此外，也可以通过将引用作为 const 类型传递，实现对于源对象的保护。

13．假设 Corporation 是基类，PublicCorporation 是派生类。再假设这两个类都定义了 head()函数，ph 是指向 Corporation 类型的指针，且指向一个 PublicCorporation 对象的地址。如果基类将 head()定义为以下两种方法，则 ph->head()将如何解释?

a．常规非虚方法。

b．虚方法。

习题解析：

与复习题 8 类似，在类的继承中，如果派生类和基类有同名的函数，通常基类的同名方法会被派生类隐藏，但是使用中基类对象和派生类对象分别调用该函数，不会发生混乱，派生类可以使用作用域解析来调用基类的隐藏函数。如果 head()是一个常规非虚方法，则 ph->head()将会根据指针的类型判断，并调用基类的方法，即 Corporation::head()；如果 head()是一个虚方法，则 ph->head()将会根据实际对象的类型调用派生类的方法，即 PublicCorporation::head()。

14．下述代码有什么问题?

```
class Kitchen
{
private:
    double kit_sq_ft;
public:
    Kitchen() {kit_sq_ft = 0.0; }
    virtual double area() const { return kit_sq_ft *kit_sq_ft; }
};
class House :public Kitchen
{
```

```
private:
    double all_sq_ft;
public:
    House() {all_sq_ft += kit_sq_ft;}
    double area(const char *s) const { cout << s; return all_sq_ft; }
};
```

习题解析：

在类的继承关系中，作为派生类，House 是没有办法直接访问 Kitchen 基类的私有数据成员的，即其 House 构造函数无法直接使用 kit_sq_ft。对于 House 中的 area()定义和 area() 的 Kitchen 版本，因为这两个方法的特征标不同，所以虚函数的功能并没有发挥，仅派生类的函数隐藏了基类的函数。

13.6 编程练习

1．以下面的类声明为基础，派生出一个 Classic 类，并添加一组 char 成员，用于存储指出 Cd 中主要作品的字符串。修改上述声明，使基类的所有函数都是虚的。如果上述定义声明的某个方法并不需要，则请删除它。使用下面的程序测试你的代码。

```
class Cd{
private:
    char performers [50];
    char label [20];
    int selections;
    double playtime; /
public:
    Cd(char *s1, char *s2, int n, double x);
    Cd(const Cd & d);
    Cd();
    ~Cd();
    void Report() const;
    Cd & operator=(const Cd & d);
};
#include <iostream>
using namespace std;
#include "classic.h"
void Bravo(const Cd & disk);
int main()
{
    Cd c1("Beatles", "Capitol", 14, 35.5);
    Classic c2 = Classic("Piano Sonata in B flat, Fantasia in C",
                         "Alfred Brendel", "Philips", 2, 57.17);
    Cd *pcd = &c1;
    cout << "Using object directly:\n";
    c1.Report();
```

```
    c2.Report();
    cout << "Using type cd *pointer to objects:\n";
    pcd->Report();
    pcd = &c2;
    pcd->Report();
    cout << "Calling a function with a Cd reference argument:\n";
    Bravo(c1);
    Bravo(c2);
    cout << "Testing assignment: ";
    Classic copy;
    copy = c2;
    copy.Report();
    return 0;
}
void Bravo(const Cd & disk)
{
    disk.Report();
}
```

```
Using object directly:
Performers: Beatles
Label: Capitol
Selections: 14
Playtime: 35.5
Performers: Piano Sonata in B flat, Fantasia in C
Label: Alfred Brendel
Selections: 2
Playtime: 57.17
Works: Philips
Using type cd *pointer to objects:
Performers: Beatles
Label: Capitol
Selections: 14
Playtime: 35.5
Performers: Piano Sonata in B flat, Fantasia in C
Label: Alfred Brendel
Selections: 2
Playtime: 57.17
Works: Philips
Calling a function with a Cd reference argument:
Performers: Beatles
Label: Capitol
Selections: 14
Playtime: 35.5
Performers: Piano Sonata in B flat, Fantasia in C
Label: Alfred Brendel
Selections: 2
Playtime: 57.17
Works: Philips
Testing assignment: Performers: Piano Sonata in B flat, Fantasia in C
```

```
Label: Alfred Brendel
Selections: 2
Playtime: 57.17
Works: Philips
```

编程分析：

题目要求在 Cd 类的基础上派生 Classic 类。原 Cd 类包含 4 个数据成员，在 Classic 类中需要添加新数据成员以描述作品属性。类的继承与派生的关键现在于构造函数和其他公用接口的重新设计。构造函数需要通过初始化列表调用基类的构造函数，其他公用接口在某些情况下也需要通过作用域运算符来调用基类的构造函数。完整代码如下。

```
/*第 13 章的编程练习 1*/
#ifndef CLASSIC_H
#define CLASSIC_H

class Cd{
private:
    char performers [50];
    char label [20];
    int selections;
    double playtime;
public:
    Cd(const char *s1, const char *s2, int n, double x);
    Cd(const Cd & d);
    Cd();
    virtual ~Cd(){};
    virtual void Report() const;
    Cd & operator=(const Cd & d);
};
class Classic : public Cd{
private:
    char works[50];
public:
    Classic();
    Classic(const Classic& c);
    Classic(const char* s1,const char* s2, const char* s3,int n,double x);
    ~Classic(){};
    virtual void Report()const ;
    Classic& operator=(const Classic& c);
};

#endif //classic.h


#include <iostream>
#include <cstring>
#include "classic.h"

using namespace std;
Cd::Cd(const char *s1, const char *s2, int n, double x)
{
```

```
        strncpy(performers,s1,50);
        if(strlen(s1)>=50) performers[40] = '\0';
        else performers[strlen(s1)] = '\0';
        strncpy(label,s2,20);
        if(strlen(s1)>=50) label[19] = '\0';
        else label[strlen(s2)] = '\0';
        selections = n;
        playtime = x;
    }
    Cd::Cd(const Cd & d)
    {
        strcpy(performers,d.performers);
        strcpy(label,d.label);
        selections = d.selections;
        playtime = d.playtime;
    }
    Cd::Cd(){
        performers[0] = '\0';
        label[0] = '\0';
        selections = 0;
        playtime = 0.0;
    }
    void Cd::Report()const
    {
        cout<<"Performers: "<<performers<<endl;
        cout<<"Label: "<<label<<endl;
        cout<<"Selecttions: "<<selections<<endl;
        cout<<"Platime: "<<playtime<<endl;
    }
    Cd & Cd::operator=(const Cd & d){
        if(this == &d)
            return *this;
        strcpy(performers,d.performers);
        strcpy(label,d.label);
        selections = d.selections;
        playtime = d.playtime;
        return *this;
    }


    Classic::Classic():Cd()
    {
        works[0] = '\n';
    }
    Classic::Classic(const Classic& c) :Cd(c)
    {
        strcpy(works,c.works);
    }
    Classic::Classic(const  char*s1,const  char*s2,const  char*s3,int  n,double  x)  :
Cd(s1,s2,n,x)
```

```
{
    strncpy(works,s3,50);
    if(strlen(s3)>=50) works[49] = '\0';
    else works[strlen(s3)] = '\0';
}

void Classic::Report()const
{
    Cd::Report();
    cout<<"Works: "<<works<<endl;
}
Classic& Classic::operator=(const Classic& c){
    if(this == &c)
        return *this;
    Cd::operator=(c);
    strcpy(works,c.works);
    return *this;
}
```

2．重做编程练习 1，使两个类使用动态内存分配而不是长度固定的数组来记录字符串。

编程分析：

原有的代码在 Cd 类中使用字符数组存储演奏者、标签等，派生类 Calssic 中新增的数据成员也使用字符数组存储。题目要求在基类和派生类中都使用动态内存分配，因此需要修改相关的成员函数。完整代码如下。

```
/*第 13 章的编程练习 2*/
#ifndef CLASSIC_H
#define CLASSIC_H

class Cd{
private:
    char* performers;
    char* label;
    int selections;
double playtime;
/*修改数据成员为指针，实现动态存储*/
public:
    Cd(const char *s1, const char *s2, int n, double x);
    Cd(const Cd & d);
    Cd();
    virtual ~Cd();
    virtual void Report() const;
    virtual Cd & operator=(const Cd & d);
};

class Classic : public Cd{
private:
    char*works;
```

```
/*修改数据成员为指针，实现动态存储*/
public:
    Classic();
    Classic(const Classic& c);
    Classic(const char* s1,const char* s2, const char* s3,int n,double x);
    ~Classic();
    virtual void Report()const ;
    Classic& operator=(const Classic& c);
};

#endif //classic.h

/*classic.cpp 文件*/
#include <iostream>
#include <cstring>
#include "classic.h"

using namespace std;
Cd::Cd(const char *s1, const char *s2, int n, double x)
{
    performers = new char[strlen(s1) + 1];
    strcpy(performers,s1);
    label = new char[strlen(s2) + 1];
    strcpy(label,s2);
    selections = n;
    playtime = x;
}
/*修改构造函数，实现动态存储*/
Cd::Cd(const Cd & d)
{
    performers = new char[strlen(d.performers) + 1];
    strcpy(performers,d.performers);
    label = new char[strlen(d.label) + 1];
    strcpy(label,d.label);
    selections = d.selections;
    playtime = d.playtime;
}
/*修改复制构造函数，实现动态存储*/
Cd::Cd(){
    performers = nullptr;
    label = nullptr;
    selections = 0;
    playtime = 0.0;
}
/*修改默认构造函数，设置空指针并初始化数据 */
Cd::~Cd(){
    if(performers != nullptr ) delete[] performers;
    if(label != nullptr ) delete[] label;
//   cout<<"Clear Cd's object."<<endl;
}
```

```
/*修改析构函数，回收存储单元*/
void Cd::Report()const
{
    if(performers == nullptr || label == nullptr){
        cout<<"Error, empty Object."<<endl;
    }else{
        cout<<"Performers: "<<performers<<endl;
        cout<<"Label: "<<label<<endl;
        cout<<"Selections: "<<selections<<endl;
        cout<<"Playtime: "<<playtime<<endl;
    }
}
Cd & Cd::operator=(const Cd & d){
    if(this == &d)
        return *this;
    performers = new char[strlen(d.performers) + 1];
    strcpy(performers,d.performers);
    label = new char[strlen(d.label) + 1];
    strcpy(label,d.label);
    selections = d.selections;
    playtime = d.playtime;
    return *this;
}

Classic::Classic():Cd()
{
    works = nullptr;
}
Classic::Classic(const Classic& c) :Cd(c)
{
    works = new char[strlen(c.works) + 1];
    strcpy(works,c.works);
}
Classic::Classic(const char* s1,const char* s2,const char* s3,int n,double x)  :
Cd(s1,s2,n,x)
{
    works = new char[strlen(s3) + 1];
    strcpy(works,s3);
}
/*修改 3 个构造函数，实现动态存储*/
Classic::~Classic(){
    delete[] works;
//   cout<<"Clear Classic's object."<<endl;
}
/*修改析构函数，回收存储单元*/

void Classic::Report()const
{
    Cd::Report();
    if(works != nullptr )
        cout<<"Works: "<<works<<endl;
}
Classic& Classic::operator=(const Classic& c){
```

```
    if(this == &c)
        return *this;
    Cd::operator=(c);
    works = new char[strlen(c.works) + 1];
    strcpy(works,c.works);
    return *this;
}
```

3．修改 baseDMA-lacksDMA-hasDMA 类的层次，使 3 个类都从一个 ABC 派生而来，然后使用与程序清单 13.10 相似的程序对结果进行测试。也就是说，它应使用 ABC 指针数组，并由用户决定要创建的对象类型。在类定义中添加 virtual View( )方法以显示数据。

编程分析：

题目需要参考程序清单 13.14 和程序清单 13.15 中的 DMA 继承关系，添加基类 ABC，并在此基础上派生其他类。

```
/*第 13 章的编程练习 3*/
//dma.h
#ifndef DMA_H_
#define DMA_H_
#include <iostream>

class ABC{
public:
    virtual ~ABC(){};
    virtual void View(){std::cout<<"This is ABC View(), it is empty.\n";};
};
/*添加基类 ABC，定义虚析构函数和虚 View()方法，表示方法的调用对象*/
// Base Class Using DMA
class baseDMA : public ABC
{
private:
    char *label;
    int rating;
public:
    baseDMA(const char *l = "null", int r = 0);
    baseDMA(const baseDMA & rs);
    virtual ~baseDMA();
    virtual void View();
    /*添加 baseDMA 类的 View() 类方法，表示方法的调用对象*/
    baseDMA & operator=(const baseDMA & rs);
    friend std::ostream & operator<<(std::ostream & os,
                                     const baseDMA & rs);
};


class lacksDMA :public baseDMA
{
private:
    enum { COL_LEN = 40};
```

```
    char color[COL_LEN];
public:
    lacksDMA(const char *c = "blank", const char *l = "null",
             int r = 0);
    lacksDMA(const char *c, const baseDMA & rs);
    virtual void View();
    /*添加 lacksDMA 类的 View() 类方法，表示方法的调用对象*/
    friend std::ostream & operator<<(std::ostream & os,
                                     const lacksDMA & rs);
};

class hasDMA :public baseDMA
{
private:
    char *style;
public:
    hasDMA(const char *s = "none", const char *l = "null",
           int r = 0);
    hasDMA(const char *s, const baseDMA & rs);
    hasDMA(const hasDMA & hs);
    ~hasDMA();
    virtual void View();
    /*添加 hasDMA 类的 View() 类方法，表示方法的调用对象*/
    hasDMA & operator=(const hasDMA & rs);
    friend std::ostream & operator<<(std::ostream & os,
                                     const hasDMA & rs);
};

#endif


//dma.cpp

#include "dma.h"
#include <cstring>

//baseDMA 的方法
baseDMA::baseDMA(const char *l, int r)
{
    label = new char[std::strlen(l) + 1];
    std::strcpy(label, l);
    rating = r;
}

baseDMA::baseDMA(const baseDMA & rs)
{
    label = new char[std::strlen(rs.label) + 1];
    std::strcpy(label, rs.label);
    rating = rs.rating;
```

```
}

baseDMA::~baseDMA()
{
    delete [] label;
}
void baseDMA::View(){
    std::cout<<"Now in baseDMA."<<std::endl;
    std::cout<<*this;
    /*通过重载运算符显示对象内容*/
}
baseDMA & baseDMA::operator=(const baseDMA & rs)
{
    if (this == &rs)
        return *this;
    delete [] label;
    label = new char[std::strlen(rs.label) + 1];
    std::strcpy(label, rs.label);
    rating = rs.rating;
    return *this;
}

std::ostream & operator<<(std::ostream & os, const baseDMA & rs)
{
    os << "Label: " << rs.label << std::endl;
    os << "Rating: " << rs.rating << std::endl;
    return os;
}


lacksDMA::lacksDMA(const char *c, const char *l, int r)
        : baseDMA(l, r)
{
    std::strncpy(color, c, 39);
    color[39] = '\0';
}

lacksDMA::lacksDMA(const char *c, const baseDMA & rs)
        : baseDMA(rs)
{
    std::strncpy(color, c, COL_LEN - 1);
    color[COL_LEN - 1] = '\0';
}
void lacksDMA::View(){
    std::cout<<"Now in lacksDMA."<<std::endl;
    std::cout<<*this;
    /*通过重载运算符显示对象内容*/

}
```

```
std::ostream & operator<<(std::ostream & os, const lacksDMA & ls)
{
    os << (const baseDMA &) ls;
    os << "Color: " << ls.color << std::endl;
    return os;
}


hasDMA::hasDMA(const char *s, const char *l, int r)
        : baseDMA(l, r)
{
    style = new char[std::strlen(s) + 1];
    std::strcpy(style, s);
}

hasDMA::hasDMA(const char *s, const baseDMA & rs)
        : baseDMA(rs)
{
    style = new char[std::strlen(s) + 1];
    std::strcpy(style, s);
}

hasDMA::hasDMA(const hasDMA & hs)
        : baseDMA(hs)  //调用基类的复制构造函数
{
    style = new char[std::strlen(hs.style) + 1];
    std::strcpy(style, hs.style);
}

hasDMA::~hasDMA()
{
    delete [] style;
}
void hasDMA::View(){
    std::cout<<"Now in hasDMA."<<std::endl;
    std::cout<<*this;
    /*通过重载运算符显示对象内容*/
}

hasDMA & hasDMA::operator=(const hasDMA & hs)
{
    if (this == &hs)
       return *this;
    baseDMA::operator=(hs);  /复制基类部分
    delete [] style;         //准备新类型
    style = new char[std::strlen(hs.style) + 1];
    std::strcpy(style, hs.style);
    return *this;
}
```

```
std::ostream & operator<<(std::ostream & os, const hasDMA & hs)
{
    os << (const baseDMA &) hs;
    os << "Style: " << hs.style << std::endl;
    return os;
}

//usebrass2.cpp
//通过 brass.cpp 进行编译
#include <iostream>
#include <string>
#include "dma.h"
const int CLIENTS = 4;

int main()
{
    using std::cin;
    using std::cout;
    using std::endl;

    ABC *p_clients[CLIENTS];
    /*创建基类 ABC 的指针数组*/
    char kind;
    for (int i = 0; i < CLIENTS; i++)
    {
        cout << "Select  1) ABC, 2) baseDMA, 3) lacksDMA, 4) hasDMA : ";

        while (cin >> kind && (kind != '1' && kind != '2' && kind != '3' && kind != '4'))
            cout <<"Enter either 1  2  3 or 4 : ";
        if (kind == '1')
             p_clients[i] = new ABC();
        else if(kind == '2'){
            while (cin.get() != '\n')
                continue;
            char l[40];
            int r;
            cout << "Enter baseDMA's label: ";
            cin.getline(l,40);
            cout << "Enter baseDMA's rating: ";
            cin >> r;
            p_clients[i] = new baseDMA(l,r);
        }
        else if(kind == '3'){
            while (cin.get() != '\n')
                continue;
            char l[40],c[40];
            int r;
            cout << "Enter lacksDMA's label: ";
            cin.getline(l,40);
            cout << "Enter lacksDMA's color: ";
```

```
            cin.getline(c,40);
            cout << "Enter lacksDMA's rating: ";
            cin >> r;
            p_clients[i] = new lacksDMA(c,l,r);
        }
        else if(kind == '4'){
            while (cin.get() != '\n')
                continue;
            char l[40],s[40];
            int r;
            cout << "Enter hasDMA's label: ";
            cin.getline(l,40);
            cout << "Enter hasDMA's style: ";
            cin.getline(s,40);
            cout << "Enter hasDMA's rating: ";
            cin >> r;
            p_clients[i] = new hasDMA(s,l,r);
        }

        while (cin.get() != '\n')
            continue;
    }
    cout << endl;
    for (int i = 0; i < CLIENTS; i++)
    {
        p_clients[i]->View();
        cout << endl;
        /*作为虚函数，View()将会依据指针指向的对象类型，匹配相应对象的View()方法*/
    }

    for (int i = 0; i < CLIENTS; i++)
    {
        delete p_clients[i];  //释放内存
    }
    cout << "Done.\n";

    return 0;
}
```

4．Benevolent Order of Programmers（BOP）用来维护瓶装葡萄酒箱。为描述它，BOP 的 PortMaster 设置了一个 Port 类，其声明如下。

```
#include <iostream>
using namespace std;
class Port {
private:
    char *brand;
    char style[20];
    int bottles;
```

```
public:
    Port(const char *br = "none", const char *st = "none", int b = 0);
    Port(const Port &p); //复制构造函数
    virtual ~Port() { delete[] brand; }
    Port &operator=(const Port &p);
    Port &operator+=(int b);
    Port &operator-=(int b);
    int BottleCount() const { return bottles; }
    virtual void Show() const;
    friend ostream &operator<<(ostream &os, const Port &p);
};
```

show()方法按下面的格式显示信息。

```
Brand: Gallo
Kind: tawny
Bottles: 20
```

operator<<()函数按下面的格式显示信息（末尾没有换行符）。

```
Gallo, tawny, 20
```

PortMaster 完成了 Port 类的方法定义后，派生了 VintagePort 类，然后被解雇——因为不小心将一瓶 45°的科佰恩酒泼到了正在准备烤肉调料的人身上。VintagePort 类如下所示。

```
class VintagePort : public Port
{
private:
    char *nickname;
    int year;
public:
    VintagePort();
    VintagePort(const char *br, int b, const char *nn, int y);
    VintagePort(const VintagePort &vp);
    ~VintagePort() { delete[] nickname; }
    VintagePort &operator=(const VintagePort &vp);
    void Show() const;
    friend ostream &operator<<(ostream &os, const VintagePort &vp);
};
```

你负责完成 VintagePort。

a．第 1 个任务是重新创建 Port 方法的定义，因为 PortMaster 在离职时销毁了方法的定义。

b．第 2 个任务是解释为什么有的方法重新定义了，而有些没有重新定义。

c．第 3 个任务是解释为何没有将 operator=()和 operator<<()声明为虚方法。

d．第 4 个任务是提供 VintagePort 中各个方法的定义。

编程分析：

题目要求在给定类声明的基础上，完成整个类的定义，完整代码如下。

```
/*第 13 章的编程练习 4*/
#ifndef VINTAGEPORT_H
#define VINTAGEPORT_H
```

```
#include <iostream>
using namespace std;
class Port {
private:
    char *brand;
    char style[20]; //即 tawny、 ruby、 vintage
    int bottles;
public:
    Port(const char *br = "none", const char *st = "none", int b = 0);
    Port(const Port &p); //复制构造函数
    virtual ~Port() { delete[] brand; }
    Port &operator=(const Port &p);
    Port &operator+=(int b);
    Port &operator-=(int b);
    int BottleCount() const { return bottles; }
    virtual void Show() const;
    friend ostream &operator<<(ostream &os, const Port &p);
};
class VintagePort : public Port
{
private:
    char *nickname;
    int year;
public:
    VintagePort();
    VintagePort(const char *br, int b, const char *nn, int y);
    VintagePort(const VintagePort &vp);
    ~VintagePort() { delete[] nickname; }
    VintagePort &operator=(const VintagePort &vp);
    void Show() const;
    friend ostream &operator<<(ostream &os, const VintagePort &vp);
};
#endif

#include <iostream>
#include "vintageport.h"

Port::Port(const char *br, const char *st, int b)
{
    brand = new char[strlen(br) + 1];
    strcpy(brand, br);
    strcpy(style, st);
    bottles = b;
}
Port::Port(const Port &p)
{
    brand = new char[strlen(p.brand) + 1];
    strcpy(brand, p.brand);
    strcpy(style, p.style);
    bottles = p.bottles;
```

```
}
Port & Port::operator=(const Port &p)
{
    if(this == &p)
       return *this;
    delete[] brand;
    brand = new char[strlen(p.brand) + 1];
    strcpy(brand, p.brand);
    strcpy(style, p.style);
    bottles = p.bottles;
    return *this;
}
Port & Port::operator+=(int b)
{
    bottles += b;
    return *this;
}
Port & Port::operator-=(int b)
{
    bottles -= b;
    return *this;
}
void Port::Show() const
{
    cout<<"Brand: "<<brand<<endl;
    cout<<"Kind: "<<style<<endl;
    cout<<"Bootles: "<<bottles<<endl;
}
ostream &operator<<(ostream &os, const Port &p)
{
    os<<p.brand<<", "<<p.style<<", "<<p.bottles;
    return os;
}

VintagePort::VintagePort(): Port()
{
    nickname = new char[5];
    strcpy(nickname,"none");
    year = 0;
}
VintagePort::VintagePort(const char *br, int b, const char *nn, int y)
       :Port(br,"none",b)
{
    nickname = new char[strlen(nn) + 1];
    strcpy(nickname, nn);
    year = y;

}
VintagePort::VintagePort(const VintagePort &vp) : Port(vp)
{
```

```
    nickname = new char[strlen(vp.nickname) + 1];
    strcpy(nickname, vp.nickname);
    year = vp.year;
}
VintagePort & VintagePort::operator=(const VintagePort &vp)
{
    if(this == &vp)
       return *this;
    Port::operator=(vp);
    delete[] nickname;
    nickname = new char[strlen(vp.nickname) + 1];
    strcpy(nickname, vp.nickname);
    year = vp.year;
    return *this;

}
void VintagePort::Show() const
{
    Port::Show();
    cout<<"NickName: "<<nickname<<endl;
    cout<<"Year: "<<year<<endl;
}
ostream & operator<<(ostream &os, const VintagePort &vp)
{
    os << (const Port &) vp;
    os<<", "<<vp.nickname<<", "<<vp.year<<endl;
    return os;
}
```

第 14 章　C++中的代码重用

本章知识点总结

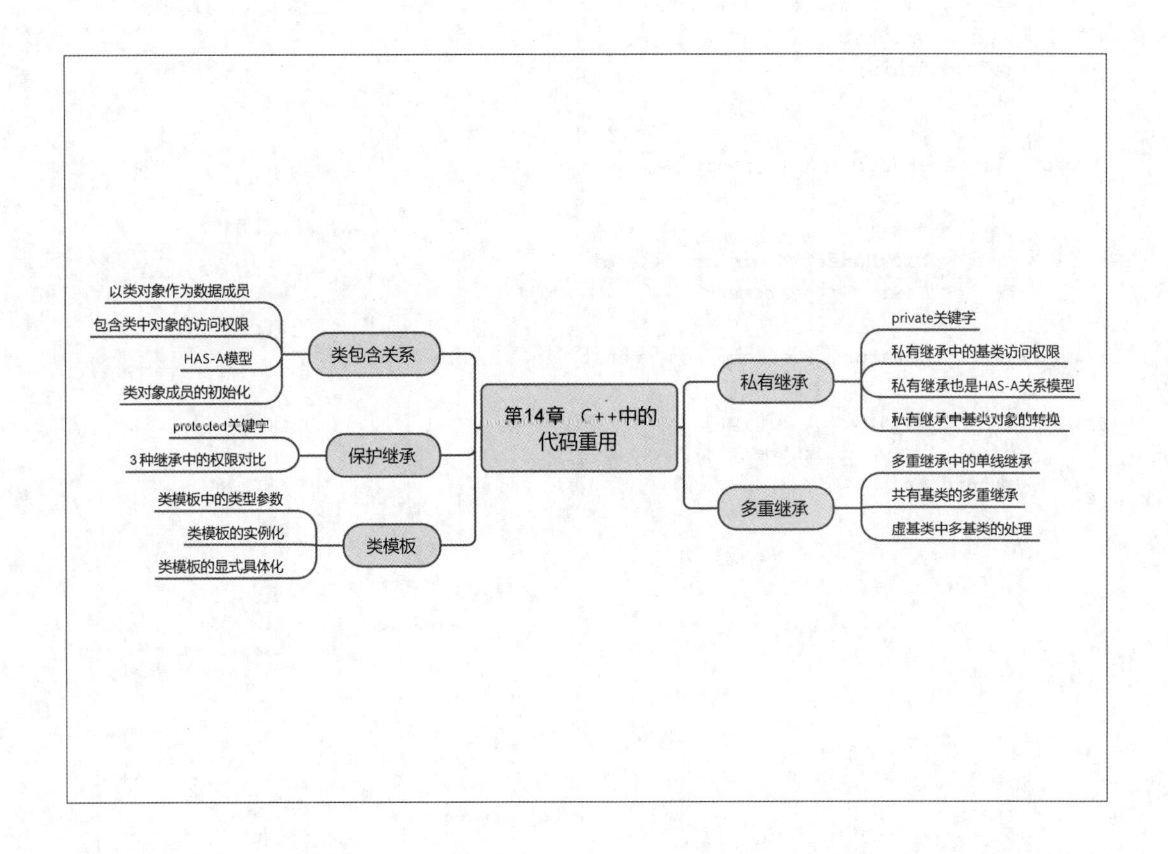

14.1　类的继承和包含关系

通过公有继承的 IS-A 模型，派生类可以继承基类的各种接口实现。除此之外，C++语言还可以通过类的包含关系和私有继承关系实现 HAS-A 模型。包含关系就是一个类的对象作为另外一个类的数据成员的形式，实现的一种两个类型的组合关系，也称为组合或者层次化。包含关系中被包含的类对象的接口不是公开的，但是可以在包含类的方法中使用被包含的类对象的方法。在使用私有继承时，只能在派生类的方法中使用基类的方法。如果需要使用基类对象，在包含关系中可以使用对象名来调用方法，私有继承则可以使用类名和作用域解析运算符来调用方法。对于公有继承，基类的公有方法将成为派生类的公有方法，这属于 IS-A 关系模型。对于私有继承，派生类不能继承基类的所有公有接口，这种继承关系和包含关系类似，属于 HAS-A 关系模型。

私有继承和包含关系都可以表示两个类的 HAS-A 关系。比较而言，包含关系通过一个对象来表述这种关系，更加直观和易于理解。私有继承关系会更加抽象，而且必须要处理同名方法。虽然私有继承关系在程序设计中较复杂，但是私有继承能够提供一些特殊的功能，主要是受保护的成员在其派生类中也可以使用。对于包含关系模型，受保护的成员则不可用。此外，在定义虚函数的情况下可以使用私有继承的形式来重新定义虚函数，但是包含关系也无法实现。简单来说，如果新类需要访问原有类中受保护的成员，或者需要重新定义虚函数，那么选择私有继承；否则，使用包含关系。

14.2　私有继承和受保护的继承

私有继承在列出基类时使用 private 关键字，基类的公有成员和受保护的成员都将成为派生类的私有成员。私有继承中基类的公有成员将成为派生类的私有成员，即不会成为派生类对象公有接口的一部分。然而，在派生类的成员函数中可以调用基类的接口方法。受保护的继承在列出基类时使用关键字 protected，在受保护的继承中基类的公有成员和保护成员都将会成为派生类受保护的成员，基类的接口在派生类中都可以使用，但是在继承结构之外均不可用。简单来说，在公有继承、私有继承和受保护的继承中，3 种数据成员的访问权限如表 14.1 所示。

表 14.1　继承中的访问权限

特征	公有继承	受保护的继承	私有继承
公有成员	变成派生类的公有成员	变成派生类受保护的成员	变成派生类的私有成员
保护成员	变成派生类受保护的成员	变成派生类受保护的成员	变成派生类的私有成员
私有成员	只能通过基类接口访问	只能通过基类接口访问	只能通过基类接口访问
能否隐式向上转换	能	能（但只能在派生类中）	不能

当以派生类作为基类再次派生时，使用私有继承的第 2 代类，在第 3 代类中所有方法均变成了私有方法；受保护的继承中第 2 代类中的公有方法成为受保护的类型，所以第 3 代类可以使用。此外，私有继承中如果需要访问基类的对象，那么需要使用强制类型转换和 this 指针，将派生类对象强制转换成基类对象。

14.3　多重继承

多重继承是指一个派生类具有多个基类的形式，声明方式如下。

```
class Derived: public Base_1,public Base_2{……};
```

如果两个基类相互独立，那么通过多重继承派生类将具备两个基类的所有数据成员和成员函数，这样可以在派生类中像单继承那样独立使用每一个基类的继承方法。派生类的多个基类之间不存在继承关系的多重继承是最简单的一种多重形式，可以简单理解成两个单重继承的叠加。如果两个基类中具有同名函数，那么在派生类中可以使用作用域运算符的形式来显式表示使用哪一个基类的函数。但是如果两个基类并非相互独立而是具有一定的派生关系，如两个基类又是从同一个基类派生的，那么第 3 代多重继承的派生类将会具备多个 2 代派生类共同的第 1 代基类。为了解决这个问题，C++语言使用虚基类技术来解决多重继承中的重复基类问题。首先虚基类使得多重继承中多个相同的基类只保留一个副本。但是当派生类混合使用虚基类和非虚基类时，派生类将包含一个表示所有虚基类的基类对象和非虚基类的多个对象。

此外，当派生类通过多种虚继承方式和非虚继承方式继承某个特定的原始基类时，该类将包含一个表示所有虚继承的基类子对象和多个分别表示各种非虚继承的基类子对象。使用虚基类将改变 C++解析二义性的方式。当使用非虚基类时，如果派生类从不同的基类继承了多个同名的数据成员，并且没有用类名进行限定，将导致二义性。但如果使用的是虚基类，派生类中的名称优先于直接或间接祖先类中的相同名称，这样即使不使用限定符，也不会导致二义性。

14.4 对象的初始化问题

包含关系是以一个类的对象作为另外一个类的数据成员的形式，因此包含关系中的对象初始化主要是数据成员的初始化，包含对象的初始化可以和基类的初始化形式一样，使用构造函数中的初始化列表来实现，但是和基类使用的类名不同，包含对象使用成员变量名进行初始化。私有继承中基类的初始化和公有继承类似，也通过派生类构造函数的初始化列表实现。在使用中需要牢记一点，初始化列表中的每一项调用都要匹配一个构造函数。

普通多重继承中基类的初始化也通过派生类的构造函数中的初始化列表实现。但是针对多重继承中的虚基类问题需要特别指出，如果不直接在初始化列表中调用第 1 代虚基类，那么第 3 代派生类中通过初始化列表调用的第 2 代构造函数将不会按照指定信息创建第 1 代虚基类的构造函数，而使用默认构造函数。因此如果对第 1 代虚基类的初始化有指定需求，则必须在第 3 代构造函数的初始化列表中显式地调用第 1 代虚基类的构造函数，相比较，普通继承则不能隔代调用基类的构造函数。

14.5 类模板（模板类）

使用模板的形式定义类，可以在类定义中实现泛型编程。类模板本质上就是类内数据成员泛型化的设计，通过对象类型的实例化可以创建对应的对象。C++语言中类模板的应用非常广泛，例如，前面章节使用的模板 array 和 vector 等都是模板类。模板类的定义方式如下。

```
template <typename T> class Class_Name{ …… };
```

其中，关键字 typename 也可以使用 class 来表示[①]。这里 typename 表明 T 是一个通用的类型说明符，在使用模板时，将使用实际的类型替换它。当调用模板时，T 将被具体的类型值取代。例如，以下代码声明了一个用 int 类型实例化的对象 TTest。

```
Class_Name< int > TTest;
```

模板的定义中也可以使用非类型（表达式）参数，例如：

```
template <typename T, int n> class Class_Name{ …… };
```

这样在模板类的实例化中就可以在 int 参数中使用非类型参数进行实例化。当然，模板类的定义中也可以包含多个类型参数，例如，多类型模板类的定义：

```
template <typename T1, typename T2 > class Class_Name{ …… };
```

在模板类的定义中也可以像函数参数一样设置类型参数的默认值。默认类型的模板参数如下。

```
template <typename T1, typename T2 = int> class Class_Name{ …… };
```

模板的显式具体化和实例化不同，显式具体化是模板类针对特定类型的一种全新的定义。即需要为特殊类型实例化模板之前，针对该特殊类型对模型进行的一种有针对性的修改。模板的显式具体化语句如下。

```
template<> class Class_Name<specialize-type>{……};
```

C++语言还允许模板类的部分具体化。

```
template <typename T1 > class Class_Name<T1, specialize-type >{ …… };
```

模板类的定义和使用是一个非常复杂的项目，需要在实践中反复地练习才能熟练掌握。

14.6　复习题

1．当以 A 栏的类为基类时，B 栏的类更合适采用公有派生还是私有派生？

习题解析：

A	B	继承类型
class Bear	class PolarBear	公有继承。北极熊是一种熊，具备熊的所有性质，因此选择公有继承，即 IS-A 模型
class Kitchen	class Home	私有继承。家和厨房的关系是包含关系，因此可以选择私有继承或者包含关系，即 HAS-A 模型
class Person	class Programmer	公有继承。程序员首先是人，因此选择公有继承
class Person	class HorseAndJockey	私有继承。马和驯马师的类可以认为是包含马和驯马师两个类型的一个组合体，因此选择私有继承或者包含关系
class Person class Automobile	class Driver	司机首先是一个人，因此 Driver 和 Person 的关系是公有继承，其次，司机驾驶机车，因此和汽车属于 HAS-A 模型，应当选择私有继承

2．假设有下面的定义。

```
class Frabjous {
```

① 在《C++ Primer Plus（第 6 版）中文版》中，模拟定义中使用 class 关键字多于 typename，因此本书中也较多使用 class 关键字来定义模板。——编者注

```
private:
    char fab[20];
public:
    Frabjous(const char *s = "C++") : fab(s) { }
    virtual void tell() { cout << fab; }
};
class Gloam {
private:
    int glip;
    Frabjous fb;
public:
    Gloam(int g = 0, const char *s = "C++");
    Gloam(int g, const Frabjous & f);
    void tell();
};
```

同时假设 Gloam 版本的 tell()应显示 glip 和 fb 的值，请为这 3 个 Gloam 方法提供定义。

习题解析：

类 Frabjous 和类 Gloam 是包含关系，其中类 Frabjous 的一个对象 fb 被定义为 Gloam 类的私有数据成员。题目内需要定义的 3 个函数包括 Gloma 的两个构造函数和 tell()函数。在构造函数内需要初始化 Frabjous 类的对象 fb。此外，tell()函数需要访问 fb 成员的数据。包含关系中，Gloam 内的方法如果需要访问 Frabjous 类，应当通过 Frabjous 类对象（即 fb 的类方法）进行访问。完整代码如下。

```
class Frabjous {
private:
    char fab[20];
public:
    Frabjous(const char *s = "C++") : fab(s) { }
    virtual void tell() { cout << fab; }
};
class Gloam {
private:
    int glip;
    Frabjous fb;
public:
    Gloam(int g = 0, const char *s = "C++");
    Gloam(int g, const Frabjous & f);
    void tell();
};

Gloam::Gloam(int g, const char *s) :glip(g),fb(s)
{
    /*在类的构造函数中，通常使用初始化列表对基类或者其他数据成员进行初始化。
     *在类内的数据成员是对象且不能使用默认构造函数时，必须使用初始化列表*/
}
Gloam::Gloam(int g, const Frabjous & f) : glip(g),fb(f)
{
}
void Gloam::tell(){
```

```
    fb.tell();
    /*类的包含关系中，Frabjous 类内的数据访问需要通过其对象和方法实现，
     *当前在 Gloam 类内无法直接访问 Frabjous 类的私有数据 fab。只能通过
     *fb.tell()函数间接访问*/
    cout<<glip<<endl;
}
```

3．假设有下面的定义。

```
class Frabjous {
private:
    char fab[20];
public:
    Frabjous(const char *s = "C++") : fab(s) {}

    virtual void tell() { cout << fab; }
};
class Gloam : private Frabjous {
private:
    int glip;
public:
    Gloam(int g = 0, const char *s = "C++");
    Gloam(int g, const Frabjous &f);
    void tell();
};
```

同时假设 Gloam 版本的 tell()应显示 glip 和 fab 的值，请为这 3 个 Gloam 方法提供定义。

习题解析：

类 Frabjous 和类 Gloam 是私有继承关系，题目要求定义的 3 个函数分别是两个构造函数和 tell()函数。同上一题的包含关系类似，构造函数内对基类的初始化只能通过初始化列表实现，但是私有继承只能通过类名 Frabjous 调用构造函数，且 tell()方法内也无法通过对象实现基类函数的调用，只能通过作用域限定符实现。完整代码如下所示。

```
class Frabjous {
private:
    char fab[20];
public:
    Frabjous(const char *s = "C++") : fab(s) {}

    virtual void tell() { cout << fab; }
};

class Gloam : private Frabjous {
private:
    int glip;
public:
    Gloam(int g = 0, const char *s = "C++");
    Gloam(int g, const Frabjous &f);
    void tell();
};
```

```
Gloam::Gloam(int g, const char *s) :glip(g),Frabjous(s)
{
     /*类的继承关系中，为了调用基类的非默认构造函数，需要通过初始化列表的形式实现。
      *区别于包含关系的是，继承关系中没有基类对象，因此必须使用类名
      *调用构造函数
      **/
}
Gloam::Gloam(int g, const Frabjous &f) : glip(g), Frabjous(f)
{
}
void Gloam::tell()
{
     Frabjous::tell();
     /*使用作用域解析的方式调用基类的 tell()函数，间接访问基类的私有数据*/
     cout<<glip<<endl;
}
```

4．假设有下面的定义，它基于程序清单 14.13 中的 Stack 模板和程序清单 14.10 中的 Woker 类。

```
Stack<Worker *> sw;
```

请写出将生成的类声明。只实现类声明，不实现非内联类方法。

习题解析：

程序清单 14.10 定义了类 Worker，程序清单 14.13 定义了 Stack 的模板类，模板声明为 template <class Type> class Stack{}。题目要求以 Worker*（即 Worker 的指针）作为模板实例化的类型，因此可以在程序清单 14.13 的基础上修改代码。要声明生成的类，需要将 Stack 模板中的类型替换成实例化的类型 Worker*。修改后的代码如下。

```
class Stack<Worker *> {
private:
     enum { MAX = 10 };
     Worker *items[MAX];
     /*替换原 Type 类型为 Worker*类型，*/
     int top; //index for top stack item
public:
     Stack();
     Boolean isempty();
     Boolean isfull();
     Boolean push(const Worker*&item);
     Boolean pop(Worker*&item);
     /*替换 push()和 pop()函数的参数类型*/
};
```

5．使用本章中的模板对下面的内容进行定义：

- string 对象数组；
- double 数组栈；
- 指向 Worker 对象的指针的栈数组。

程序清单 14.18 生成了多少个模板类定义?

习题解析：

- 程序清单 14.17 内定义了数组的模板 ArrayTP，其声明格式为

```
template<class T , int n> class ArrayTP;
```

其中，int 类型的非类型参数表示数组的长度，在此基础上可以定义 string 对象的数组，格式如下。

```
ArrayTP<string, 100> sa;
```

其中，整型数据可以根据需要指定。

- 程序清单 14.13 定义了模板 Stack，其声明格式为

```
template <class Type> class Stack;
```

题目要求创建一个 double 数组的栈，因此需要结合 ArrayTP 模板，首先定义 double 数组类型，再定义栈。通常程序设计中为了简化可以使用 typedef，这里可以定义为

```
StackTP<ArrayTP<double,100> > stack_arr_d;
```

- 指向 Worker 对象的指针的栈数组表明数据是一个关于栈的数组，因此首先定义 Worker 对象的指针的栈，即 StackTP<Worker *>，再定义关于该栈的数组形式，最后的格式如下。

```
ArrayTP< StackTP<Worker *>, 100> arr_stk_wpr;
```

程序清单 14.18 生成了 4 个模板。

- ArrayTP<int, 10>sums：int 类型数据，长度为 10 的数组。
- ArrayTP<double, 10>saves：double 类型数据，长度为 10 的数组。
- ArrayTP<int, 5>，int 类型数据，长度为 5 的数组。
- Array<ArrayTP <int, 5>, 10>twodee：以上一个数组为类型定义一个长度为 10 的数组。其本质就是一个 5×10 的二维数组。

6．指出虚基类与非虚基类之间的区别。

习题解析：

非虚基类和虚基类的差别体现在当派生类的多个基类有共同的祖先时，非虚基类的继承关系中，多个基类的共同祖先会存在多个相同的副本，而虚基类中共同的祖先只会保存一个副本。此外，当使用非虚基类时，派生类从不同的基类那里继承了多个同名的数据成员，如果没有用类名进行限定，将导致二义性。但如果使用的是虚基类，派生类中的名称优先于直接或间接祖先类中的相同名称，这样即使不使用限定符也不会导致二义性。

14.7　编程练习

1．Wine 类有一个 string 类的对象成员（参见第 4 章）和一个 Pair 对象（参见本章）。其中，前者用于存储葡萄酒的名称，而后者有两个 valarray<int>对象（参见本章），这两个 valarray<int>对象分别保存了葡萄酒的酿造年份和该年份生产的瓶数。例如，Pair 的第 1 个

valarray<int>对象可能表示 1988 年、1992 年和 1996 年，第 2 个 valarray<int>对象可能表示 24 瓶、48 瓶和 144 瓶。Wine 最好有一个 int 成员，用于存储年数。另外，一些 typedef 可能有助于简化编程工作。

```
typedef std::valarray<int> ArrayInt;
typedef Pair<ArrayInt, ArrayInt> PairArray;
```

这样，PairArray 表示的是类型 Pair<std::valarray<int>, std::valarray<int> >。使用包含关系来实现 Wine 类，并用一个简单的程序对其进行测试。Wine 类应该有一个默认构造函数以及如下构造函数。

```
//初始化标签为 1, 初始化年数为 y,
//初始化 vintage 的年数为 yr[],初始化瓶数为 bot[]
Wine(const char *1, int y, const int yr[], const int bot[]);
//初始化标签为 1, 初始化年数为 y,
//创建关于长度的数组对象
Wine(const char *1, int y);
```

Wine 类应该有一个 GetBottles( )方法，它根据 Wine 对象能够存储几个年份，提示用户输入年份和瓶数。方法 Label( )返回一个指向葡萄酒名称的引用。sum( )方法返回 Pair 对象的第 2 个 valarray<int>对象中的瓶数总和。

测试程序应提示用户输入葡萄酒名称、元素个数以及每个元素存储的年份和瓶数等信息。程序将使用这些数据来构造一个 Wine 对象，然后显示对象中保存的信息。

下面是一个简单的测试程序。

```
//pe14-1.cpp
#include <iostream>
#include "winec.h"
int main ( void )
{
    using std::cin;
    using std::cout;
    using std::endl;

    cout << "Enter name of wine: ";
    char lab[50];
    cin.getline(lab, 50);
    cout << "Enter number of years: ";
    int yrs;
    cin >> yrs;
    Wine holding(lab, yrs);
    holding.GetBottles();
    holding.Show() ;

    const int YRS = 3;
    int y[YRS] = {11993, 1995, 1998};
    int b[YRS] = { 48, 60, 72};
    Wine more("Gushing Grape Red",YRS, y, b);
    more.Show();
    cout << "Total bottles for " << more.Label() //use Label() method
    << ":; " << more.sum() << endl; //use sum() method
    cout << "Bye\n";
```

```
    return 0;
}
```

下面是该程序的运行情况。

```
Enter name of wine: Gully Wash
Enter number of years: 4
Enter Gully Wash data for 4 year(s):
Enter year: 1988
Enter bottles for that year: 42
Enter year: 1994
Enter bottles for that year: 58
Enter year: 1998
Enter bottles for that year: 122
Enter year: 2001
Enter bottles for that year: 144
Wine: Gully Wash
      Year Bottles
      1988 42
      1994 58
      1998 122
      2001 144
Wine: Gushing Grape Red
      Year Bottles
      1993 48
      1995 60
      1998 72
Total bottles for Gushina Grape Red: 180
Bye
```

编程分析：

Pair 模板类的定义可以参考程序清单 14.19，并在此基础上使用 typedef 简化定义。

```
typedef std::valarray<int> ArrayInt;
typedef Pair<ArrayInt, ArrayInt> PairArray;
```

定义 PairArray 类型，并在此基础上定义类 Wine。PairArray 类型首先是两个 ArrayInt 组成的 a、b 两个数据成员的配对，而 ArrayInt 类型又是 C++语言预定义的数值类型，因此 PairArray 就是由两个 int 类型数组配对之后形成的一个数据结构。类 Wine 中需要使用 string 对象和 PairArray 对象描述 Wine 的多个属性，从程序的输出可以判断，string 对象表示 Wine 的标签名称，Pair 对象表示 Wine 的年份和对应数量。从其模型和结构上可以得出 Wine 类型和 string、PairArray 类的关系属于 HAS-A 模型（无法构成 IS-A 模型），因此可以选择和使用包含关系与私有继承。其中题目中已经提供了 Wine 类的构造函数的原型，结合程序的输出，可以推断成员函数的基本功能和参数。本题可以先使用包含关系，编程练习 2 将要求使用私有继承来实现 Wine 类。

```
/*第 14 章的编程练习 1*/
#ifndef WINEC_H
#define WINEC_H
```

```
#include <iostream>
#include <string>
#include <valarray>
using namespace std;

template <class T1, class T2> class Pair;
/*Pair 模板类的声明*/
typedef std::valarray<int> ArrayInt;
typedef Pair<ArrayInt,ArrayInt> PairArray;
/*PairArray 中两个类型参数相同，均为 ArrayInt*/

template <class T1, class T2>
class Pair
{
private:
    T1 a;
    T2 b;
public:
    T1 & first();
    T2 & second();
    T1 first() const { return a; }
    T2 second() const { return b; }
    Pair(const T1 & aval, const T2 & bval) : a(aval), b(bval) { }
    Pair() {}
};

class Wine
{
private:
    string label;
    PairArray info;
   /*Pair 模板类对象存储 Wine 的年份和对应数量*/
     int year;
public:
    Wine(const char*l, int y,const int yr[],const int bot[]);
    Wine(const char*l, int y);
    void GetBottles();
    const string& Label() const;
    int sum() const;
    void Show();
};
#endif

#include "winec.h"

template<class T1, class T2>
T1 & Pair<T1,T2>::first()
{
    return a;
}
```

```
template<class T1, class T2>
T2 & Pair<T1,T2>::second()
{
    return b;
}
/*Pair 模板类的成员函数的定义*/

Wine::Wine(const char* l, int y,const int yr[],const int bot[])
        : label(l), year(y), info(ArrayInt(yr,y),ArrayInt(bot,y))
{
}
Wine::Wine(const char* l, int y)
        :label(l),year(y),info(ArrayInt(0,0),ArrayInt(0,0))
{
}
void Wine::GetBottles()
{
    cout << "Enter " << label << " data for " << year << " year(s):\n";
    info.first().resize(year);
    info.second().resize(year);
    for (int i = 0; i < year; i++)
    {
       cout << "Enter year: ";
       cin >> info.first()[i];
       cout << "Enter bottles for that year: ";
       cin >> info.second()[i];
    }
}
const string& Wine::Label() const
{
    return label;
}
int Wine::sum() const
{
    return info.second().sum();
}
void Wine::Show()
{
    cout << "Wine: " << label << endl;
    cout << "   Year    Bottles" << endl;
    for (int i = 0; i < year; i++)
    {
       cout << "   " << info.first()[i]
            << "    " << info.second()[i] << endl;
    }
}
/*Wine 类内成员函数的定义*/
```

2．采用私有继承而不是包含来完成编程练习 1。同样，一些 typedef 可能会有所帮助，另外，你可能还需要考虑诸如下面这样的语句的含义。

```
PairArray::operator=(PairArray(ArrayInt(),ArrayInt()));
cout<<(const string& )(*this);
```

你设计的类应该可以使用编程练习 1 中的测试程序进行测试。

编程分析：

题目要求在编程练习 1 的基础上，将包含关系改写为私有继承。Wine 类的基本方法已经在编程练习 1 中简要表述。要改写为私有继承，在编写代码时需要注意私有继承中构造函数的初始化方法，需要在初始化列表中直接使用类名进行初始化。

```
Wine::Wine(const char*l, int y) :string(l)...
```

此外，作为私有继承，因为派生类无法访问私有继承的数据成员，所以在某些特定数据的操作过程中需要使用类型强制转换，例如，通过以下语句将 wine 对象转换成 string 对象并通过标准输入/输出进行输出。完整代码如下。

```
cout << "Wine: " << (const string&) (*this) << endl;

/*第 14 章的编程练习 2*/
#ifndef WINEC_H
#define WINEC_H

#include <iostream>
#include <string>
#include <valarray>
using namespace std;

template <class T1, class T2> class Pair;
typedef std::valarray<int> ArrayInt;
typedef Pair<ArrayInt,ArrayInt> PairArray;

template <class T1, class T2>
class Pair
{
private:
    T1 a;
    T2 b;
public:
    T1 & first();
    T2 & second();
    T1 first() const { return a; }
    T2 second() const { return b; }
    Pair(const T1 & aval, const T2 & bval) : a(aval), b(bval) { }
    Pair() {}
};

class Wine: private PairArray, private string
/*把原有的两个数据成员修改为私有继承*/
{
private:
```

```
    int year;
public:
    Wine(const char*l, int y,const int yr[],const int bot[]);
    Wine(const char*l, int y);
    void GetBottles();
    const string& Label() const;
    int sum() const;
    void Show();
};
#endif

#include "winec.h"

template<class T1, class T2>
T1 & Pair<T1,T2>::first()
{
    return a;
}
template<class T1, class T2>
T2 & Pair<T1,T2>::second()
{
    return b;
}

Wine::Wine(const char*l, int y,const int yr[],const int bot[])
: string(l), year(y), PairArray(ArrayInt(yr,y),ArrayInt(bot,y))
{
}
/*私有继承中，基类在初始化列表中初始化*/
Wine::Wine(const char* l, int y) :string(l),year(y),PairArray(ArrayInt(0,0),ArrayInt(0,0))
{
}
/*私有继承中，基类在初始化列表中初始化*/
void Wine::GetBottles() {
    cout << "Enter " << (const string&) (*this) << " data for " << year << " year(s):\n";
    this->first().resize(year);
    this->second().resize(year);
    for (int i = 0; i < year; i++)
    {
        cout << "Enter year: ";
        cin >> this->first()[i];
        cout << "Enter bottles for that year: ";
        cin >> this->second()[i];
    }
}
const string& Wine::Label() const
{
    return (const string&) (*this);
}
```

```
/*私有继承中，基类数据成员通过转换访问，即先转换成基类，再访问其成员*/
int Wine::sum() const
{
    return this->second().sum();
}
void Wine::Show()
{
    cout << "Wine: " << (const string&) (*this) << endl;
    cout << "   Year    Bottles" << endl;
    for (int i = 0; i < year; i++)
    {
        cout << "    " << this->first()[i]
             << "     " << this->second()[i] << endl;
    }
}
```

3．定义一个 QueueTp 模板。然后在一个类似于程序清单 14.12 的程序中创建一个指向 Worker 的指针队列（参见程序清单 14.10 中的定义），并使用该队列来测试它。

编程分析：

Queue 是一个队列类型，其定义参见程序清单 12.10。该类的主要的数据成员是队列内的元素，程序清单 12.10 使用数组的形式表示，重要的成员函数包括判空、判满、入队、出队和相关构造函数。题目要求修改成模板形式，因此首先需要修改原有类的定义。修改时需要重点关注对数据成员的操作函数。最后实现 Worker 的指针队列，并进行测试。完整代码如下。

```
/*第 14 章的编程练习 3*/
//queuetp.h
#ifndef QUEUETP_H_
#define QUEUETP_H_
#include <iostream>
#include <string>
using namespace std;

class Worker
{
private:
    std::string fullname;
    long id;
protected:
    void Data() const;
    void Get();
public:
    Worker() : fullname("no one"), id(0L) {}
    Worker(const std::string & s, long n) : fullname(s), id(n) {}
    ~Worker(){};
    void Set();
    void Show() const;
```

```
};
/*原有的 Worker 类保持不变*/

template<class T> class QueueTp
{
private:
    enum {Q_SIZE = 10};
    struct Node{T item; Node*next;};
/*把 Node 结构体修改成参数类型*/

    Node*front;
    Node*rear;
    int items;
    const int qsize;
    QueueTp (const QueueTp & q) : qsize(0) { }
    QueueTp & operator=(const QueueTp & q) { return *this; }
public:
    QueueTp(int qs = Q_SIZE): qsize(qs)
    {
        front = rear = nullptr;
        items = 0;
    }
    ;
    ~QueueTp()
    {
        Node *temp;
        while(front != nullptr)
        {
            temp = front;
            front = front->next;
            delete temp;
        }
    };
    bool isempty() const;
    bool isfull() const;
    int queuecount () const;
    bool enqueue(const T &item);
    bool dequeue (T &item) ;
    /*入队、出队函数修改其参数类型*/
};
#endif

#include "queuetp.h"

void Worker::Set()
{
    cout<<"Enter worker's name: ";
    getline(cin,fullname);
    cout<<"Enter worker's ID: ";
    cin>>id;
```

```
    while(cin.get()!='\n')
        continue;
}
void Worker::Show() const
{
    cout<<"Name: "<<fullname<<endl;
    cout<<"Employee ID: "<<id<<endl;
}
/*Worker 类的成员函数的实现*/

template <class T>
QueueTp <T>::QueueTp(int qs):qsize(qs)
/*模板类的构造函数*/
{
   front = rear = nullptr;
   items = 0;
}

template <class T>
QueueTp<T>::~QueueTp()
/*模板类的析构函数*/
{
   Node *temp;
   while(front != nullptr)
   {
       temp = front;
       front = front->next;
       delete temp;
   }
}

template <class T>
bool QueueTp<T>::isempty() const
{
    return items == 0;
}

template <class T>
bool QueueTp<T>::isfull() const
{
    return items == qsize;
}

template <class T>
int QueueTp<T>::queuecount() const
{
    return items;
}

template <class T>
```

```
bool QueueTp<T>::enqueue(const T &item)
{
    if(isfull())
        return false;
    Node *temp = new Node;
    temp->item = item;
    temp->next = nullptr;
    items++;
    if(front == nullptr)
        front = temp;
    else
        rear->next = temp;
    rear = temp;
    return true;
}
/*定义模板类的入队函数*/

template <class T>
bool QueueTp<T>::dequeue(T &item)
{
    if(isempty())
        return false;
    item = front->item;
    items--;
    Node *temp = front;
    front = front->next;
    delete temp;
    if(items == 0)
        rear = nullptr;
    return true;
}
/*定义模板类的出队函数*/


#include <iostream>
#include <string>
using namespace std;

int main()
{
    QueueTp<Worker> lolas;
    Worker w1;
    w1.Set();
    lolas.enqueue(w1);
    Worker w2;
    lolas.dequeue(w2);
    w2.Show();
    cout << "Bye.\n";
    return 0;
}
```

4．Person 类保存人的名和姓。除构造函数外，它还有 Show()方法，用于显示名和姓。Gunslinger 类以 Person 类为虚基类派生而来，它包含一个 Draw()成员，该方法返回一个 double 值，表示枪手的拔枪时间。这个类还包含一个 int 成员，表示手枪上的刻痕数。最后，这个类还包含一个 Show()函数，用于显示所有这些信息。PokerPlayer 类以 Person 类为虚基类派生而来。它包含一个 Draw()成员，该函数返回一个介于 1～52 的随机数，用于表示扑克牌的值（也可以定义一个 Card 类，其中包含花色和面值成员，然后由 Draw()返回一个 Card 对象）。PokerPlayer 类使用 Person 类的 show()函数。BadDude 类从 Gunslinger 和 PokerPlayer 类派生而来。它包含 Gdraw()成员（返回枪手拔枪的时间）和 Cdraw()成员（返回下一张扑克牌），还有一个合适的 Show()函数。请定义这些类和方法以及其他必要的方法（如用于设置对象值的方法），并使用一个类似于程序清单 14.12 的简单程序对它们进行测试。

编程分析：

题目要求实现一系列的派生关系类。首先基类是 Person，成员函数和 Show()方法的基本功能已经给定；由 Person 直接派生的类包括 Gunslinger 类和 PokerPlayer 类；在 Gunslinger 中需要添加 int 类型成员以表示刻痕，添加 Draw() 方法以表示拔枪时间，隐藏基类的 Show()方法；为 PokerPlayer 类添加 Draw()方法。BadDude 类是 Person 类的第 3 代派生类，从 Gunslinger 和 PokerPlayer 类派生。本题中的继承关系是 IS-A 模型，因此选用公有继承。此外，在多重继承的类模型中，需要注意的主要问题如下：

- 虚基类的设计；
- 继承过程中同名函数的隐藏关系；
- 作用域解析符的使用；
- 派生类的构造函数设计（初始化列表）。

此外，还需要定义 Card 类，设计 PorkPlayer 的相关成员函数。完整代码如下。

```
/*第 14 章的编程练习 4*/
#ifndef PERASON_H
#define PERASON_H

#include <iostream>
#include <string>

using namespace std;
class Person
{
private:
    string fname;
    string lname;
public:
    Person():fname("no name"),lname("no name"){};
    Person(string f,string l);
    virtual ~Person(){};
    virtual void Show() const;
};
```

```
/*基类 Person 的声明*/

class Gunslinger:virtual public Person
{
private:
    int nick;
/*新增数据成员*/
public:
    Gunslinger():Person(),nick(0){ };
    Gunslinger(string f,string l,int n);
    ~Gunslinger(){};
    double Draw();
    void Show()const;
/*新增成员函数，隐藏 Show()函数*/
};
/*派生类 Gunslinger，继承自 Person 类*/


struct Card{
    enum SUITE {SPADE,HEART,DIAMOND,CLUB};
    SUITE suite;
    int number;
};
/*Card 结构体的定义*/

class PokerPlayer:virtual public Person
{
public:
    ~PokerPlayer(){};
    Card Draw() const;
    /*新增成员函数*/
};
/*派生类 PokerPlayer，继承自 Person 类*/

class BadDude: public Gunslinger,public PokerPlayer
{
public:
    double GDraw() const;
    int CDraw() const;
    void Show() const;
/*新增成员函数，隐藏 Draw()函数*/
};
/*派生类 BadDude，继承自 PokerPlayer、Gunslinger 类*/
#endif //


#include <iostream>
#include <cstdlib>
#include <ctime>
#include "perason.h"
```

```
Person::Person(string f,string l): fname(f),lname(l)
{
}
void Person::Show() const
{
    cout<<fname<<"."<<lname<<endl;
}
Gunslinger::Gunslinger(string f, string l, int n): Person(f,l), nick(n)
{
}
double Gunslinger::Draw(){
    srand(time(0));
    return rand() % 60;
}
void Gunslinger::Show()const{
    Person::Show();
    cout<<"Nick: "<<nick<<endl;
}
Card PokerPlayer::Draw() const{
    Card temp;
    srand(time(0));
    temp.number = rand() % 52;
    temp.suite =  Card::SUITE (rand() % 4);
    return temp;
}
int main(){
    Person person("Jakey","Slong");
    person.Show();
    /*测试person对象的创建，输出信息*/
    Gunslinger gl("Tidy","White",12);
    gl.Show();
    cout<<"Gunslinger's nick is "<<gl.Draw()<<endl;
    /*测试Gunslinger对象的创建，输出信息*/
    PokerPlayer pokerplayer;
    pokerplayer.Show();
    /*测试pockerplayer的默认构造函数*/
}
/*在main()函数内进行基本功能测试*/
```

5．下面是一些类声明。

```
//emp.h
#include <iostream>
#include <string>
class abstr_emp
{
private:
    std::string fname;
    std::string lname;
    std::string job;
```

```
public:
    abstr_emp();
    abstr_emp(const std::string & fn, const std::string & ln,
              const std::string & j);
    virtual void ShowAll() const;
    virtual void SetAll();
    friend std::ostream &
    operator<<(std::ostream & os, const abstr_emp & e);
    virtual ~abstr_emp() = 0;
};
class employee : public abstr_emp
{
public:
    employee () ;
    employee(const std::string & fn, const std::string & ln,
             const std::string & j);
    virtual void ShowAll() const;
    virtual void SetAll();
};
class manager: virtual public abstr_emp
{
private:
    int inchargeof;
protected:
    int InChargeOf() const { return inchargeof; }
    int & InChargeOf(){ return inchargeof; }
public:
    manager () ;
    manager(const std::string & fn, const std::string & ln,
            const std::string & j, int ico = 0);
    manager(const abstr_emp & e, int ico);
    manager (const manager & m);
    virtual void ShowAll() const;
    virtual void SetAll();
};
class fink: virtual public abstr_emp
{
private:
    std::string reportsto;
protected:
    const std::string ReportsTo() const { return reportsto; }
    std::string & ReportsTo(){ return reportsto; }
public:
    fink();
    fink(const std::string & fn, const std::string & ln,
        const std::string & j, const std::string & rpo);
    fink(const abstr_emp & e, const std::string & rpo);
    fink(const fink & e);
    virtual void ShowAll() const;
    virtual void SetAll();
```

```
};
class highfink: public manager, public fink
{
public:
    highfink();
    highfink(const std::string & fn, const std::string & ln,
             const std::string & j, const std::string & rpo, int ico);
    highfink(const abstr_emp & e, const std::string & rpo, int ico);
    highfink(const fink & f, int ico);
    highfink(const manager & m, const std::string & rpo);
    highfink(const highfink & h);
    virtual void ShowAll() const;
    virtual void SetAll();
};
```

注意，该类层次结构使用了带虚基类的 MI，所以要牢记这种情况下用于构造函数初始化列表的特殊规则。还需要注意的是，有些方法被声明为受保护的。这可以简化一些 highfink 方法的代码（例如，如果 highfink::ShowAll()只调用 fink::ShowAll()和 manager::ShowAll()，则它将调用 abstr_emp::ShowAll()两次）。请提供类方法的实现，并在一个程序中对这些类进行测试。下面是一个小型测试程序。

```
//pe14-5.cpp
//useempl.cpp
#include <iostream>
using namespace std;
#include "emp.h"
int main(void)
{
    employee em("Trip", "Harris", "Thumper") ;
    cout << em << endl;
    em.ShowAll() ;
    manager ma("Amorphia", "Spindragon", "Nuancer", 5);
    cout << ma << endl;
    ma.ShowAll() ;

    fink fi("Matt", "Oggs", "Oiler", "Juno Barr");
    cout << fi << endl;
    fi.ShowAll ();
    highfink hf(ma, "Curly Kew"); //recruitment?
    hf.ShowAll();
    cout << "Press a key for next phase:\n";
    cin.get();
    highfink hf2;
    hf2.SetAll();
    cout << "Using an abstr_emp *pointer:\n";
    abstr_emp *tri[4] = {&em, &fi, &hf, &hf2};
    for (int i = 0; i < 4; i++)
        tri[i]->ShowAll();
```

```
    return 0;
}
```

- 为什么没有定义赋值运算符？
- 为什么要将 ShowAll()和 SetAll()定义为虚的？
- 为什么要将 abstr_emp 定义为虚基类？
- 为什么 highfink 类没有数据部分？
- 为什么只需要一个 operator<<()版本？
- 如果使用下面的代码替换程序的结尾部分，将会发生什么情况？

编程分析：

本题是一个关系较复杂的多重继承关系，基类是 abstr_emp。注意，由于该类的析构函数被设置称为纯虚函数，因此该类是抽象类，但是纯虚函数形式的析构函数也必须实现，这一点在本章理论部分有讨论。第 1 代派生类为 employee、manager 和 fink，分别为它们添加了相应的数据成员，也需要重新定义部分成员函数。第 2 代派生类为 highfink，从 fink 和 manager 派生。本题的重点和难点在于 highfink 类的设计上，如题目所述，如果 highfink::ShowAll() 只调用 fink::ShowAll() 和 manager::ShowAll()，则它将调用 abstr_emp::ShowAll()两次)，因此要合理利用 manager 类中受保护的方法 ReportsTo()。通过该方法既可以避免二次调用基类方法，又可以实现所有数据的访问。题目中几个问题的参考答案如下。

- 为什么没有定义赋值运算符？

对于类内的所有数据成员，没有分配动态存储，且 string 类型的对象也实现了赋值运算，使用默认的赋值运算符能够正确工作，因此可以不重新定义赋值运算符。

- 为什么要将 ShowAll()和 SetAll()定义为虚的？

在类的派生过程中，ShowAll()和 SetAll()两个函数在每一个派生类中实现的功能都有所不同，为了能够更好地依据实际对象来查找函数，必须使用虚函数来实现，例如，对于 main()函数中的以下语句，如果没有定义虚函数，那么 tri 指针数组的成员函数将无法正常调用。

```
abstr_emp *tri[4] = {&em, &fi, &hf, &hf2};
```

- 为什么要将 abstr_emp 定义为虚基类？

本题中类 highfink 实现了多重继承，且两个基类又有共同的基类 abstr_emp，因此需要使用虚基类的形式来简化 highfink 的两个基类中的 abstr_emp 副本。多重继承中单一基类保留多份副本，占用较多的存储空间，还容易产生命名冲突，因此通常需要使用虚基类进行简化处理。

- 为什么 highfink 类没有数据部分？

highfink 类完全继承和使用 manager 类与 fink 类的数据成员，完全能够描述本身的属性，因此不需要再添件新的数据成员。

- 为什么只需要一个 operator<<()版本？

abstr_emp 中的运算符重载 operator<<()是通过友元形式实现的，友元关系无法继承。ShowAll()函数通过继承实现相关功能，在形式上更加简单，因此没有必要通过运算符的形式实现输出功能。

- 如果使用下面的代码替换程序的结尾部分，将会发生什么情况？

```
abstr_emp tri[4] = {em,fi,hf,hf2};
for(int i = 0; i< 4 ;i++)
  tri[i].ShowAll();
```

作为抽象类，abstr_emp 是无法创建对象数组的，因此 abstr_emp tri[4]声明语句错误。如果修正该错误，将 abstr_emp 改为普通类，程序可以运行，进行对象类型的强制转换，但是 ShowAll()函数将会只显示姓名和职务，无法实现虚函数原有的功能。

```
/*第 14 章的编程练习 5*/
/*使用题目给定的 emp.h 和 useemp.cpp，下面仅为 emp.cpp*/

#include "emp.h"
using namespace std;
class highfink: public manager, public fink //management fink
{
public:
    highfink();
    highfink(const std::string & fn, const std::string & ln,
             const std::string & j, const std::string & rpo, int ico);
    highfink(const abstr_emp & e, const std::string & rpo, int ico);
    highfink(const fink & f, int ico);
    highfink(const manager & m, const std::string & rpo);
    highfink(const highfink & h);
    virtual void ShowAll() const;
    virtual void SetAll();
};

abstr_emp::abstr_emp() :lname("none"),fname("none"),job("none")
{
}
abstr_emp::abstr_emp(const std::string & fn, const std::string & ln,
                     const std::string & j):fname(fn),lname(ln),job(j)
{
}
abstr_emp::~abstr_emp(){}

void abstr_emp::ShowAll() const
{
    cout<<"NAME: "<<fname<<"."<<lname<<endl;
    cout<<"JOB TITLE: "<<job<<endl;
}
void abstr_emp::SetAll()
{
    cout<<"Entenr the first name: ";
    getline(cin,fname);
    cout<<"Enter the last name: ";
    getline(cin,lname);
    cout<<"Enter th job title: ";
    getline(cin,job);
}
```

```
std::ostream &operator<<(std::ostream & os, const abstr_emp & e){
    os<<"NAME: "<<e.fname<<"."<<e.lname<<endl;
    os<<"JOB TILTE: "<<e.job<<endl;
    return os;
}

employee::employee () :abstr_emp()
{
}
employee::employee(const std::string & fn, const std::string & ln,
                   const std::string & j): abstr_emp(fn,ln,j)
{
}
void employee::ShowAll() const{
    abstr_emp::ShowAll();
}
void employee::SetAll()
{
    abstr_emp::SetAll();
}

manager::manager ():abstr_emp(),inchargeof(0)
{
}
manager::manager(const std::string & fn, const std::string & ln,
                 const std::string & j, int ico ): abstr_emp(fn,ln,j),inchargeof(ico)
{
}
manager::manager(const abstr_emp & e, int ico): abstr_emp(e)
{
    inchargeof = ico;
}
manager::manager (const manager & m): abstr_emp(m)
{
    inchargeof = m.inchargeof;
}
void manager::ShowAll() const{
    abstr_emp::ShowAll();
    cout<<"IN CHARGE OF: "<<inchargeof <<endl;
}
void manager::SetAll()
{
    abstr_emp::SetAll();
    cout<<"Enter the number of in charge: ";
    cin>>inchargeof;
    while(cin.get() != '\n')
        continue;
}

fink::fink():abstr_emp(),reportsto("none")
```

```
{
}

fink::fink(const std::string & fn, const std::string & ln,
          const std::string & j, const std::string & rpo): abstr_emp(fn,ln,j),
          reportsto(rpo)
{
}
fink::fink(const abstr_emp & e, const std::string & rpo):abstr_emp(e),reportsto(rpo)
{
}
fink::fink(const fink & e):abstr_emp(e)
{
    reportsto = e.reportsto;
}
void fink::ShowAll() const
{
    abstr_emp::ShowAll();
    cout<<"REPORT TO: "<<reportsto<<endl;
}
void fink::SetAll()
{
    abstr_emp::SetAll();
    cout<<"Enter the reports to whom: ";
    getline(cin,reportsto);
}

highfink::highfink()
{
}
highfink::highfink(const std::string & fn, const std::string & ln,
                  const std::string & j, const std::string & rpo, int ico):
       abstr_emp(fn,ln,j),manager(fn,ln,j,ico),fink(fn,ln,j,rpo)
{
}
highfink::highfink(const abstr_emp & e, const std::string & rpo, int ico):
       abstr_emp(e),manager(e,ico),fink(e,rpo)
{
}
highfink::highfink(const fink & f, int ico):
       abstr_emp(f),fink(f),manager(f,ico)
{
}
highfink::highfink(const manager & m, const std::string & rpo):
       abstr_emp(m), manager(m),fink(m,rpo)
{


}
highfink::highfink(const highfink & h):abstr_emp(h),manager(h),fink(h)
{
```

```
}
void highfink::ShowAll() const
{
    manager::ShowAll();
    cout << "Reportsto: " << ReportsTo() << endl;
    cout << endl;

}
void highfink::SetAll()
{
    manager::SetAll();
    cout << "Enter the reportsto: ";
    getline(cin, fink::ReportsTo());

}
```

第 15 章　友元、异常和其他

本章知识点总结

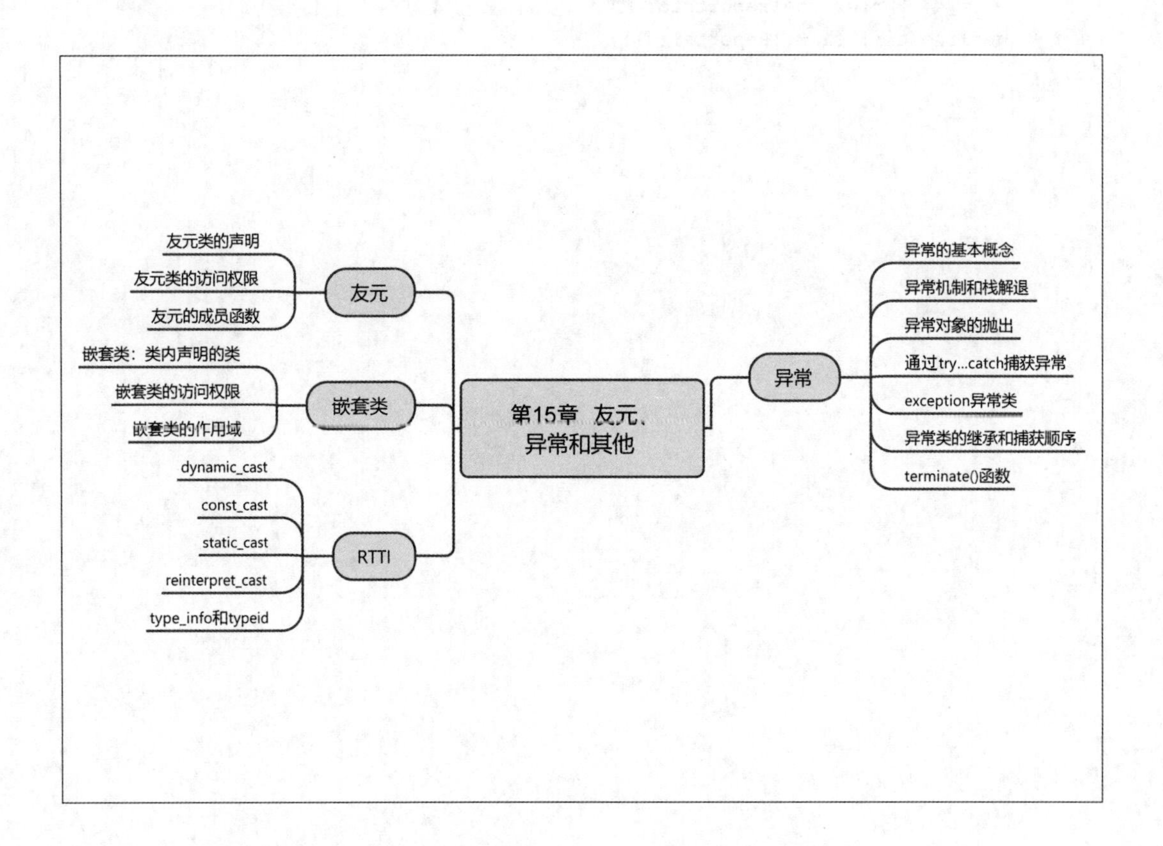

15.1　友元类和类的嵌套

友元类和友元函数的形式相似，只是将整个类作为另外一个类的友元，这样友元类的所有方法都能够访问原始类的私有成员和保护成员。如果不需要将这个类作为原始类的友元类，也可以选择类内的部分方法，将其作为友元。友元类的定义方式如下。

```
class A{
…
friend class B;
}
```

这样类 B 将作为类 A 的友元类，B 内的所有方法都可以访问 A 的私有成员和受保护的成员。如果只需要使 B 类的指定函数作为友元来访问类 A，那么可以定义为：

```
class B;
class A{
…
friend void B::f();
}
```

将声明语句 class B;放置在最前面是为了告诉编译器 B 是一个自定义类，避免编译器的编译错误。实际应用中，为了使一个类能够访问其他两个类的私有对象，可以将这个类作为其他两个类的友元，这样能够更加灵活地处理类之间的数据交换。

C++中如果一个类的声明和定义在另外一个类当中，被包含的类称为嵌套类。包含类的成员函数可以创建和使用嵌套类的对象。嵌套类的基本结构如下。

```
class A{
…
        class B{
        }
}
```

这样类 B 就是存在于 A 类当中的嵌套类，如果嵌套类定义在私有部分，那么只能在 A 类当中使用嵌套类 B；如果嵌套类 B 定义在保护部分，则嵌套类对于 A 类是可见的，对外不可见，对于 A 类的派生类也可见；如果嵌套类在定义公有部分，类外可以创建和使用嵌套类。当嵌套类可见之后，嵌套类的数据成员的访问控制也会限制外部对嵌套类的访问，这些访问控制和常规类相同。

15.2　异常与异常处理

C++语言中的异常机制是指程序在运行过程中出现的一些不可预期的运行错误，例如，运算中不适当地输入数值、I/O 读取失败、网络连接失败、内存空间非法访问等。这些错误会导致程序的运行错误，程序设计者应该在程序设计中考虑到如何正确处理这类可能会发生的错误，从而在这些意外情况发生时正确发现并处理，这样会增强整个程序的容错性和健壮性。C++语言中使用异常和异常处理的形式来处理这些程序运行时的异常情况。

异常处理分为两个阶段——异常产生阶段和异常处理阶段。在异常产生阶段，程序需要抛出异常信号，详细说明当前异常的情况和状态。使用 throw 语句抛出异常，throw 语句类似于返回语句，它将终止函数的执行，但 throw 不将控制权返回给调用程序，而会导致程序沿函数调用序列后退，直到被异常处理代码捕获。C++语言中的异常处理有专门的异常处理代码。在异常处理阶段，通过 try...catch 语句捕获上一阶段抛出的异常信号，并分析和处理被捕获的异常。其中抛出异常信息的语句使用 throw 关键字和特定的异常对象。异常对象内能够保存一些异常的基本信息，以便捕获代码处理。异常的捕获使用 try...catch 语句块，其基本格式如下。

```
try{
    //抛出异常的代码
} catch( Exception){
```

```
        //处理异常
    }
    catch( Other_Exception){
        //处理异常
    }
```

C++在函数调用中通常将信息放在栈中来处理。栈是一种先入后出（FILO）的数据存储模型，即程序将调用函数的指令的地址、函数的参数、函数中自动存储的变量等数据放到栈中，如果被调用的函数调用了另一个函数，则后者的信息将被添加到栈中，并以此类推，不断叠加到栈的顶部，最先进入的数据在栈的底部。当函数结束时，程序将返回被调用时的状态并释放栈顶的相关数据。当函数由于出现异常而终止时，程序释放栈中数据的方式不同，程序不会在释放栈的第 1 个返回地址（即函数的嵌套调用中的最后一个函数调用）后停止，而是继续释放栈顶数据，直到找到一个位于 try 块中的返回地址，并将控制权转交给块尾的异常处理程序，这个过程称为栈解退。这种栈解退过程将自动释放 try 块和 throw 之间整个函数调用序列中的所有对象，从而避免了由于程序的异常而造成函数调用中栈内数据的回收错误。

15.3 异常类和异常规范

在异常处理过程中，一旦发现异常应立即处理，否则将会导致整个程序的异常退出。意外异常（不与任何异常规范匹配的异常）在默认情况下和未捕获异常类似。在异常发生但未处理时，程序首先调用函数 terminate()，该函数负责调用 abort()函数终止程序。因此，如果我们希望程序不退出，可以修改 terminate() 函数调用的函数指针，从而避免调用 terminate()函数，标准格式如下。

```
set_terminate(My_Function);
```

如果函数中引发了异常规范中没有的异常，程序将会调用以下 unexpected()函数，该函数将调用 terminate()函数。

```
set_unexpected(My_Function);
```

早期 C++版本中，异常规范表示函数定义中标识的函数可能抛出异常的代码，用如下格式表示。

```
void my_function() throw(expection_thing);
```

其中 throw()表示函数 f()可能会抛出的异常类型，但是这样的异常规范可能会影响整个代码的编译和运行效率，因此 C++11 摒弃了函数内的异常规范，这一点在编程中需要注意。但是 C++11 中可以使用以下语句表示函数不会引发异常。

```
void my_function() noexpect;
```

C++异常的主要目的是为设计容错程序提供语言支持。为了更好地实现异常处理功能，C++定义了 exception 类作为程序员自定义异常类型的基类。exception 类中有一个名为 what()的虚函数，该函数会返回一个字符串来描述异常信息。此外，C++系统库还定义了多个异常类，如 logic_error、runtime_error、bad_alloc 等，用户在编程过程中可以选择和使用。如果有一个异常类继承层次结构，应将捕获位于层次结构最下面的异常类的 catch 语句放在最前面，将捕获基类异常的 catch 语句放在最后面。

15.4　运行阶段类型识别

C++中由于类的派生和继承，因此对象指针在程序运行中指向的对象需要进行类型识别，才能正确识别和处理指针指向的具体派生类对象。运行阶段类型识别（Run-Time Type Identification，RTTI）的功能就是让程序在程序的运行阶段动态检测对象的实际类型。C++语言中有 3 个支持 RTTI 的元素——dynamic_cast 运算符、typeid 运算符、type_info 运算符。其中 typeid 返回的 type_info 对象，通过 type_info 可以确定对象是否为特定的类型和关于对象的信息。

对于类型转换，C++语言中主要使用如下 4 个类型转换运算符（程序中应当选择合适的运算符进行数据类型的转换）。

- dynamic_cast：用于将派生类的指针转换为基类的指针，即在类层次结构中进行向上转换。使用方法如下。

```
dynamic_cast<type_name> (expression);
```

如果能够安全转换该指针，则返回转换后的指针；否则，返回空指针。

- static_cast：通常用于低风险的转换，仅当两个类型能够进行隐式类型转换时才可以使用，如整型和浮点型、字符型之间的互相转换，且 static_cast 不能用于在不同类型的指针之间互相转换。
- const_cast：仅用于进行去除 const 属性的转换，即改变值为 const 或 volatile。
  从基类到派生类的转换在不进行显式类型转换的情况下是非法的，因此可以使用 static_cast 来向下转换。
- reinterpret_cast：该运算符功能强大，可以用来处理无关类型之间的转换，因此也会带来很大的风险；运行中它会产生一个新的值，这个值与原始参数有完全相同的二进制数据。

15.5　复习题

1．下面建立友元的尝试有什么错误?

a．class snap {

```
        friend clasp;
        ...
};
class clasp {…};
```

b．class cuff {

```
public:
        void snip(muff &) {…}
        ...
};
class muff {
```

```
        friend void cuff::snip(muff &);
        ...
};
```

c．class muff {

```
        friend void cuff::snip(muff &) ;
        ...
};
class cuff {
public:
        void snip(muff &) {…}
        ...
};
```

习题解析：

a．声明一个友元类需要使用以下格式。

```
friend class clasp;
```

题目内没有使用关键字 class，从而造成编译错误。

b．类 cuff 内的公有函数 sniff 使用了类 muff 作为函数的参数，但是在类 cuff 前并没有 muff 的声明，因此需要在此之前进行 muff 类的前向声明，即在 class cuff 语句前添加 `class muff;`语句，完整形式可以如下。

```
class muff;
class cuff {
public:
        void snip(muff &) {…}
        ...
};
class muff {
        friend void cuff::snip(muff &);
        ...
};
```

c．题目给出的两个类内中，类 muff 使用类 cuff 的成员函数作为友元，类 cuff 内的成员函数使用 muff 类作为参数，因此两个类互为前导，编译器将无法正确识别相关语句。应当将 cuff 类前置，前导声明类 muff，后置类 muff，并声明友元函数，这样编译器才能正确识别。完整代码如下。

```
class muff;
class cuff {
public:
        void snip(muff &) {…}
        ...
};

class muff {
        friend void cuff::snip(muff &) ;
        ...
};
```

2．了解如何建立互为友元的关系后，你能够创建一种更严格的友元关系（即类 B 只有部分成员是类 A 的友元，而类 A 只有部分成员是类 B 的友元）吗？请解释原因。

习题解析：

部分互为友元的关系是无法建立的。为了使 A 拥有一个本身为类 B 的成员函数的友元，B 的声明必须位于 A 的声明之前，因此需要多个前向声明以表示 B 是一个类，且要表示 B 类的多个成员方法。同样，如果 B 以 A 的一个成员函数作为友元，则在 B 之前也需要多个前向声明。因此，A、B 的前向声明均无法实现，互为部分友元关系目前在语法上无法实现。

3．下面的嵌套类声明中可能存在什么问题？

```
class Ribs
{
private:
        class Sauce
        {
                int soy;
                int sugar;
        public:
        Sauce(int s1, int s2) : soy(s1), sugar(s2) { }
        };
...
};
```

习题解析：

题目提供的代码在语法上是正确的，但是我们设计和使用嵌套类的主要目的是在隐藏部分细节的基础上，在类内部正常访问该类对象。但是 Sauce 类定义在私有部分，且 Sauce 的两个数据成员也是私有数据，因此在 Ribs 类内没有足够的接口，用于访问 Sauce 对象内的数据。解决办法是为 Sauce 添加相应接口函数，或者将成员设置为公有成员，便于在类外直接访问。

4．throw 和 return 之间的区别何在？

习题解析：

throw 和 return 语句都会造成当前函数终止并返回，但是具体在实现上，尤其在函数数据栈的处理上是有差别的。假设函数 `f1()` 调用函数 `f2()`，`f2()` 中的返回语句导致程序会自动返回 `f1()` 函数，并且开始在函数 `f1()` 中调用函数 `f2()` 后面的一条语句。

throw 语句导致程序沿函数调用的当前序列回溯，直到找到直接或间接包含对 `f2()` 的调用的 try 语句块为止。return 语句可能在并不在 `f1()` 中，因此会持续回退，出现栈解退，直到找到这样的 try 语句块后，才执行下一个匹配的 catch 语句块，从而影响程序原本功能的实现。

5．假设有一个从异常基类派生出来的异常类层次结构，则应按什么顺序放置 catch 块？

习题解析：

异常类的派生关系可以使用户更加细致和针对性地处理程序设计中的异常问题，继承关系使用户可以使用捕获基类对象的方式一次性捕获所有派生类的异常。因此，为了更好地实现异常处理，catch 块应当首先捕获继承中的派生类异常对象，随后逐渐从继承关系向上，捕获基类的异常，这样才能兼顾异常处理的针对性和全面性。

6．对于本章定义的 Grand 类、Superb 类和 Magnificient 类，假设 pg 为 Grand *指针，并将其中某个类对象的地址赋给了它，而 ps 为 Superb *指针，则下面两个代码示例的行为有什么不同?

```
if (ps = dynamic _cast<Superb *>(pg) )
        ps->say(); //示例 1
if (typeid(*pg) == typeid(Superb) )
        (Superb *) pg)->say(); //示例 2
```

习题解析：

本章定义的 Grand 类、Superb 类和 Magnificent 类是单继承关系，因此题目中的 pg 为基类指针，ps 为 Grand 的派生类（Magnificent 的基类）指针。

对于示例 1，if 语句的条件表达式 `ps = dynamic _cast<Superb *>(pg)`是一个赋值语句，因此 pg 指向一个 Superb 对象或从 Superb 派生而来的任何类的对象，能够正确进行类型转换，赋值成功后 if 语句的条件为真。

对于示例 2，if 语句的条件表达式 `typeid(*pg) == typeid(Superb)`判断两个数据对象的 typeid 是否相等，因此仅当指针 pg 指向 Superb 对象时，逻辑表达式的结果才为真，如果指向的是其他类的对象，则 typeid 不相等，逻辑表达式的结果为假。

7．static_cast 运算符与 dynamic_cast 运算符有什么不同?

习题解析：

dynamic_cast 运算符用于将派生类的指针转换为基类的指针，即允许沿类层次结构向上转换；而 static_cast 仅当两个类型能够进行隐式类型转换时才可以使用，它允许向上转换和向下转换。static_cast 运算符还允许枚举类型和整型之间以及数值类型之间的转换。

15.6 编程练习

1．对 Tv 和 Remote 类做如下修改：

a．让它们互为友元；

b．在 Remote 类中添加一个状态变量成员，该成员描述遥控器是处于常规模式还是互动模式；

c．在 Remote 中添加一个显示模式的方法；

d．在 Tv 类中添加一个对 Remote 中新成员进行切换的方法，该方法应仅当电视处于打开状态时才能运行。

编写一个小程序来测试这些新特性。

编程分析：

程序清单 15.1 和程序清单 15.2 定义了 Tv 类与 Remote 类，并将 Remote 类作为 Tv 类的友元类。本题要求修改程序清单，在建立互为友元的关系上，需要注意类的前向声明；在添加数据成员和方法时，可以通过枚举类型实现。完整代码如下。

```
/*第 15 章的编程练习 1*/
//tv.h —— Tv 类和 Remote 类
#ifndef TV_H_
#define TV_H_
#include <iostream>
using namespace std;
class Remote;
/*互为友元类，需要在前面声明 Remote 类*/
class Tv
{
public:
    friend class Remote;
    enum {Off, On};
    enum {MinVal,MaxVal = 20};
    enum {Antenna, Cable};
    enum {TV, DVD};

    Tv(int s = Off, int mc = 125) : state(s), volume(5),
                                  maxchannel(mc), channel(2), mode(Cable), input(TV) {}
    void onoff() {state = (state == On)? Off : On;}
    bool ison() const {return state == On;}
    bool volup();
    bool voldown();
    void chanup();
    void chandown();
    void set_mode() {mode = (mode == Antenna)? Cable : Antenna;}
    void set_input() {input = (input == TV)? DVD : TV;}
    void settings() const;
    void set_Rmode(Remote & r);

private:
    int state;
    int volume;
    int maxchannel;
    int channel;
    int mode;
    int input;
};
```

```
class Remote
{
public:
    enum {Normal, InterActive};
    /*添加状态类型*/
private:
    int mode;
    int work_mode;
    /*添加状态变量*/
public:
    friend class Tv;
    Remote(int m = Tv::TV) : mode(m) {}
    bool volup(Tv & t) { return t.volup();}
    bool voldown(Tv & t) { return t.voldown();}
    void onoff(Tv & t) { t.onoff(); }
    void chanup(Tv & t) {t.chanup();}
    void chandown(Tv & t) {t.chandown();}
    void set_chan(Tv & t, int c) {t.channel = c;}
    void set_mode(Tv & t) {t.set_mode();}
    void set_input(Tv & t) {t.set_input();}
    int show_mode()const{return work_mode;}
    /*添加状态查询函数*/
};
#endif

#include "tv.h"

bool Tv::volup()
{
    if (volume < MaxVal)
    {
        volume++;
        return true;
    }
    else
        return false;
}
bool Tv::voldown()
{
    if (volume > MinVal)
    {
        volume--;
        return true;
    }
    else
        return false;
}

void Tv::chanup()
{
```

```
    if (channel < maxchannel)
        channel++;
    else
        channel = 1;
}

void Tv::chandown()
{
    if (channel > 1)
        channel--;
    else
        channel = maxchannel;
}

void Tv::settings() const
{
    using std::cout;
    using std::endl;
    cout << "TV is " << (state == Off? "Off" : "On") << endl;
    if (state == On)
    {
        cout << "Volume setting = " << volume << endl;
        cout << "Channel setting = " << channel << endl;
        cout << "Mode = "
             << (mode == Antenna? "antenna" : "cable") << endl;
        cout << "Input = "
             << (input == TV? "TV" : "DVD") << endl;
    }
}

inline void Tv::set_Rmode(Remote& r)
{
    r.work_mode = (r.work_mode == Normal) ? InterActive : Normal;
}
/*添加状态查询函数的定义*/

#include <iostream>
#include "tv.h"

//use_tv.cpp —— 使用 Tv 类和 Remote 类


int main()
{
    using std::cout;
    Tv s42;
    cout << "Initial settings for 42\" TV:\n";
    s42.settings();
    s42.onoff();
    s42.chanup();
```

```
    cout << "\nAdjusted settings for 42\" TV:\n";
    s42.settings();

    Remote grey;

    grey.set_chan(s42, 10);
    grey.volup(s42);
    grey.volup(s42);
    cout << "\n42\" settings after using remote:\n";
    s42.settings();

    Tv s58(Tv::On);
    s58.set_mode();
    grey.set_chan(s58,28);
    cout << "\n58\" settings:\n";
    s58.settings();
    return 0;
}
```

2．修改程序清单 15.11，使两种异常类型都是从头文件<stdexcept>提供的 logic_error 类派生出来的类。使每个 what()方法都报告函数名和问题的性质。异常对象不用存储错误的参数值，而只用支持 what()方法。

编程分析：

程序清单 15.11 定义了两个相互独立的类 bad_hmean 和 bad_gmean，并自定义了 mesg()函数，以实现信息输出功能。题目要求修改之前的代码，使两个类都从 logic_error 类派生。使用异常类派生的优势之一就是可以实现相对统一的错误消息提示功能。可以在构造函数内通过 what_arg 成员初始化，实现 what()函数的错误消息输出，从而简化类的定义。题目要求删除 bad_hmean 和 bad_gmean 类内存储错误参数值的数据成员。完整代码如下，重点在于继承中构造函数的初始化。

```
/*第 15 章的编程练习 2*/
#include <iostream>
#include <cmath>
#include <stdexcept>
using namespace std;

/*修改程序清单 15.10，直接从 logic_error 派生 bad_hmean 和 bad_gmean
 *利用构造函数初始化 waht_arg 参数，并通过 what()函数输出数据， 由于程序比较简单，
 *因此在一个文件内编译*/
class bad_hmean : public std::logic_error
{
public:
    bad_hmean(const string what_arg = "HMean, Invalid argument ") : logic_error(what_arg) {}
};
class bad_gmean : public std::logic_error
{
public:
```

```
    bad_gmean(const string what_arg = "GMean, Invalid argument ") : logic_error(what_arg) {}
};

double hmean(double a, double b);
double gmean(double a, double b);

int main()
{
    double x, y, z;

    cout << "Enter two numbers: ";
    while (cin >> x >> y)
    {
        try {
            z = hmean(x,y);
            cout << "Harmonic mean of " << x << " and " << y
                << " is " << z << endl;
            cout << "Geometric mean of " << x << " and " << y
                << " is " << gmean(x,y) << endl;
            cout << "Enter next set of numbers <q to quit>: ";
        }
        catch (bad_hmean & bg)
        {
            bg.what();
            cout << "Try again.\n";
            continue;
        }
        catch (bad_gmean & hg)
        {
            cout << hg.what();
            //cout << "Values used: " << hg.v1 << ", "
            // << hg.v2 << endl;
            cout << "Sorry, you don't get to play any more.\n";
            break;
        }
    }
    cout << "Bye!\n";
    return 0;
}

double hmean(double a, double b)
{
    if (a == -b)
        throw bad_hmean();
    return 2.0 *a *b / (a + b);
}

double gmean(double a, double b)
{
    if (a < 0 || b < 0)
```

```
        throw bad_gmean();
    return std::sqrt(a *b);
}
```

3．这个编程练习与编程练习 2 相同，但异常类是从一个这样的基类派生而来的，该基类是从 logic_error 派生而来的，并存储两个参数值。异常类应该有一个方法，用于报告这些值以及函数名。程序使用一个 catch 块来捕获基类的异常，其中任何一种从该基类的异常派生而来的异常都将导致循环结束。

编程分析：

在编程练习 2 的基础上，题目要求修改类的继承关系，bad_hmean 和 bad_gmean 两个类需要从同一个基类派生（该基类又是 logic_error 类的派生类），并添加两个数据成员存储错误参数，添加一个方法用于显示错误参数的信息。另外，题目要求使用一个 catch 块捕获所有的基类异常，因此应当使用基类对象捕获异常，并根据对象类型来进行具体异常的分析。即使用 RTTI 方法判断对象类型，并通过 dynamic_cast 进行类型转换，完成对应异常的处理工作。完整代码如下。

```
/*第 15 章的编程练习 3*/
#include <iostream>
#include <cmath>
using namespace std;

class unexpected_mean : public std::logic_error
{
private:
    double v1;
    double v2;
public:
    unexpected_mean(double a = 0, double b = 0, std::string s = "mean error")
          : v1(a), v2(b), logic_error(s) {}
    virtual void mesg() = 0;
    //virtual ~unexpected_mean() {}
};
/*首先定义一个基类 unexpected_mean ，从 logic_error 派生，添加两个
 *数据变量以描述错误消息，把成员函数 mesg()设置为虚函数，可以使用默认析构函数*/

class bad_hmean : public unexpected_mean
{
private:
public:
    bad_hmean(double a = 0, double b = 0, std::string s = "HMean")
            : unexpected_mean(a,b,s) {}
    void mesg();
};
/*定义 bad_hmean 类*/
class bad_gmean : public unexpected_mean
{
```

```
public:
    bad_gmean(double a = 0, double b = 0,std::string s = "GMean")
            : unexpected_mean(a,b,s){}
    void mesg();
};
/*定义 bad_gmean 类*/

inline void unexpected_mean::mesg()
{
    cout<<v1<<" "<<v2<<endl;
}


inline void bad_hmean::mesg()
{
    std::cout << "bad_HMean() now!" << std::endl;
    std::cout << what() << "\n";
    std::cout << "Hmean invalid arguments\n";
    unexpected_mean::mesg();
}
/*定义 mesg()函数，输出错误消息*/

inline void bad_gmean::mesg()
{
    std::cout << "bad_GMean() now!" << std::endl;
    std::cout << what() << "\n";
    std::cout << "Gmean invalid arguments\n";
    unexpected_mean::mesg();
}
/*定义 bad_gmean 类，定义 mesg()函数，输出错误消息*/

double hmean(double a, double b);
double gmean(double a, double b);
int main()
{
    double x, y, z;

    cout << "Enter two numbers: ";
    while (cin >> x >> y)
    {
        try {
            cout << "Harmonic mean of " << x << " and " << y
                 << " is " << hmean(x,y) << endl;
            cout << "Geometric mean of " << x << " and " << y
                 << " is " << gmean(x,y) << endl;
            cout << "Enter next set of numbers <q to quit>: ";
        }// try 块结束
        catch (unexpected_mean & bg)
        {
            if(typeid (bg) == typeid (bad_hmean))
```

```
            {
                bad_hmean*ph = dynamic_cast<bad_hmean*> (&bg);
                ph->mesg();
            }else if(typeid (bg) == typeid (bad_gmean))
            {
                bad_gmean*pg = dynamic_cast<bad_gmean*> (&bg);
                pg->mesg();
            }
            cout << "Try again.\n";
            continue;
        }
    }
    cout << "Bye!\n";
    return 0;
}

double hmean(double a, double b)
{
    if (a == -b)
        throw bad_hmean(a,b);
    return 2.0 *a *b / (a + b);
}

double gmean(double a, double b)
{
    if (a < 0 || b < 0)
        throw bad_gmean(a,b);
    return std::sqrt(a *b);
}
```

4. 程序清单 15.16 在每个 try 块后面都使用两个 catch 块，以确保 nbad_index 异常导致方法 label_val()被调用。请修改该程序，在每个 try 块后面只使用一个 catch 块，并使用 RTTI 来确保正确地调用 label_val()。

编程分析：

程序清单 15.16 通过嵌套类，在 Sales 类内定义了嵌套的异常类 bad_index；在 Sales 的派生类 LabeledSales 中通过嵌套类定义了异常类 nbad_index。nbad_index 又是 bad_index 的派生类，因此在 try...catch 语句中需要使用两个语句块分别捕获两个嵌套类的对象。由于两个类是派生关系，因此通过捕获基类对象并使用 RTTI 方法判断对象的具体类型，可以简化代码并正确调用 label_val()函数。完整代码如下。

```
/*第 15 章的编程练习 4*/
//sales.h
#include <stdexcept>
#include <string>

class Sales
{
```

```
public:
    enum {MONTHS = 12};
    class bad_index : public std::logic_error
    {
    private:
        int bi;
    public:
        explicit bad_index(int ix,
                           const std::string & s = "Index error in Sales object\n");
        int bi_val() const {return bi;}
        virtual ~bad_index() throw() {}
    };
    explicit Sales(int yy = 0);
    Sales(int yy, const double *gr, int n);
    virtual ~Sales() { }
    int Year() const { return year; }
    virtual double operator[](int i) const;
    virtual double & operator[](int i);
private:
    double gross[MONTHS];
    int year;
};

class LabeledSales : public Sales
{
public:
    class nbad_index : public Sales::bad_index
    {
    private:
        std::string lbl;
    public:
        nbad_index(const std::string & lb, int ix,
                   const std::string & s = "Index error in LabeledSales object\n");
        const std::string & label_val() const {return lbl;}
        virtual ~nbad_index() throw() {}
    };
    explicit LabeledSales(const std::string & lb = "none", int yy = 0);
    LabeledSales(const std::string & lb, int yy, const double *gr, int n);
    virtual ~LabeledSales() { }
    const std::string & Label() const {return label;}
    virtual double operator[](int i) const;
    virtual double & operator[](int i);
private:
    std::string label;
};

//sales.cpp

#include "sales.h"
using std::string;
```

```
Sales::bad_index::bad_index(int ix, const string & s )
        : std::logic_error(s), bi(ix)
{
}

Sales::Sales(int yy)
{
    year = yy;
    for (int i = 0; i < MONTHS; ++i)
        gross[i] = 0;
}

Sales::Sales(int yy, const double *gr, int n)
{
    year = yy;
    int lim = (n < MONTHS)? n : MONTHS;
    int i;
    for (i = 0; i < lim; ++i)
        gross[i] = gr[i];
    //for i > n and i < MONTHS
    for ( ; i < MONTHS; ++i)
        gross[i] = 0;
}

double Sales::operator[](int i) const
{
    if(i < 0 || i >= MONTHS)
        throw bad_index(i);
    return gross[i];
}

double & Sales::operator[](int i)
{
    if(i < 0 || i >= MONTHS)
       throw bad_index(i);
    return gross[i];
}

LabeledSales::nbad_index::nbad_index(const string & lb, int ix,
                                     const string & s ) : Sales::bad_index(ix, s)
{
    lbl = lb;
}

LabeledSales::LabeledSales(const string & lb, int yy)
        : Sales(yy)
{
    label = lb;
}
```

```
LabeledSales::LabeledSales(const string & lb, int yy, const double *gr, int n)
        : Sales(yy, gr, n)
{
    label = lb;
}

double LabeledSales::operator[](int i) const
{   if(i < 0 || i >= MONTHS)
        throw nbad_index(Label(), i);
    return Sales::operator[](i);
}

double & LabeledSales::operator[](int i)
{
    if(i < 0 || i >= MONTHS)
        throw nbad_index(Label(), i);
    return Sales::operator[](i);
}

//
//use_sales.cpp ——嵌套的异常
#include <iostream>
#include "sales.h"

int main()
{
    using std::cout;
    using std::cin;
    using std::endl;

    double vals1[12] =
        {
        1220, 1100, 1122, 2212, 1232, 2334,
        2884, 2393, 3302, 2922, 3002, 3544
        };

    double vals2[12] =
        {
        12, 11, 22, 21, 32, 34,
        28, 29, 33, 29, 32, 35
        };

    Sales sales1(2011, vals1, 12);
    LabeledSales sales2("Blogstar",2012, vals2, 12 );

    cout << "First try block:\n";
    try
    {
      int i;
```

```
    cout << "Year = " << sales1.Year() << endl;
    for (i = 0; i < 12; ++i)
    {

        cout << sales1[i] << ' ';
        if (i % 6 == 5)
            cout << endl;
    }
    cout << "Year = " << sales2.Year() << endl;
    cout << "Label = " << sales2.Label() << endl;
    for (i = 0; i <= 12; ++i)
    {

        cout << sales2[i] << ' ';
        if (i % 6 == 5)
            cout << endl;
    }
    cout << "End of try block 1.\n";
}
catch(Sales::bad_index & bad)
/*捕获基类的对象*/
{
    if(typeid(LabeledSales::nbad_index)  == typeid(bad))
    {
        LabeledSales::nbad_index*p = dynamic_cast<LabeledSales::nbad_index*> (&bad);
    /*通过typeid()识别捕获对象的类型，并特殊化处理*/
        cout << p->what();
        cout << "Company: " << p->label_val() << endl;
        cout << "bad index: " << p->bi_val() << endl;
    }
    else if(typeid(Sales::bad_index) == typeid(bad))
    {
    /*通过typeid()识别捕获对象的类型，并特殊化处理*/
        cout << bad.what();
        cout << "bad index: " << bad.bi_val() << endl;
    }

}
cout << "\nNext try block:\n";
try
{
    sales2[2] = 37.5;
    sales1[20] = 23345;
    cout << "End of try block 2.\n";
}
catch(Sales::bad_index & bad)
{
    if(typeid(LabeledSales::nbad_index)  == typeid(bad))
    {
        LabeledSales::nbad_index*p = dynamic_cast<LabeledSales::nbad_index*> (&bad);
```

```
            cout << p->what();
            cout << "Company: " << p->label_val() << endl;
            cout << "bad index: " << p->bi_val() << endl;
        }
        else if(typeid(Sales::bad_index) == typeid(bad))
        {
            cout << bad.what();
            cout << "bad index: " << bad.bi_val() << endl;
        }

    }
    cout << "done\n";
    return 0;
}
```

第 16 章　string 类和标准模板库

本章知识点总结

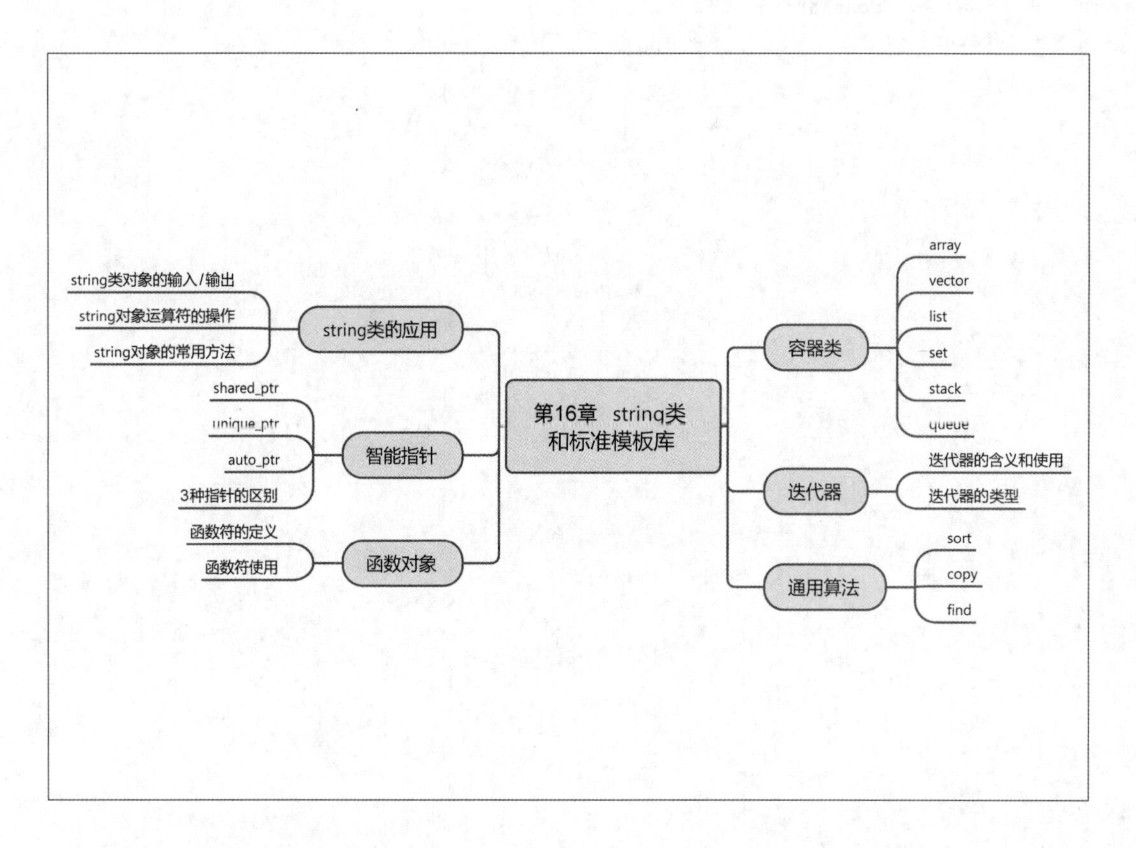

本章主要介绍 C++中预定义的标准模板库（Standard Template Library，STL），熟练掌握这些标准库能够为常见编程问题提供一个安全快速的解决方案。

16.1　C++中的 string 类

与 C 风格字符串不同，string 类为将字符串作为对象处理提供了一种方便的方法，实现了自动内存管理功能以及众多处理字符串的方法和函数。通过 string 类的对象能够实现字符串合并、判断相等、字符串插入、字符串反转、字符搜索或子串查找等操作。对比 C 风格字符串，string 类在 I/O 读取过程中，需要使用系统中的 getline()函数，而不能直接使用 cin

对象的 getline()方法。此外，使用 string 类对象的重要优势是在类的定义中包含 string 对象（作为数据成员），简化使用字符指针的各种动态存储管理的复杂操作。在程序设计中可以通过手册查询 string 的相关功能。下面是 string 类一些常用的成员函数。

- find()：从前往后查找子串或字符出现的位置。
- rfind()：从后往前查找子串或字符出现的位置。
- replace()：对 string 对象中的子串进行替换。
- insert()：在 string 对象中插入另一个字符串。
- erase()：删除 string 对象中的子串。

16.2　智能指针模板类

智能指针是行为类似于指针的类对象。C++标准库中定义了 3 个智能指针模板——auto_ptr、unique_ptr 和 shared_ptr。这 3 个智能指针在使用中与普通指针类似，直接通过赋值获得 new 创建的对象的地址。但是和普通指针需要手动使用 delete 释放对象的存储空间不同，当智能指针过期时，其析构函数将自动使用 delete 来释放该指针指向对象的内存。因此，在使用智能指针时无须手动释放内存，这样在各类异常发生时智能指针能够保证系统内存不发生泄露。智能指针的基本使用方式如下。

```
#include <memory> //智能指针的头文件
auto_ptr<type_name> pa(new type_name);
unique_ptr<type_name> pu(new type_name);
shared_ptr<type_name> ps(new type_name);
```

其中，auto_ptr 在 C++11 中已摒弃，因此在使用 C++11 进行编译时应尽量使用 unioque_ptr 和 shared_ptr 两种智能指针。

智能指针的差别在于，如果程序要使用多个指向同一个对象的指针，应选择 shared_ptr（可以理解为多个指针共享一个对象）。常见的应用场景包括：

- 有一个指针数组，并使用一些辅助指针来标识特定的元素；
- 两个对象都包含指向第 3 个对象的指针；
- STL 容器包含指针。

如果程序不需要多个指向同一个对象的指针，则推荐使用 unique_ptr（可以理解为一个指针对应一个对象）。本质上说，智能指针是建立在所有权概念的基础上的。对于特定的对象，只能有一个智能指针拥有它，这样只有拥有对象的智能指针的构造函数会删除该对象。系统可以通过赋值操作转让所有权。auto_ptr 和 unique_ptr 使用这种相对严格的操作策略。shared_ptr 指针通过跟踪引用特定对象的智能指针数来记录指向该对象的指针个数，并在计数为 0 时回收该对象。

16.3　STL 中的容器类

STL 是 C++语言提供的一组标准化模板库，它主要提供了容器类模板、迭代器类模板、函数对象模板和算法函数模板的集合。程序设计中可以直接使用 STL 工具来解决编程问题，也可以使用它们作为基本部件来解决程序中的其他问题。

STL 中的容器是存储相同类型数据对象的一个大型存储单元（容器），单元内的数据对象按照一定的规则存储。通过相应的容器类方法，用户可以对容器内的数据进行管理和维护。和容器相关的迭代器是一个广义上的指针对象，通过它能够像普通指针那样访问容器内的数据对象。STL 使得用户能够快速构造各种数据类型的容器（包括数组、队列和链表），通过容器和迭代器能够执行各种数据操作（包括搜索、排序和随机排列）。STL 定义了多种容器类模板——vector、deque、list、set、multiset、map、multimap 和 bitset，还定义了适配器类模板——queue、priority_queue 和 stack。此外，为了更好地实现数据的操作，STL 还定义了大量的通用算法。STL 通用算法将迭代器作为容器和通用算法之间的接口，从而不必为每种容器提供一个专用操作方法。通过不同数据类型的容器和相关的迭代器实现通用算法。以 vector 模板类为例，以下语句定义了一个 double 类型的矢量容器 scores，并获得了一个指向第 1 个元素的迭代器。

```
vector<double> scores;
vector<double>::iterator pd = scores.begin();
```

使用这个迭代器，可以对整个容器内的数据进行各类操作。例如，通过以下语句，可以遍历容器内的元素。

```
for(pd = scores.begin();pd != scores.end();pd++){}
```

各种通用函数也可以根据输入不同容器的迭代器参数进行相应处理。例如，以下语句将会完成容器内元素的排序，并通过函数指针实现每个元素的对应操作。

```
sort(scores.begin(), scores.end());
for_each(scores.begin(), scores.end(),MyFunction);
```

STL 中的容器类模板较多，其中常用的方法如表 16.1 所示。

表 16.1　容器类模板常用的方法

方法	说明
begin()	返回一个指向第 1 个元素的迭代器
end()	返回一个指向超尾的迭代器
size()	返回元素数目
swap()	交换两个容器的内容

针对每一个特定容器类还有特有的方法，如 insert()、resize()等，可以在应用中查询相关手册。

16.4 STL 中的迭代器和通用算法

STL 中定义了 5 种迭代器，分别是输入迭代器、输出迭代器、正向迭代器、双向迭代器和随机访问迭代器。每一种迭代器都可以执行解除引用的操作（即通过*运算符访问指向对象），也可执行判断相等的操作。

- 输入迭代器：可用来读取容器中的信息。即输入迭代器解除引用将使程序能够读取容器中的值，但不一定能够使程序修改值。
- 输出迭代器：与输入迭代器相似，只是输出迭代器解除引用使程序能够修改容器值，而不能读取。
- 正向迭代器：只使用++运算符来遍历容器，所以它每次沿容器向前移动一个元素。正向迭代器可以读取和修改数据，也可以只读取数据。
- 双向迭代器：同时支持递增和递减运算符。
- 随机访问迭代器：能够直接跳转到容器中的任何一个元素，随机访问迭代器具有双向迭代器的所有特性，同时支持随机访问。

各种迭代器的类型并不是确定的，而只是一种概念性描述，需要在应用中灵活处理。STL 中每个容器类都定义了一个类级 typedef 名称——iterator，例如，vector<int>类的迭代器类型为 vector<int> :: iterator，list<int>类的迭代器类型为 list<int> :: iterator。但是程序设计中很多时候直接使用 auto 类型来进行迭代器的声明。

STL 预定义了处理容器的非成员函数，包括 sort()、copy()、find()等。使用这些通用算法模板提供泛型，使用迭代器来标识要处理的容器内的数据。容器类和迭代器的设计使得不同类型的容器可以实现跨容器的数据操作，例如，可以使用 copy()将 vector 对象中的值复制到 list 对象中。STL 将算法库主要分为非修改式序列操作、修改式序列操作、排序和通用数字运算 4 种。STL 还提供了函数对象（函数符），函数对象是重载了()运算符（即 operator()()方法）的类。函数符可以使用函数表示法来调用这种类对象。由于 STL 模板库中容器和算法众多，因此可以在程序设计中查询相关手册。

16.5　复习题

1．考虑下面的类声明，若将它转换为使用 string 对象的声明，哪些方法不再需要显式定义?

```
class RQ1 {
private:
    char *st;
public:
    RQ1() {st = new char[1];strcpy(st, "");}
    RQ1(const char *s) {
        st = new char[strlen(s) + 1];strcpy(st, s);}
    RQ1(const RQ1 &rq)
    {st = new char[strlen(rq.st) + 1];strcpy(st, rq.st);}
    ~RQ1() { delete[] st };
    RQ1 &operator=(const RQ1 &rq);

};
```

习题解析：

本题要求将字符指针实现的字符串转换成 string 类型的字符串，由于 string 类已经定义了相关构造函数，实现了动态的存储管理，重载了赋值运算符，完成了析构函数的存储空间回收，因此修改为 string 类型之后，类 RQ1 不需要用户自定义复制构造函数、析构程序和赋值运算符。可以修改为如下代码。

```
class RQ1 {
private:
    string st;
public:
    RQ1() :st(""){ }
    RQ1(const char *s):st(s) { }
    ~RQ1() { };

};
```

2．在易于使用方面，指出 string 对象至少两个优于 C 风格字符串的地方。

习题解析：

string 类是 C++语言中定义的字符串类，相比 C 风格字符串，string 类对象通过类内的构造函数、析构函数、赋值运算符以及相关的大量公有接口，能够让用户更简单地使用字符串，而不用考虑动态存储管理、类内部的深拷贝等问题。例如，可以将一个 string 对象通过赋值运算符 "=" 直接赋给另一个 string 对象；string 对象提供了自己的内存管理功能，所以一般不需要担心字符串超出存储容量。

3．编写一个函数，用 string 对象作为参数，将 string 对象转换为全部大写。

习题解析：

string 类内没有定义转换大小写的公有接口，因此必须通过修改 string 对象的每一个字符实现。要访问 string 对象的单个字符，可以使用字符数组的形式实现，即通过运算符"[]"获取和修改 string 对象的单个字符。程序还需要使用 cctype 头文件中的字符转换函数 toupper()。完整代码如下。

```
#include <iostream>
#include <cctype>
void ToUpper(string& st)
{
    for(int i = 0; i < st.size(); i++ )
        st[i] = =toupper(st[i])
}
```

4．从概念上或语法上说，下面哪个不是正确使用 auto_ptr 的方法（假设已经包含了所需的头文件）？

- `auto_ptr<int> pia(new int[20]);`
- `auto_ptr<string> (new string) ;`
- `int rigue = 7;`

```
auto_ptr<int>pr (&rigue) ;
```

- `auto_ptr dbl (new double) ;`

习题解析：

- `auto_ptr<int> pia(new int(20));`

模板 auto_ptr 使用 delete 而不是 delete []，因此只能与 new 一起使用，而不能与 new [] 一起使用。

- `auto_ptr<string> ps (new string) ;`

缺少智能指针变量名，因此添加智能指针变量名 ps。

- ```
  int rigue = 7;
  auto_ptr<int>pr (&rigue) ;
  ```

智能指针 auto_ptr 必须指向 new 生成的新的内存区域，不能使用已有变量的存储地址。

- `auto_ptr<double>dbl (new double) ;`

模板中缺少表示 double 类型的"<>"，修改为 `auto_ptr<double>dbl (new double);`。

5．如果可以生成一个存储高尔夫球棍（而不是数字）的栈，为何它（从概念上说）是一个坏的高尔夫袋子？

习题解析：

栈的首要特性是先入后出（FILO）。先入后出表示首先入栈的数据会最后从栈内取出，利用该特性存储高尔夫球棍，意味着可能会反复使用栈顶部（最后入栈）的部分球杆。如果需要使用最先入栈的球棍，也必须拿出前面的所有球棍，这样的处理过程非常不适合高尔夫球运动。

6．若逐洞记录高尔夫成绩，为什么说 set 容器是糟糕的选择？

习题解析：

STL 标注模板库中的集合（set）容器的重要特点是不保存重复数据，即集合的多个相同数据将会只保存一个副本，因此高尔夫球记分系统中相同的分数将无法体现，set 容器并不适合保存成绩，但是很多时候我们也可以利用该特性，实现重复数据的合并功能。

7．既然指针是一个迭代器，为什么 STL 设计人员没有简单地使用指针来代替迭代器呢？

习题解析：

C++语言的 STL 中，指针可以作为一个迭代器进行运算。也可以认为迭代器是一种广义上的指针，但是并不能简单认为可以用指针代替迭代器来设计 STL。这是因为 STL 中容
  ```

器类和迭代器的主要目的是实现一种泛型编程，指针无法适应多种容器类内部的多种复杂数据类型和数据的组织形式，并且特定类型的指针和泛型编程的目标相违背。

8. 为什么 STL 设计人员仅定义了迭代器基类，而使用继承来派生其他迭代器类型的类，并根据这些迭代器类来表示算法？

习题解析：

C++语言的继承功能和 RTTI 技术使得迭代器基类与其他派生类型的迭代器在通用算法中能够实现完整的功能，因此可以将 STL 函数用于指向常规数组的常规指针以及指向 STL 容器类的迭代器，进行统一化处理，提高整个 STL 模板的通用性。

9．给出 vector 对象比常规数组方便的 3 个例子。

习题解析：

STL 中容器类的重要特点是可以创建多种类型数据的统一化容器，并通过公有接口进行数据的维护和管理，例如，可以将一个 vector 对象通过赋值语句赋值给另一个 vector 对象，vector 模板能够自己管理的内存，可以通过公有接口将元素直接插入向量中，vector 会自动调整长度不用担心数组越界。

10．如果程序清单 16.9 是使用 list（而不是 vector）实现的，则该程序的哪些部分将是非法的？非法部分能够轻松修复吗？如果可以，如何修复呢？

习题解析：

程序清单 16.9 使用 vector 类型的数据来存储书籍的基本信息。程序中使用 sort()函数与 for_each()函数进行排序和遍历，可以使用 list 容器的成员函数替换，但是 random_shuffle()函数要求使用随机访问迭代器，而 list 对象只有双向迭代器，因此无法实现这种乱序访问。如果使用 list 容器，也可以通过容器复制，将 list 复制到 vector 进行数据的 random_shuffle()处理，再重新将数据复制回 list，这样也可以间接实现该功能。

11．假设有程序清单 16.15 所示的函数符 TooBig，下面的代码有何功能？赋给 bo 的是什么值？

```
bool bo = TooBig<int>(10)(15);
```

习题解析：

原 TooBig 模板类的定义如下。

```
template<class T>  //定义 operator()() 的函数类
class TooBig
{
private:
    T cutoff;
public:
```

```
    TooBig(const T & t) : cutoff(t) {}
    bool operator()(const T & v) { return v > cutoff; }
};
```

TooBig<int>(10)语句首先生成一个匿名函数对象，其初始化数据成员 cutoff 为 10，随后该对象调用了 operator()方法，参数是 15，该方法通过 **return** v > cutoff 语句判断输入参数 15 大于 cutoff，因此返回 true，所以布尔类型变量 bo 通过赋值得到 true。

16.6　编程练习

1. 回文指的是顺读和逆读都一样的字符串。例如，“tot”和“otto”都是简短的回文。编写一个程序，要求用户输入字符串，并将字符串引用传递给一个布尔函数。如果字符串是回文，该函数将返回 true；否则，返回 false。此时，不要担心诸如大小写、空格和标点符号这些复杂的问题。即这个简单的版本将拒绝“Otto”和“Madam，I'm Adam”。请查看《C++ Primer Plus（第 6 版）中文版》附录 F 中的字符串方法列表，以简化这项任务。

编程分析：

要判断字符串是否是回文，分析方法是将字符串翻转，并判断翻转前后字符串是否相等。如果相等，就是回文字符串；否则，不是。题目不要求考虑大小写字符和标点符号，因此该算法符合题目要求。字符串的翻转可以使用 STL 函数中的 reverse()函数实现（也可以通过自定义的循环进行翻转，两者效果相同）。完整代码如下。

```
/*第 16 章的编程练习 1*/
#include <iostream>
using namespace std;

bool palindromic(string& s);

int main()
{
    string st;
    cout<<"Enter the string to test: ";
    getline(cin, st);
    cout<<st<<endl;
    cout<<palindromic(st);
    return 0;
}

bool palindromic(string& s)
{
    string temp = s;
    reverse(temp.begin(), temp.end());
    /*翻转输入字符串 s，并进行比较，如果需要自己实现字符串翻转，
     *可以使用循环，交换前半部分与后半部分的字符*/
    return (s == temp);
```

```
    /*翻转前后相等，返回 true*/
}
```

2．与编程练习 1 中给出的问题相同，但要考虑诸如大小写、空格和标点符号这样的复杂问题。即“Madam，I'm Adam”不是回文。例如，测试函数可能会将字符串缩写为“madamimadam”，然后测试倒过来是否一样。不要忘了 cctype 库，你可能从中找到几个有用的 STL 函数，尽管不一定非要使用它们。

编程分析：

与编程练习 1 比较，本题需要考虑字符的大小写和标点符号。在使用字符串时，首先需要过滤标点符号，并转换大小写。基本算法是可以通过字符串处理生成不包含标点符号且统一大小写的临时字符串，并使用编程练习 1 的算法比较。也可以通过回文的特点，通过比较头尾字符来实现。下面使用头尾字符比较的算法进行回文的判断。

```
/*第 16 章的编程练习 2*/
#include <iostream>
#include <cctype>
using namespace std;

bool palindromic(string& s);

int main()
{
    string st;
    cout<<"Enter the string to test: ";
    getline(cin, st);
    cout<<"String "<<st<<" is ";
    if(palindromic(st))
        cout<<"a palindromic string. "<<endl;
    else
        cout<<"not a palindromic string."<<endl;

}
bool palindromic(string& s)
{
    /*该算法忽略了全部非字母的情况，如果要排除这种情况，需要添加一个判断标识*/
    //std::string::iterator phead, ptail;
    auto phead = s.begin();
    auto ptail = s.end();
    /*可以使用两种方式定义头尾的迭代器*/
    while(ptail > phead){
        if(!isalpha(*phead))
        {
            phead++;
            continue;
        }
        /*忽略头部的非字母字符*/
        if(!isalpha(*ptail))
```

```
        {
           ptail--;
           continue;
        }
        /*忽略尾部的非字母字符*/
        if(toupper(*phead) == toupper(*ptail))
        {
            phead++;
            ptail--;
        }else
        {
            return false;
        }
    }
    return true;
}
```

3．修改程序清单 16.3，使之从文件中读取单词。一种方案是使用 vector<string>对象，而不是 string 数组。这样便可以使用 push_back()将数据文件中的单词复制到 vector<string>对象中，并使用 size()来确定单词列表的长度。由于程序应该每次从文件中读取一个单词，因此应使用运算符>>，而不是 getline()。文件中包含的单词应该用空格、制表符或换行符分隔。

编程分析：

在从文件中读取数据时，getline()读取到换行符，运算符>>读取到第 1 个空白字符，因此，要读取单词，应当使用运算符>>。题目要求将读取到的数据存储到 vector 中，并通过 vector 判断文件的单词数量。在程序清单 16.3 的基础上添加相应功能，完整代码如下。

```
/*第 16 章的编程练习 3*/
//hangman.cpp —— string 对象的一些方法
#include <iostream>
#include <string>
#include <cstdlib>
#include <ctime>
#include <cctype>
#include <fstream>
#include <vector>
/*添加相应的头文件 fstream 和 vector*/
const int NUM = 26;

int main()
{
    using std::cout;
    using std::cin;
    using std::tolower;
    using std::endl;
    using std::string;
/*原来使用 using 声明，可以修改为 using 编译指令*/
```

```
/*添加文件读取功能，创建文件对象，从文件内读取
 *相应的单词 ，存入 string 的 vector 对象内*/
    std::ifstream fin;
    fin.open("wordlist.txt", std::ifstream::in);
    if(!fin.is_open())
    {
        cerr<<"Can't open file wordlist.txt."<<endl;
        exit(EXIT_FAILURE);
    }
    string word;
    std::vector<string> wordlist;
    if(fin.good())
    {
        while(fin >> word)
            wordlist.push_back(word);
    }
    int length = wordlist.size();
    fin.close();
/*数据读取完*/
    std::srand(std::time(0));
    char play;
    cout << "Will you play a word game? <y/n> ";
    cin >> play;
    play = tolower(play);
    while (play == 'y')
    {
        string target = wordlist[std::rand() % NUM];
        int length = target.length();
        string attempt(length, '-');
        string badchars;
        int guesses = 6;
        cout << "Guess my secret word. It has " << length
             << " letters, and you guess\n"
             << "one letter at a time. You get " << guesses
             << " wrong guesses.\n";
        cout << "Your word: " << attempt << endl;
        while (guesses > 0 && attempt != target)
        {
            char letter;
            cout << "Guess a letter: ";
            cin >> letter;
            if (badchars.find(letter) != string::npos
                || attempt.find(letter) != string::npos)
            {
                cout << "You already guessed that. Try again.\n";
                continue;
            }
            int loc = target.find(letter);
            if (loc == string::npos)
            {
```

```
                cout << "Oh, bad guess!\n";
                --guesses;
                badchars += letter; //add to string
            }
            else
            {
                cout << "Good guess!\n";
                attempt[loc]=letter;
                //check if letter appears again
                loc = target.find(letter, loc + 1);
                while (loc != string::npos)
                {
                    attempt[loc]=letter;
                    loc = target.find(letter, loc + 1);
                }
            }
            cout << "Your word: " << attempt << endl;
            if (attempt != target)
            {
                if (badchars.length() > 0)
                    cout << "Bad choices: " << badchars << endl;
                cout << guesses << " bad guesses left\n";
            }
        }
        if (guesses > 0)
            cout << "That's right!\n";
        else
            cout << "Sorry, the word is " << target << ".\n";

        cout << "Will you play another? <y/n> ";
        cin >> play;
        play = tolower(play);
    }

    cout << "Bye\n";
    return 0;
}
```

4．编写一个具有老式接口的函数，其原型如下。

```
int reduce(long ar[], int n);
```

实参应是数组名和数组中的元素个数。该函数对数组进行排序，删除重复的值，返回缩减后数组中的元素数目。请使用 STL 函数编写该函数（如果决定使用通用的 unique()函数，请注意它将返回结果区间的结尾）。使用一个小程序测试该函数。

编程分析：

reduce()函数处理数组参数，对其进行排序并删除重复数据，如果使用普通算法进行排序并删除数据，排序可以在原数组中进行，删除数据则可能需要使用临时数组，整体算法和程序设计比较复杂。如果使用 STL，可以使用 list 容器进行排序，使用 unique( )公用接口删除重复数据，这样程序会更加简单。完整代码如下。

```
/*第 16 章的编程练习 4*/
#include <iostream>
#include <list>

int reduce(long ar[], int n);
int main()
{
    long ar1[5] = {45000, 3400, 45000, 100000, 2500};
    int resize = reduce(ar1, 5);
    std::cout << "array1: \n";
    for (int i = 0; i < resize; i++)
    {
        std::cout << ar1[i] << " ";
    }
    return 0;
}

int reduce(long ar[], int n)
{
    std::list<long> ls;
    ls.insert(ls.end(),ar,ar + n);
    /*将数组的内容复制到 list 末尾*/
    ls.sort();
    ls.unique();
    /*利用 list 的排序和删除重复数据的函数*/
    auto pd = ls.begin();
    for (int i = 0; i < ls.size(); i++, pd++)
       ar[i] = *pd;
    /*将处理完的数据复制回数组中，注意，list 长度已改变*/
    return ls.size();
}
```

5．问题与编程练习 4 相同，但要编写一个模板函数。

```
template <class T>
int reduce(T ar[], int n);
```

在一个使用 long 实例和 string 实例的小程序中测试该函数。

编程分析：

题目要求实现一个模板函数，处理数组的重复元素并实现排序功能。针对函数模板的设计需求，使用编程练习 4 中的算法实现，即使用 list 容器和迭代器实现模板函数。完整代码如下。

```
/*第 16 章的编程练习 5*/
#include <iostream>
#include <list>
/*模板函数的定义*/
template <typename T>
int reduce(T ar[], int n)
```

```
{
    std::list<T> ls;
    ls.insert(ls.end(), ar, ar + n);
    /*创建 T 类型的 list，并将数据复制到 list 中*/
    ls.sort();
    ls.unique();
    /*排序，删除重复数据*/
    auto pd = ls.begin();
    for (int i = 0; i < ls.size(); i++, pd++)
        ar[i] = *pd;
    return ls.size();
}
int main()
{
    long ar1[5] = {45000, 3400, 45000, 100000, 2500};
    int resize = reduce(ar1, sizeof(ar1)/sizeof(long));
    std::cout << "array1: \n";
    int i;
    for (i = 0; i < resize; i++)
    {
        std::cout << ar1[i] << " ";
    }
    std::string ar2[6] = {"it", "aboard", "it", "zone", "quit", "aa"};
    resize = reduce(ar2, sizeof(ar2)/sizeof(std::string));
    std::cout << "\narray2: \n";
    for (i = 0; i < resize; i++)
    {
        std::cout << ar2[i] << " ";
    }
    return 0;
}
```

6. 使用 STL queue 模板类而不是第 12 章的 Queue 类，重写程序清单 12.12。

编程分析：

程序清单 12.12 的主要功能是用 Queue 类模拟银行排队等候任务。在该程序中 Queue 的主要功能是模拟实现指定长度的入队、出队功能，主要用到的函数是 enqueue()、dequeue()，及队列的判空和判满函数。STL 中 queue 模板类完全兼容原有 Queue 类的所有功能，只需要在函数上稍做替换。替换后完整代码如下。

```
/*第 16 章的编程练习 6*/
//bank.cpp
//通过 queue.cpp 进行编译
#include <iostream>
#include <cstdlib>
#include <ctime>
#include <queue>

const int MIN_PER_HR = 60;
```

```
class Customer
{
private:
    long arrive;
    int processtime;
public:
    Customer() : arrive(0), processtime (0){}
    void set(long when);
    long when() const { return arrive; }
    int ptime() const { return processtime; }
};
void Customer::set(long when)
{
    processtime = std::rand() % 3 + 1;
    arrive = when;
}

typedef Customer Item;
/*声明 Customer 为 Item 类型*/
/*使用 queue 模板，因此可以删除自定义 queue 类型的头文件和 cpp*/
bool newcustomer(double x);

int main()
{
    using std::cin;
    using std::cout;
    using std::endl;
    using std::ios_base;
    std::srand(std::time(0));

    cout << "Case Study: Bank of Heather Automatic Teller\n";
    cout << "Enter maximum size of queue: ";
    int qs;
    cin >> qs;
    /*应用 queue 模板定义变量 line、元素 item 的类型*/
    std::queue<Item>line ;

    cout << "Enter the number of simulation hours: ";
    int hours;
    cin >> hours;

    long cyclelimit = MIN_PER_HR *hours;

    cout << "Enter the average number of customers per hour: ";
    double perhour;
    cin >> perhour;
    double min_per_cust;
    min_per_cust = MIN_PER_HR / perhour;

    Item temp;
```

```
long turnaways = 0;
long customers = 0;
long served = 0;
long sum_line = 0;
int wait_time = 0;
long line_wait = 0;

for (int cycle = 0; cycle < cyclelimit; cycle++)
{
    if (newcustomer(min_per_cust))
    {
        if (line.size() >= qs)
            /*调用 queue 的函数 size() ，判断其元素数是否大于最大数量 qs，
             *不需要使用原来自定义 isfull()函数来判断*/
            turnaways++;
        else
        {
            customers++;
            temp.set(cycle);    //cycle = time of arrival
            /*push()函数加数据入队列*/
            line.push(temp);
        }
    }
    if (wait_time <= 0 && !line.empty())
    {
        /*pop()函数 将队列元素取出*/
        line.pop();
        wait_time = temp.ptime();
        line_wait += cycle - temp.when();
        served++;
    }
    if (wait_time > 0)
        wait_time--;
    sum_line += line.size();
}


if (customers > 0)
{
    cout << "customers accepted: " << customers << endl;
    cout << "  customers served: " << served << endl;
    cout << "         turnaways: " << turnaways << endl;
    cout << "average queue size: ";
    cout.precision(2);
    cout.setf(ios_base::fixed, ios_base::floatfield);
    cout << (double) sum_line / cyclelimit << endl;
    cout << " average wait time: "
        << (double) line_wait / served << " minutes\n";
}
else
```

```
        cout << "No customers!\n";
    cout << "Done!\n";
    return 0;
}


bool newcustomer(double x)
{
    return (std::rand() *x / RAND_MAX < 1);
}
```

7. 彩票卡是一个常见的游戏。卡片上是带编号的圆点，其中一些圆点被随机选中。编写一个 lotto()函数，它接受两个参数。第 1 个参数是彩票卡上圆点的个数，第 2 个参数是随机选择的圆点个数。该函数返回一个 vector<int>对象，其中包含（按排列后的顺序）随机选择的号码。例如，可以按下面这样使用该函数：

```
vector<int> winners;
winners = Lotto(51,6);
```

这样将把一个向量赋给 winner，该向量包含 1～51 中随机选定的 6 个数字。注意，仅仅使用 rand()无法完成这项任务，因为它会生成重复的值。提示：由函数创建一个包含所有可能值的向量，使用 random_shuffle()，然后通过打乱后的向量的第 1 个值来获取值。编写一个小程序来测试这个函数。

编程分析：

题目要求设计一个彩票抽奖函数，函数返回值为 vector<int>类型的对象，参数是彩票卡上的圆点个数和随机选择的圆点数量。lotto()函数的原型如下。

```
std::vector<int> Lotto(int, int);
```

返回的向量中包含 6 个随机数字，为了保证数字的随机性，要求在函数内部多次使用随机生成的向量数据，再通过 random_shuffle()函数进行乱序，并依次选择第 1 个数据来生成最后的返回值。完整代码如下。

```
/*第 16 章的编程练习 7*/
#include <iostream>
#include <vector>
#include <cstdlib>
#include <ctime>
using namespace std;

vector<int> Lotto(int dot, int sdot);
/*声明彩票函数，返回值是 vector<int>对象*/
int main()
{
    vector<int> winners;
    winners = Lotto(51,6);
    vector<int>::iterator pd;
    cout << "winners: \n";
    for (pd = winners.begin(); pd != winners.end(); pd++)
```

```
        cout << *pd << " ";
    return 0;
}

vector<int> Lotto(int dot, int sdot)
{
    vector<int> result, temp;
    /*定义两个 vector<int>对象，一个用于生成临时数据*/
    srand(time(0));
    for(int i = 0; i < sdot; i++)
    {
        for(int j = 0; j < dot; j++)
            temp.push_back(rand()%dot);
        random_shuffle(temp.begin(), temp.end());
        /*temp 用于存储临时数据，乱序后，取出第一个元素，转储到 result 中，
         *从而实现更加优质的随机数据*/
        result.push_back(*temp.begin());
    }
    return result;
}
```

8. Mat 和 Pat 希望邀请他们的朋友来参加派对。他们要编写一个程序来完成下面的任务。

- 让 Mat 输入他朋友的姓名列表。首先把姓名存储在一个容器中，然后按排列后的顺序显示出来。
- 让 Pat 输入她朋友的姓名列表。首先把姓名存储在另一个容器中，然后按排列后的顺序显示出来。
- 创建第 3 个容器，将两个列表合并，删除重复的部分，并显示这个容器的内容。

编程分析：

Mat 和 Pat 的好友名单分别存储在一个容器中，该名单需要排序并显示。合并后的总名单中需要删除重复数据，因此可以选择 list 或者 set 容器类存储名单。下面的例子使用 set 容器的主要原因是 set 中的元素是无重复的，因此可以不使用 list 的 unique()函数进行重复元素的删除。

```
/*第 16 章的编程练习 8*/
#include <iostream>
#include <set>
#include <string>

using namespace std;
int main()
{
    set<string> Mat_set, Pat_set, Guest_set;
    /*使用集合模板定义 3 个对象，分别用于存储个人数据和合并后的数据*/
    cout << "Enter Mat's friends(q to quit): ";
    string name;
```

```
    while(getline(cin, name) && name != "q")
    {
        Mat_set.insert(name);
        cout<<name<<" add to Mat's list. (q to quit):";
    }
    /*添加姓名到 Mat 的名单中*/
    cout << "\nMat's friends are: \n";
    for(auto pd = Mat_set.begin(); pd != Mat_set.end(); pd++)
         cout << *pd << " ";

    cout << "Enter Pat's friends(q to quit): ";
    while(getline(cin, name) && name != "q")
    {
        Pat_set.insert(name);
        cout<<name<<" add to Pat's list. (q to quit):";
    }
    /*添加姓名到 Pat 的名单中*/

    cout << "\nPat's friends are: \n";
    for(auto pd = Pat_set.begin(); pd != Pat_set.end(); pd++)
         cout << *pd << " ";

    Guest_set.insert(Mat_set.begin(),Mat_set.end());
    Guest_set.insert(Pat_set.begin(),Pat_set.end());
    /*通过 instcr()函数，合并两个个人名单至总名单*/
    cout << "\nAll friends are: \n";
    for(auto pd = Guest_set.begin(); pd != Guest_set.end(); pd++)
         cout << *pd << " ";
    return 0;
}
```

9．相对于数组，在链表中添加和删除元素更容易，但排序速度更慢。这就产生了一种可能性：相对于使用链表算法进行排序，将链表复制到数组中，对数组进行排序，再将排序后的结果复制到链表中的速度可能更快；但这种方式可能占用更多的内存。请使用如下方法检验上述假设。

a．创建大型 vector<int>对象 vi0，并使用 rand()给它提供初始值。

b．创建 vector<int>对象 vi 和 list<int>对象 li，它们的长度和初始值都与 vi0 相同。

c．计算使用 STL 算法 sort()对 vi 进行排序所需的时间，再计算使用 list 的方法 sort()对 li 进行排序所需的时间。

d．将 li 重置为排序后 vi0 的内容，并计算执行如下操作所需的时间：将 li 的内容复制到 vi 中，对 vi 进行排序，并将结果复制到 li 中。

要计算这些操作所需的时间，可使用 ctime 库中的 clock()。如程序清单 5.14 所示，可使用下面的语句来获取开始时间。

```
clock_t start = clock();
```

再在操作结束后使用下面的语句获取经过了多长时间。

```
clock_t end = clock();
```

```
cout<<(double)(end-start)/CLOCKS_PER_SEC;
```

这种测试并非绝对可靠，因为结果取决于很多因素，如可用内存量、是否支持多处理以及数组（列表）的长度（随着要排序的元素数增加，数组相对于列表的效率将更明显）。另外，如果编译器提供了默认生成方式和发布生成方式，请使用发布生成方式。由于当今计算机的速度非常快，要获得有意义的结果，可能需要使用尽可能大的数组。例如，可尝试包含 100 000 个、1 000 000 个和 10 000 000 个元素。

编程分析：

本题主要目的是通过对比，分析数组和链表两种数据结构的排序时间，其本质是讨论和研究计算机科学中时间复杂度的问题。算法的时间复杂度是一个函数，它定性描述该算法的运行时间，通常用大 O 符号表示。本题只是一个简单的时间对比，程序整体结构比较简单。完整代码如下。

```
/*第 16 章的编程练习 9*/
#include <iostream>
#include <ctime>
#include <vector>
#include <list>

using namespace std;
const int LENGTH = 1000000;
int main()
{
    vector<int> vi0;
    /*定义 vector<int>对象 vi0*/
    srand (time(0));
    for(int i = 0; i < LENGTH; i++)
        vi0.push_back(rand()%1000);
    /*在 vi0 对象内存入 LENGTH 个随机数*/
    vector<int> vi(vi0);
    /*利用 vi0 复制构造 vector<int>对象 vi*/
    list<int> li(vi0.begin(), vi0.end());
    /*定义 list<int>对象 li，将 vi0 数据复制到 li 中*/
    clock_t time = clock();
    sort(vi.begin(), vi.end());
    time = clock() - time;
    /*对 vi 排序并计时*/
    cout << "Time used sort by vector.sort(): ";
    cout << (double)(time) / CLOCKS_PER_SEC << " second"<<endl;
    /*输出 vi 的排序时间*/
    time = clock();
    li.sort();
    time = clock() - time;
    /*记录 list 的排序时间*/
    cout << "Time used sort by list.sort(): ";
    cout << (double)(time) / CLOCKS_PER_SEC << " second"<<endl;
    /*输出 li 的排序时间*/
```

```
    li.assign(vi0.begin(), vi0.end());
    /*重置 li 对象的数据*/
    time = clock();
    vi.assign(li.begin(), li.end());
    sort(vi.begin(), vi.end());
    li.assign(vi.begin(), vi.end());
    time = clock() - time;
    /*统计“把 list 数据复制到 vector 中、通用排序 vector、vector 复制回 list”这三步操作的用时记录*/
    cout << "Time used by generic sort : ";
    cout << (double)(time) / CLOCKS_PER_SEC << " second"<< "\n";
    return 0;
}
```

10．请按如下方式修改程序清单 16.9（vect3.cpp）。

a．在结构体 Review 中添加成员 price。

b．不使用 vector<Review>来存储输入，而使用 vector<shared_ptr<Review>>。注意，必须使用 new 返回的指针来初始化 shared_ptr。

c．在输入阶段结束后，使用一个循环让用户选择如下方式之一显示书籍，包括按原始顺序显示，按字母表顺序显示，按评级升序显示，按评级降序显示，按价格升序显示，按价格降序显示。

下面是一种可能的解决方案：获取输入后，创建一个 shared_ptr 向量，并使用原始数组初始化它。定义一个对指向结构体的指针进行比较的 operator<()函数，并使用它对第 2 个向量进行排序，使其中的 shared_ptr 按其指向的对象中的书名排序。重复上述过程，创建按 rating 和 price 排序的 shared_ptr 向量。请注意，通过使用 rbegin()和 rend()，可避免创建按相反的顺序排列的 shared_ptr 向量。

编程分析：

程序清单 16.9 的主要功能是利用 vector 容器进行图书信息管理，题目要求在 Review 类中添加数据成员 price，并使用指向 Review 的指针作为 vector 的元素类型，创建一个图书管理软件。由于 vector 容器内是指向 Review 对象的智能指针，因此在比较排序函数上需要修改原有算法。可以按照题目给出的解决方案进行间接排序。完整代码如下。

```
/*第 16 章的编程练习 10*/
#include <iostream>
#include <string>
#include <vector>
#include <algorithm>
#include <memory>

using namespace std;

struct Review {
    string title;
    int rating;
    int price;
    /*添加 price 成员*/
```

```
};

bool operator<(const shared_ptr<Review> & p1, const shared_ptr<Review> & p2);
bool FillReview(Review & rr);
void ShowReview(const shared_ptr<Review> & p);

bool worseThan(const shared_ptr<Review> & p1, const shared_ptr<Review> & p2);
bool expenThan(const shared_ptr<Review> & p1, const shared_ptr<Review> & p2);
/*修改原有的 worseThan()函数，把参数修改为智能指针，添加 expenThan()函数*/
int main()
{
    using namespace std;

    vector<shared_ptr<Review> > books;
    Review temp;
    while(FillReview(temp))
    {
       shared_ptr<Review> pd (new Review(temp));
       /*智能指针需要使用 new 动态存储一个新的对象*/
       books.push_back(pd);
    }
    if (books.size() > 0)
    {
       cout << "Choose the way to sort: "
            << "r: rate, s: rate r, p: price, d: price r, q: quit\n";
       char choice;
       while(cin >> choice && choice != 'q')
       {
          switch(choice)
          {
             case 'r': sort(books.begin(), books.end(), worseThan);
                 break;
             case 's': sort(books.rbegin(), books.rend(), worseThan);
                 break;
             case 'p': sort(books.begin(), books.end(), expenThan);
                 break;
             case 'd': sort(books.rbegin(), books.rend(), expenThan);
                 break;
             default:  sort(books.begin(), books.end());
                 break;
          }
          for_each(books.begin(), books.end(), ShowReview);
          cout << "Please choose the way to sort: "
               << "r: rate, s: rate r, p: price, d: price r, q: quit\n";
       }
    }
    return 0;
}

bool FillReview(Review & rr)
```

```
{
    cout << "Enter book title (quit to quit): ";
    getline(cin, rr.title);
    if (rr.title == "quit" || rr.title == "")
       return false;
    cout << "Enter book rating: ";
    cin >> rr.rating;
    if (!cin)
       return false;
    cout << "Enter book price: ";
    cin >> rr.price;
    if (!cin)
       return false;
    while (cin.get()!='\n')
       continue;
    return true;
}
/*对于相关函数，需要将原 Review 对象的方式修改为智能指针，注意成员运算符*/
void ShowReview(const shared_ptr<Review> & p)
{
    cout << p->rating << "\t" << p->title << "\t" << p->price << endl;
}

bool operator<(const shared_ptr<Review> & p1, const shared_ptr<Review> & p2)
{
    if (p1->title < p2->title)
        return true;
    else if (p1->title == p2->title && p1->rating < p2->rating)
        return true;
    else
        return false;
}

bool worseThan(const shared_ptr<Review> & p1, const shared_ptr<Review> & p2)
{
    if (p1->rating < p2->rating)
        return true;
    else
        return false;
}

bool expenThan(const shared_ptr<Review> & p1, const shared_ptr<Review> & p2)
{
    if (p1->price < p2->price)
        return true;
    else
        return false;
}
```

第 17 章　输入、输出和文件

本章知识点总结

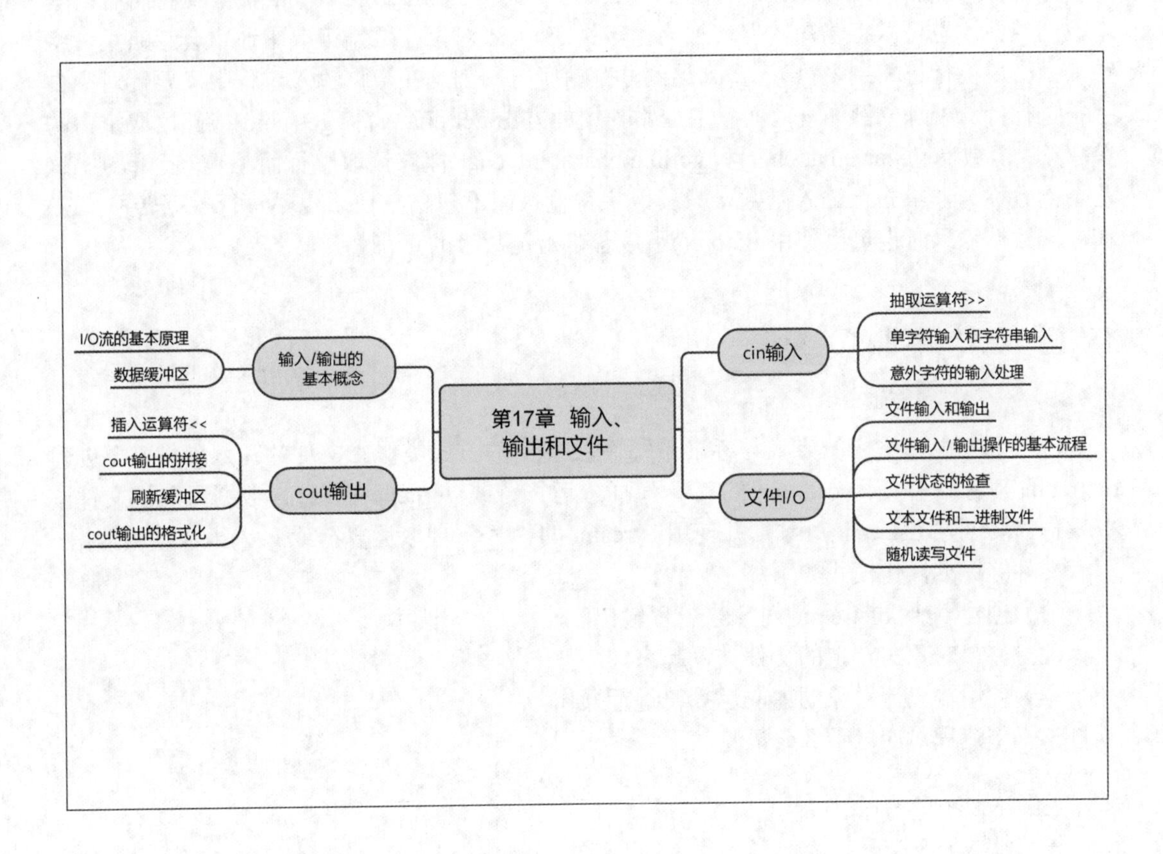

17.1　C++中的I/O流

C++语言把数据的输入和输出看作数据信息以字节流的形式实现的传输模式。程序在数据的输入源和数据的输出源中间通过流这种形式进行了数据的传输。输入时程序从输入流中抽取数据；输出时程序将字节插入输出流中。而数据输入源和数据输出源则可以是外部设备、磁盘文件等形式。C++语言中通过流的形式读写数据的整体模型更加简单，程序设计中只需要创建相应的对象，并连接对应的输入/输出源，就可以调用相应的读写函数进行操作。

系统内部在数据的读写操作中使用缓冲区可以提高输入和输出的运行效率。缓冲区是用作中介的内存块，它是将信息从设备传输到程序或从程序传输到设备的临时的中间存储区域。通常缓冲区通过暂时存储待读写的数据，帮助同步不同设备之间的信息传输速率来提高读写效率。C++中通过预定义的 I/O 类及其流对象进行数据的 I/O 操作，最常用的类包括 istream、ostream 和 iostream，及 iostream 内的 cin、cout 等。

cout 对象通过重载的“<<”重定向符可以支持所有基本数据类型的直接输出。cout 的输出格式可以通过相关的控制符、setf()函数和格式常量进行设置，其中主要包括输出的计数系统、字段宽度、浮点型数据的显示精度等。

cin 对象通过对输入流的“>>”重定向操作可以读取相应数据。 cin 能够支持所有的基本数据类型，也可以将输入数据转换为指定格式。cin 对象在数据读取中通常会忽略空白字符（空格、换行符和制表符），直到遇到非空白字符，因此在数据输入过程中需要针对空白字符进行特殊处理。例如，前面章节通常使用的方法 get(char&)和 get(void)能够读取空白的单字符；函数 get(char*, int, char)和 getline(char*, int, char)能够读取整行而不是一个单词。这几个函数称为非格式化输入函数，只读取字符输入而不忽略空白，也不进行数据转换。cin 在读取数据输入时也可以使用 ignore()函数忽略指定数量的字符。

17.2 文件 I/O

C++文件输入/输出的方式与标准系统输入/输出的方式类似。文件的写入操作需要创建 ofstream 对象，并使用 ostream 的“<<”插入运算符或 write()函数进行数据写入；文件读取操作需要创建 ifstream 对象，并使用 istream 的“>>”抽取运算符或 get()函数读取数据。写入文件的标准操作模式如下。

（1）创建一个 ofstream 对象来管理输出流。

（2）将该对象与特定的文件关联起来。

（3）使用输出流对象的抽取运算符进行操作。

该操作流程对应以下 3 条语句。

```
ofstream fout;
fout.open("jar.txt");
fout<<"Hello World!";
```

相应地，读取文件的标准操作模式如下。

（1）创建一个 ifstream 对象来管理输入流。

（2）将该对象与特定的文件关联起来。

（3）使用输入流对象的抽取运算符进行操作。

该操作流程对应以下 3 条语句。

```
ifstream fin;
fin.open("jar.txt");
fin>>ch;
```

在读写操作完成后，应当调用 close()函数关闭文件。实际应用中还应在 open()函数中通过 ios_base 的文件模式常量控制文件的读写操作的具体方式。

系统文件在存储中分为文本格式或二进制格式。虽然所有的数据在计算机内都是以二进制形式存储的，但是文件的文本格式是指系统将所有内容都转换成文本字符的形式，这样更加易于字处理程序读取和编辑文本文件，也利于文本文件的传输和转换。而文件的二进制格式将所有内容直接以二进制编码存储（通常用于非文本文件，如机器语言编码、图像、音乐等）。文本模式中系统会自动按照文本编码的特点（针对不同系统平台的差异和字符编码差异）对文本文件做一些字符转换，在不同操作系统平台下可能会有不同的转换方式，因此会导致文本格式的文件存在一定的转换差异。通常二进制格式占用的空间较小，且二进制格式中可以直接读取文件的每个字节，因此不会有转换误差或舍入误差。C++中，可以将文件模式设置为 ios_base::binary 常量来指定二进制格式。要以二进制格式存储数据，可以使用 write()成员函数。通过这种方法将内存中指定数目的字节复制到文件中，不进行任何转换。

当文件使用 get()、put()等方法进行数据的输入/输出时，从打开文件的起始位置开始，直到文件的末尾，顺序进行读写。文件的随机存取指的是可以从文件的任何位置开始读写操作，不需要按照顺序进行读写，这样程序可以更加灵活地查找和跳转到某个指定位置，存取该位置的数据。fstream 类有两个方法——seekg()和 seekp()，前者将输入指针移动到文件指定的位置，后者将输出指针移动到文件指定的位置。如果要检查文件指针的当前位置，可以使用 tellg()方法（输入流）或者 tellp()方法（输出流），它们都返回一个表示当前位置的 streampos 值，通过 streampos 值可以实现文件的快速定位和移动。

17.3　复习题

1．iostream 文件在 C++ I/O 中扮演何种角色？

习题解析：

C++语言把数据的输入/输出以字节信息流的形式进行处理。通过类的继承关系，C++中预定义了大量用于数据输入/输出的类，其中 iostream 文件的核心是 istream 和 ostream 两个类的定义，分别表示输入流与输出流中用于管理输入和输出的类、常量以及运算符，是系统 I/O 中最基本的库文件。该文件还创建了一些标准对象（cin、cout、cerr 和 clog 以及对应的宽字符对象），用于处理与每个程序相连的标准输入流和输出流。

2．对于键入的数字（如 121），为什么要求程序进行转换？

习题解析：

系统的标准化输入工具首先将从键盘输入的信息按照一系列字符的形式读取进数据缓冲区。例如，输入 121 将生成 3 个字符，分别是 1、2、1，按照系统的规定每个字符使用 1 字节的二进制数据表示。C++程序中如果通过标准 I/O 表示输入的数据是 int 类型的数值，那么程序将会自动将这 3 个字符转换为 int 类型的数据 121，并按照 int 类型的方式以 32 位数据的形式存储为二进制数据。

3．标准输出与标准错误之间有什么区别？

习题解析：

在默认情况下，标准输出和标准错误都将连接标准输出设备（通常为显示器），因此两者的差别不大。但是如果要求操作系统将输出重定向到文件，则标准输出将与文件（而不是显示器）相连，但标准错误仍与显示器相连。

4. 在不为每个类型提供明确指示的情况下，为什么 cout 仍能够显示不同的 C++类型？

习题解析：

C++语言中预定义的标准输入/输出中的 ostream 类已经为 C++基本类型（如各种 int 类型、浮点类型、字符类型）定义了 operator <<() 运算符重载函数，因此当输出不同类型的数据时“<<”运算符能够查找相应的重载函数。但是对于用户自定义类型，还需要在类定义中手动重载相应的运算符，才能够正确使用运算符“<<”输出。

5．输出方法的定义中的哪个特征能够用来拼接输出？

习题解析：

当使用标准输出流进行数据输出时，对于用户自定义类型，需要重载“<<”运算符。为了实现输出语句的拼接，即 `cout<<"Hello "<<"world."<<endl` 的形式，在重载运算符时，必须返回 ostream &类型的对象。这样，当通过一个对象调用方法时，将返回该对象，返回对象可以继续调用序列中的下一个方法，从而拼接输出。

6．编写一个程序，要求用户输入一个整数，然后以十进制、八进制和十六进制显示该整数。在宽度为 15 个字符的字段中显示每种形式，并将它们显示在同一行上，同时使用 C++中的数基前缀。

习题解析：

在 cout 对象进行标准输出时，可以使用 setf()、setw()等一系列函数和标准控制符设置输出数据的格式与类型。具体代码如下。

```
#include <iostream>
#include <iomanip>
int main()
{
    using namespace std;
    cout << "Enter an integer: ";
    int n;
    cin >> n;
    cout << setw(15) << "base ten" << setw(15)
        << "base sixteen" << setw(15) << "base eight" << "\n";
```

```
    cout.setf(ios::showbase) ; //或者 cout << showbase;
    cout << setw(15) << n << hex << setw(15) << n
         << oct << setw(15) << n << "\n";
    return 0;
}
```

7. 编写一个程序，请求用户输入下面的信息，并按下面的格式显示它们。

```
Enter your name: Billy Gruff
Enter your hourly wages: 12
Enter number of hours worked: 7.5
First format:
                   Billy Gruff: $     12.00:  7.5
Second format:
Billy Gruff                    : $12.00     :7.5
```

习题解析：

依据题目提供的显示信息，首先通过系统的输入获取用户姓名（使用字符串存储），以及薪水、工作时间（用 float 类型变量存储）。输出信息首先右对齐，然后左对齐。完整代码如下。

```
#include <iostream>
#include <iomanip>
int main()
{
    using namespace std;
    char name[20] ;
    float hourly;
    float hours;
    cout << "Enter your name: ";
    cin.get (name, 20).get();
    cout << "Enter your hourly wages: ";
    cin >> hourly;
    cout << "Enter number of hours worked: ";
    cin >> hours;
    cout.setf(ios::showpoint) ;
    cout.setf(ios::fixed, ios::floatfield) ;
    cout.setf(ios::right, ios::adjustfield) ;
    /*调用 serf()函数设置输出右对齐*/
    cout << "First format:\n";
    cout << setw(30) << name << ": $" << setprecision(2)
         << setw(10) << hourly << ":" << setprecision(1)
        << setw(5) << hours << "\n";
    /*通过 setw()函数设置输出宽度*/
    cout << "Second format:\n";
    cout.setf(ios::left, ios::adjustfield) ;
    /*调用 serf()函数设置输出左对齐*/
    cout << setw(30) << name << ": $" << setprecision(2)
         << setw(10) << hourly << ":" << setprecision(1)
         << setw(5) << hours << "\n";
```

```
    /*setw()函数设置输出宽度*/
    return 0;
}
```

8．对于下面的程序，回答下面的问题。

```
//rq117-8.cpp
#include <iostream>
int main()
{
    using namespace std;
    char ch;
    int ct1 = 0;
    cin >> ch;
    while (ch != 'q')
    {
        ct1++;
        cin >> ch;
    }
    int ct2 = 0;
    cin.get (ch) ;
    while (ch != 'q')
    {
        ct2++;
        cin.get (ch) ;
    }
    cout << "ct1 = " << ct1 << "; ct2 = " << ct2 << "\n";
    return 0;
}
```

如果输入如下，该程序将输出什么内容?

```
I see a q<Enter>
I see a q<Enter>
```

其中，<Enter>表示按回车键。

习题解析：

程序运行后，显示如下结果。

```
I see a q
I see a q
ct1 = 5; ct2 = 9
```

该程序的前半部分使用 cin >> ch;读取标准输入，并通过 ctl 计数，该方法会忽略空格和换行符。后半部分使用 cin.get(ch);读取标准输入，并通过 ctl 计数包括空白在内的所有字符数据。此外，程序的后半部分从第 1 个 q 后面的换行符开始读取，会将换行符计算在内，因此显示 ct2=9。

9．下面的两条语句都读取并丢弃行尾之前的所有字符（包括行尾）。这两条语句的行为有哪些不同?

```
while (cin.get() != '\n')
```

```
        continue;
    cin.ignore(80, '\n');
```

习题解析：

cin.get()方法会一次读取一个字符，并判断是否为换行符，如果不是，则丢弃重新读取下一个数据，因此可以持续循环。通过 ignore(80,'\n')函数读取数据并丢弃有两个条件，一是数据不超过 80 个字符，二是遇见第 1 个换行符，因此在这种情况下，ignore()函数最多将跳过前 80 个字符。

17.4　编程练习

1. 编写一个程序，统计输入流中第 1 个$之前的字符数目，并将$留在输入流中。

编程分析：

题目要求计算程序能够读取系统标准输入，并对$字符前的所有字符进行计数，因此应当使用 cin.get()函数，并要求$符号保留在输入流的缓冲区中，因此应当使用 peek()函数预先判断下一个字符是否是$，随后再决定读取还是保留。完整代码如下。

```
/*第 17 章的编程练习 1*/
#include <iostream>

int main()
{
    char input;
    int count = 0;
    std::cout << "Enter a phase: ";
    while(std::cin.peek() != '$')
    /*通过 peek()函数检查，并未清空缓冲区*/
    {
        std::cin.get(input);
        count++;
    }
    std::cout << count << " chars.\n";
    return 0;
}
```

2. 编写一个程序，将键盘输入（直到模拟的文件末尾）复制到通过命令行指定的文件中。

编程分析：

题目要求首先通过命令行 argv 获取指定文件名，随后把通过标准输入读取的数据保存至该文件中，直到用户输入 EOF（模拟文件末尾）为止。程序结构比较简单，完整代码如下。

```
/*第 17 章的编程练习 2*/
```

```
#include <iostream>
#include <fstream>

using namespace std;
int main(int argc, char *argv[])
{
    if(argc == 1)
    {
        cout<<"Usage: "<<argv[0]<<" filename[s]"<<endl;
        exit(EXIT_FAILURE);
    }
    /*命令行参数的数量检查*/
    char ch;
    ofstream fout(argv[1],ios_base::out);
    if(fout.is_open())
    {
        cout << "Inpue the data:\n";
        while(cin.get(ch) && ch != EOF)
            fout << ch;
    /*循环写入文件，直到输入EOF*/
    }
    else
    {
        cout << "error to create the file!";
        exit(EXIT_FAILURE);
    }
    fout.close();
    return 0;
}
```

3．编写一个程序，将一个文件复制到另一个文件中。使程序通过命令行获取文件名。如果文件无法打开，程序将指出这一点。

编程分析：

题目要求通过命令行获取源文件和目标文件的文件名，因此首先判断 argc 参数的值。文件的复制过程需要一个 ifsteam 对象和一个 ofstream 对象，前者负责数据的读取，后者负责数据的写入。下列代码使用 get()函数，每次只能读取一个字符，运行效率较低。可以使用 read()和 write()函数，通过较大的数据缓冲区一次读取多个数据，提高运行效率。

```
/*第 17 章的编程练习 3*/
#include <iostream>
#include <fstream>

using namespace std;

int main(int argc, char *argv[])
{
    if(argc < 3)
    {
```

```
        cout<<"Usage: "<<argv[0]<<" srcfile desfile"<<endl;
        exit(EXIT_FAILURE);
    }
    /*检查命令行参数*/
    char ch;

    ifstream fin(argv[1],ios_base::in);
    ofstream fout(argv[2],ios_base::out);
    if(!fin.is_open())
    {
        cout << "Can't open the file " << argv[1] << " !"<<endl;
        exit(EXIT_FAILURE);
    }
    if(!fout.is_open())
    {
        cout << "Can't open the file " << argv[2] << " !"<<endl;
        exit(EXIT_FAILURE);
    }
    /*检查文件是否打开*/
    while(fin.get(ch)) fout << ch;
    /*文件循环读取、复制*/
    fin.close();
    fout.close();
    return 0;
}
```

4. 编写一个程序，它打开两个文本文件进行输入，打开一个文本文件进行输出。该程序将两个输入文件中对应的行拼接起来，并用空格分隔，然后将结果写入输出文件中。如果一个文件比另一个短，则将较长文件中余下的几行直接复制到输出文件中。例如，假设第 1 个输入文件的内容如下。

```
eggs kites donuts
balloons hammers
stones
```

而第 2 个输入文件的内容如下。

```
zero lassitude
finance drama
```

则得到的文件内容将如下。

```
eggs kites donuts zero lassitude
balloons hammers finance drama
stones
```

编程分析：

根据题目提供的文件输出可以判断文件的读取使用了 getline()函数，每次读取一行数据，两个文件的同一行合并，并在合并文件同一行后输出换行符。程序设计中需要注意输入文件中的任何一个文件到文件末尾后，都需要复制另一个文件的剩余内容。完整代码如下。

```
/*第 17 章的编程练习 4*/
#include <iostream>
```

```
#include <fstream>

using namespace std;

int main()
{
    string line;

    ifstream fin1("file1.txt",ios_base::in);
    ifstream fin2("file2.txt",ios_base::in);
    ofstream fout("CombFile.txt",ios_base::out);
    /*打开 3 个文件，准备读写处理*/
    if(fin1.is_open() && fin2.is_open() && fout.is_open())
    {
    /*判断文件是否正常打开*/
        while(!fin1.eof() || !fin2.eof())
        {
             if(getline(fin1, line) && line.size() > 0)
                 fout << line;
             if(getline(fin2, line) && line.size() > 0)
                 fout << line;
             /*while 循环的结束条件为两个输入文件均到末尾，否则循环读写*/
            fout << endl;
            /*写入文件中的一行，输出换行符*/
        }
    }
    else
    {
        cout << "Can't open the file!\n";
        exit(EXIT_FAILURE);
    }
    fin1.close();
    fin2.close();
    fout.close();
    return 0;
}
```

5．Mat 和 Pat 想邀请他们的朋友来参加派对，就像第 16 章中的编程练习 8 那样，但现在他们希望在程序中使用文件。请编写一个完成下述任务的程序。

- 从文本文件 mat.dat 中读取 Mat 的朋友名单，其中每行表示一个朋友。首先把姓名存储在容器中，然后按顺序显示出来。
- 从文本文件 pat.dat 中读取 Pat 的朋友名单，其中每行表示一个朋友。首先把姓名存储在容器中，然后按顺序显示出来。
- 合并两个清单，删除重复的条目，并将结果保存在文件 matpat.dat 中，其中每行表示一个朋友。

编程分析：

在第 16 章的编程练习 8 的基础上，将标准输入/输出重定向到文件，并使用 set 容器进行数据的合并和排序，最后输出到文件 matpat.dat 中。完整代码如下。

```
/*第 17 章的编程练习 5*/
#include <iostream>
#include <fstream>
#include <set>

using namespace std;

int main()
{
    ifstream fin_mat("mat.txt", ios_base::in);
    ifstream fin_pat("pat.txt", ios_base::in);
    /*定义输入文件*/
    string guest;
    set<string> mat_guest, pat_guest, guest_set;
    /*定义 set<string> 对象，存储 Mat 的朋友名单、Pat 的朋友名单和总名单*/

    if(!fin_mat.is_open() || !fin_pat.is_open())
    {
        cout<<"Error open files."<<endl;
        exit(EXIT_FAILURE);
    }
    while(getline(fin_mat, guest) && guest.size() > 0)
        mat_guest.insert(guest);
    cout << "\nMat's friends are: \n";
    for(auto pd = mat_guest.begin(); pd != mat_guest.end(); pd++)
        cout << *pd << " ";
    /*从输入文件读取 Mat 的朋友名单，并存储在 set 内*/
    while(getline(fin_pat, guest) && guest.size() > 0)
        pat_guest.insert(guest);
    cout << "\nPat's friends are: \n";
    for(auto pd = pat_guest.begin(); pd != pat_guest.end(); pd++)
        cout << *pd << " ";
    /*从输入文件读取 Pat 的朋友名单，并存储在 set 内*/

    fin_pat.close();
    fin_mat.close();
    /*关闭文件流*/
    guest_set.insert(mat_guest.begin(),mat_guest.end());
    guest_set.insert(pat_guest.begin(),pat_guest.end());
    /*合并名单*/
    ofstream fout("guest.txt", ios_base::out);
    if(!fout.is_open())
    {
        cout<<"Error open files."<<endl;
        exit(EXIT_FAILURE);
    }

    for(auto pd = guest_set.begin(); pd != guest_set.end(); pd++)
        fout << *pd << " ";
```

```
    fout.close();
    /*文件输出*/

    ifstream fin("guest.txt", ios_base::in);
    if(!fin.is_open())
    {
        cout<<"Error open files."<<endl;
        exit(EXIT_FAILURE);
    }
    /*测试合并后的名单，并输出*/
    cout << "\nAll Guest list : \n";
    while(getline(fin, guest))
        cout << guest << " ";
    cout << endl;
    fin.close();
    return 0;
}
```

6．考虑第 14 章的编程练习 5 中的类定义。首先需要完成该练习，然后完成下面的任务。

编写一个程序，它使用标准 C++ I/O、文件 I/O 以及第 14 章的编程练习 5 中定义的 employee、manager、fink 和 highfink 类型的数据。该程序应包含程序清单 17.17 中的代码行，即允许用户将新数据添加到文件中。首次运行该程序时，将要求用户输入数据，然后显示所有的数据，并将这些信息保存到一个文件中。当该程序再次运行时，将首先读取并显示文件中的数据，然后要求用户添加数据，并显示所有的数据。差别之一是，通过一个指向 employee 类型的指针数组来处理数据。这样，指针可以指向 employee 对象，也可以指向从 employee 派生出来的其他 3 种对象中的任何一种。数组较小有助于检查程序，例如，可以将数组限定为最多包含 10 个元素。

```
const int MAX = 10; //不超过 10 个元素
employee *pc [MAX] ;
```

为通过键盘输入，程序应使用一个菜单，供用户选择要创建的对象类型。菜单将使用一个 switch，以便使用 new 来创建指定类型的对象，并将它的地址赋给 pc 数组中的一个指针。然后该对象可以使用虚函数 setall()来提示用户输入相应的数据:

```
pe[i]->setall(); //根据对象的类型调用函数
```

为将数据保存到文件中，应设计一个虚函数 writeall()。

```
for (1 = 0; i < index; i++)
pe [i] ->writeall(fout);//把 fout ofstream 连接到输出文件
```

注意，对于这个练习，应使用文本 I/O，而不是二进制 I/O（遗憾的是，虚对象包含指向虚函数指针表的指针，而 write()将把这种信息复制到文件中。当使用 read()读取文件的内容以填充对象时，函数指针的值将为乱码，这将扰乱虚函数的行为）。可使用换行符将字段分隔开，这样在输入时将很容易识别各个字段。也可以使用二进制 I/O，但不能将对象作为一个整体写入，而应该分别对每个类成员应用 write()和 read() 的类方法，这样，程序将只把所需的数据保存到文件中。

比较难处理的部分是使用文件恢复数据。程序如何才能知道接下来要恢复的项是 employee 对象、manager 对象、fink 对象还是 highfink 对象？一种方法是，在把对象的数据写入文件时，在数据前面加上一个指示对象类型的整数，这样，在向文件写入时，程序便可以读取该整数，并使用 switch 语句创建一个适当的对象来接收数据。

```
enum classkind{Employee, Manager, Fink, HighFink};//在类的头文件中
......
int classtype
while((fin>>classtype).get(ch)){//用换行符分隔字段
        switch(classtype){
        case Employee: pc[i] = new employee;
                      :break;
```

然后便可以使用指针调用虚函数 getall()来读取信息。

```
pc[i++]->getall();
```

编程分析：

题目需要将第 14 章编程练习 5 中使用标准输入/输出实现的多重继承的程序修改为文件数据输入和输出。首先添加 writeall()函数，实现文件写入功能，按照题意使用文本写入方式写入信息，并实现虚函数在继承中的多态性设计。此外，在文件恢复阶段要考虑不同对象在输出时如何保存对象的类别信息，这样才能保证在数据读取中匹配正确的类对象。

在通过文件存储不同类型的对象时有很多种方式。最简单的解决方案是修改类的数据成员，添加一个标志位，并在 writeall()函数内写入该标志位。在读取文件时依据该标志判断其类型，这种解决方案容易理解，但是需要对类定义进行多处改动，操作较为烦琐。下面的代码在数据存储阶段（main()函数内）根据 RTTI 判断对象类型，在调用对象的 writeall()函数前先存储标志数据。在读取时重新根据标志位恢复数据，这样的方式需要修改主程序，添加并使用多条条件语句。RTTI 方法维护了类定义的完整性，但是并未很好地实现数据和操作的类内封装；修改类数据成员的方法需要在设计阶段对数据的功能和性质进行更加全面的设计。实际应用中可以根据特定情况选择使用。RTTI 方法的完整代码如下。

```
/*第 17 章的编程练习 6*/
#ifndef EMP_H
#define EMP_H
//emp.h ——abstr_emp 类及其子类的头文件
#include <iostream>
#include <string>
#include <fstream>
using namespace std;

class abstr_emp
{
private:
    std::string fname;
    std::string lname;
    std::string job;
public:
    abstr_emp();
    abstr_emp(const std::string & fn, const std::string & ln,
            const std::string & j);
    virtual void ShowAll() const;
    virtual void SetAll();
```

```
    friend std::ostream &
    operator<<(std::ostream & os, const abstr_emp & e);
    virtual ~abstr_emp() = 0;
    virtual void writeall(ofstream& fout) const;
    virtual void getall(ifstream& fin);
};
class employee : public abstr_emp
{
public:
    employee () ;
    employee(const std::string & fn, const std::string & ln,
          const std::string & j);
    virtual void ShowAll() const;
    virtual void SetAll();
    virtual void writeall(ofstream& fout) const;
    virtual void getall(ifstream& fin);

};
class manager: virtual public abstr_emp
{
private:
    int inchargeof;
protected:
    int InChargeOf() const { return inchargeof; }
    int & InChargeOf(){ return inchargeof; }
public:
    manager () ;
    manager(const std::string & fn, const std::string & ln,
          const std::string & j, int ico = 0);
   manager(const abstr_emp & e, int ico);
   manager (const manager & m);
   virtual void ShowAll() const;
   virtual void SetAll();
   virtual void writeall(ofstream& fout) const;
   virtual void getall(ifstream& fin);

};
class fink: virtual public abstr_emp
{
private:
    std::string reportsto; //to whom fink reports
protected:
    const std::string ReportsTo() const { return reportsto; }
    std::string & ReportsTo(){ return reportsto; }
public:
    fink();
    fink(const std::string & fn, const std::string & ln,
        const std::string & j, const std::string & rpo);
    fink(const abstr_emp & e, const std::string & rpo);
    fink(const fink & e);
    virtual void ShowAll() const;
    virtual void SetAll();
```

```
    virtual void writeall(ofstream& fout) const;
    virtual void getall(ifstream& fin);
};
class highfink: public manager, public fink
{
public:
    highfink();
    highfink(const std::string & fn, const std::string & ln,
             const std::string & j, const std::string & rpo, int ico);
    highfink(const abstr_emp & e, const std::string & rpo, int ico);
    highfink(const fink & f, int ico);
    highfink(const manager & m, const std::string & rpo);
    highfink(const highfink & h);
    virtual void ShowAll() const;
    virtual void SetAll();
    virtual void writeall(ofstream& fout) const;
    virtual void getall(ifstream& fin);
};

#endif

#include "emp.h"

abstr_emp::abstr_emp() :lname("none"),fname("none"),job("none")
{
}
abstr_emp::abstr_emp(const std::string & fn, const std::string & ln,
                     const std::string & j):fname(fn),lname(ln),job(j)
{
}
abstr_emp::~abstr_emp()
{
}
void abstr_emp::ShowAll() const
{
    cout<<"NAME: "<<fname<<"."<<lname<<endl;
    cout<<"JOB TITLE: "<<job<<endl;
}
void abstr_emp::SetAll()
{
    cout<<"Entenr the first name: ";
    getline(cin,fname);
    cout<<"Enter the last name: ";
    getline(cin,lname);
    cout<<"Enter th job title: ";
    getline(cin,job);
}
std::ostream &operator<<(std::ostream & os, const abstr_emp & e){
    os<<"NAME: "<<e.fname<<"."<<e.lname<<endl;
    os<<"JOB TILTE: "<<e.job<<endl;
    return os;
}
```

```
void abstr_emp::writeall(ofstream& fout) const
{
    fout << fname << endl;
    fout << lname << endl;
    fout << job << endl;
}
void abstr_emp::getall(ifstream& fin)
{
    getline(fin,fname);
    getline(fin,lname);
    getline(fin,job);
}

employee::employee () :abstr_emp()
{
}
employee::employee(const std::string & fn, const std::string & ln,
                   const std::string & j): abstr_emp(fn,ln,j)
{
}
void employee::ShowAll() const{
    abstr_emp::ShowAll();
}
void employee::SetAll()
{
    abstr_emp::SetAll();
}
void employee::writeall(ofstream& fout) const
{
    abstr_emp::writeall(fout);
}
void employee::getall(ifstream& fin)
{
    abstr_emp::getall(fin);
}

manager::manager ():abstr_emp(),inchargeof(0)
{
}
manager::manager(const std::string & fn, const std::string & ln,
                 const std::string & j, int ico):abstr_emp(fn,ln,j),inchargeof(ico)
{
}
manager::manager(const abstr_emp & e, int ico): abstr_emp(e)
{
    inchargeof = ico;
}
manager::manager (const manager & m): abstr_emp(m)
{
    inchargeof = m.inchargeof;
}
void manager::ShowAll() const{
```

```
        abstr_emp::ShowAll();
        cout<<"IN CHARGE OF: "<<inchargeof<<endl;
    }
    void manager::SetAll()
    {
        abstr_emp::SetAll();
        cout<<"Enter the number of in charge: ";
        cin>>inchargeof;
        while(cin.get() == '\n')
            continue;
    }
    void manager::writeall(ofstream& fout) const
    {
        abstr_emp::writeall(fout);
        fout<<inchargeof<<endl;
    }
    void manager::getall(ifstream& fin)
    {
        abstr_emp::getall(fin);
        fin>>inchargeof;
        fin.get();
    }


    fink::fink():abstr_emp(),reportsto("none")
    {
    }

    fink::fink(const std::string & fn, const std::string & ln,
               const std::string & j, const std::string & rpo): abstr_emp(fn,ln,j),
reportsto(rpo)
    {
    }
    fink::fink(const abstr_emp & e, const std::string & rpo):abstr_emp(e),reportsto(rpo)
    {
    }
    fink::fink(const fink & e):abstr_emp(e)
    {
        reportsto = e.reportsto;
    }
    void fink::ShowAll() const
    {
        abstr_emp::ShowAll();
        cout<<"REPORT TO: "<<reportsto<<endl;
    }
    void fink::SetAll()
    {
        abstr_emp::SetAll();
        cout<<"Enter the reports to whom: ";
        getline(cin,reportsto);
    }
```

```
void fink::writeall(ofstream& fout) const
{
    abstr_emp::writeall(fout);
    fout<<reportsto<<endl;
}
void fink::getall(ifstream& fin)
{
    abstr_emp::getall(fin);
    getline(fin,reportsto);
}

highfink::highfink() :abstr_emp(),manager(), fink()
{
}
highfink::highfink(const std::string & fn, const std::string & ln,
                   const std::string & j, const std::string & rpo, int ico):
       abstr_emp(fn,ln,j),manager(fn,ln,j,ico),fink(fn,ln,j,rpo)
{
}
highfink::highfink(const abstr_emp & e, const std::string & rpo, int ico):
       abstr_emp(e),manager(e,ico),fink(e,rpo)
{
}
highfink::highfink(const fink & f, int ico):
       abstr_emp(f),fink(f),manager(f,ico)
{
}
highfink::highfink(const manager & m, const std::string & rpo):
       abstr_emp(m), manager(m),fink(m,rpo)
{
}
highfink::highfink(const highfink & h):abstr_emp(h),manager(h),fink(h)
{
}
void highfink::ShowAll() const
{
     manager::ShowAll();
     cout << "Reportsto: " << ReportsTo() << endl;
     cout << endl;

}
void highfink::SetAll()
{
     manager::SetAll();
     cout << "Enter the reportsto: ";
     getline(cin, fink::ReportsTo());

}
void highfink::writeall(ofstream& fout) const
{
     manager::writeall(fout);
     fout<<fink::ReportsTo()<<endl;
```

```
}
void highfink::getall(ifstream& fin)
{
    manager::getall(fin);
    getline(fin,fink::ReportsTo());
}


//pe14-5.cpp
//useempl.cpp
#include <iostream>
#include <fstream>
#include "emp.h"

using namespace std;

const int MAX = 4;
enum ClassType{ Employee, Manager, Fink, HighFink};

int main(void)
{
   employee em("Trip", "Harris", "Thumper") ;
   cout << em << endl;
   em.ShowAll() ;
   manager ma("Amorphia", "Spindragon", "Nuancer", 5);
   cout << ma << endl;
   ma.ShowAll() ;

   fink fi("Matt", "Oggs", "Oiler", "Juno Barr");
   cout << fi << endl;
   fi.ShowAll ();
   highfink hf(ma, "Curly Kew"); //recruitment?
   hf.ShowAll();
   cout << "Press a key for next phase:\n";
   cin.get();

   ofstream fout("EMP.TXT");
   if(!fout.is_open()){exit(EXIT_FAILURE);}
   cout << "Using an abstr_emp *pointer to write file.\n";
   abstr_emp *tri[MAX] = {&em, &ma, &fi, &hf};
/*
 *以上为第 14 章的编程练习 5 的主程序部分，下面开始修改原有程序，测试本章的文件读写部分。
 *读写部分使用 RTTI 方法实现，在写入文件时，RTTI 判断 tri[i]的对象类别，并依照类别写入
 *标志位，基本形式如下
 **/
 for (int i = 0; i < MAX; i++)
 {
    if(typeid(*tri[i]) == typeid(employee))
    {
       fout << Employee << endl;
       tri[i]->writeall(fout);
    }
```

```
    else if(typeid(*tri[i]) == typeid(manager))
    {
        fout << Manager << endl;
        tri[i]->writeall(fout);
    }
    else if(typeid(*tri[i]) == typeid(fink))
    {
        fout << Fink << endl;
        tri[i]->writeall(fout);
    }
    else if(typeid(*tri[i]) == typeid(highfink))
    {
        fout << HighFink << endl;
        tri[i]->writeall(fout);
    }else{
        fout << -1 << endl;
        tri[i]->writeall(fout);
    };
 }
fout.close();

abstr_emp*pc[MAX];
int classtype;
int i = 0;
ifstream fin("EMP.TXT");
if(!fin.is_open()){exit(EXIT_FAILURE);}
/*文件读取部分，首先读取文件内写入的标志位，并通过对比标志位，判断类型，
 *switch 语句内调用不同对象的方法，进行数据恢复
 **/
   while((fin>>classtype))
   {
       fin.get();
       switch(classtype)
       {
           case Employee:
               pc[i] = new employee;
               pc[i++]->getall(fin);
               break;
          case Manager:
               pc[i] = new manager;
               pc[i++]->getall(fin);
               break;
          case Fink:
               pc[i] = new fink;
               pc[i++]->getall(fin);
               break;
          case HighFink:
               pc[i] = new highfink;
               pc[i++]->getall(fin);
               break;
       }
   }
```

```
    for (i = 0; i < MAX; i++) pc[i]->ShowAll();
    fin.close();
    return 0;
}
```

7. 下面是某个程序的部分代码。该程序将键盘输入读取到一个由 string 对象组成的 vector 中，将字符串内容（而不是 string 对象）存储到一个文件中，然后把该文件的内容复制到另一个由 string 对象组成的 vector 中。

```
int main()
{
    using namespace std;
    vector<string> vostr;
    string temp;

    cout << "Enter strings (empty line to quit):\n";
    while (getline(cin,temp) && temp[0] != '\0')
        vostr.push_ back (temp) ;
    cout << "Here is your input.\n";
    for_each(vostr.begin(), vostr.end(), ShowStr) ;

    ofstream fout("strings.dat", ios base::out | ios base::binary) ;
    for_each(vostr.begin(), vostr.end(), Store(fout) );
    fout.close();


    vector<string> vistr;
    ifstream fin("strings.dat", ios base::in | ios _base::binary) ;
    if (!fin.is_open())
    {
        cerr << "Could not open file for input.\n";
        exit (EXIT_FAILURE) ;
    }
    GetStrs(fin, vistr);
    cout << "\nHere are the strings read from the file:\n";
    for_each(vistr.begin(), vistr.end(), ShowStr) ;
    return 0;
}
```

该程序以二进制格式打开文件，并想使用 read()和 write()来完成 I/O。余下的工作如下所述。

- 编写函数 void ShowStr(const string&)，它显示一个 string 对象，并在显示完后换行。
- 编写函数符 Store，它将字符串信息写入文件中。Store 的构造函数应接受一个指定 ifstream 对象的参数，而重载的 operator()(const string &)应指出要写入文件中的字符串。一种可行的计划是，首先将字符串的长度写入文件中，然后将字符串的内容写入文件中。例如，如果 len 存储了字符串的长度，可以这样做。

```
os.write((char*)&len,sizeof(std::size_t)); //存储字符串的长度
os.write(s.data(),len);//存储字符
```

成员函数 data()返回一个指针，该指针指向一个存储字符串字符的数组。它类似于成员函数 c_str()，只是后者在数组末尾加上了一个空字符。

- 编写函数 GetStrs()，它根据文件恢复信息。该函数可以使用 read()来获得字符串的长度，然后使用一个循环从文件中读取相应数量的字符，并将它们附加到一个原来为空的临时 string 末尾。由于 string 的数据是私有的，因此必须使用 string 类的方法来将数据存储到 string 对象中，而不能直接存储。

编程分析：

题目要求实现二进制文件的读取和写入操作。二进制文件的读取和写入主要使用 write()函数，在写入字符串时需要写入字符串长度和字符串内容，具体用法可以参考题目中给出的范例。程序的另一个难点是使用 Store 函数符来实现数据读写，因此需要重载 Store 类的 operator()(const string &)运算符。最后程序还需要使用 GetStrs()函数恢复文件信息，恢复函数应当首先读取字符串长度，随后按照指定长度利用 read()函数读取字符串内容。完整代码如下。

```
/*第 17 章的编程练习 7*/
#include <iostream>
#include <vector>
#include <string>
#include <fstream>

using namespace std;
const int LIMIT = 50;

void ShowStr(const string & str);
void GetStrs(ifstream & fin, vector<string> & v);

class Store
{
private:
    string str;
    ofstream*fout;
public:
    Store(ofstream &out):fout(&out){ }
    bool operator()(const string & str);
    ~Store() {}
};
/*定义 Store 函数符*/

void ShowStr(const string & str)
{
    cout << str << endl;
}

void GetStrs(ifstream & fin, vector<string> & v)
{
    unsigned int len;
    char*p;
    while(fin.read((char *)&len, sizeof len))
    {
```

```
        p = new char[len];
        fin.read(p, len);
        v.push_back(p);
    }
}
/*读取字符串，先读取字符串长度数据，再按照该长度读取指定长度的字符串*/

bool Store::operator()(const string & str)
{
    unsigned int len = str.length();
    if (fout->is_open())
    {
        fout->write((char *)&len, sizeof len);
        fout->write(str.data(), len);
        return true;
    }
    else return false;
}
/*函数符将字符串写入文件，使用题目中的 write()函数，
 *先写入字符串长度，再写入字符串*/

int main()
{
    using namespace std;
    vector<string> vostr;
    string temp;

    cout << "Enter strings (empty line to quit):\n";
    while (getline(cin, temp) && temp[0] != '\0')
        vostr.push_back(temp);
    cout << "Here is your input.\n";
    for_each(vostr.begin(), vostr.end(), ShowStr);


    ofstream fout("strings.txt", ios_base::out | ios_base::binary);
    for_each(vostr.begin(), vostr.end(), Store(fout));
    fout.close();

    vector<string> vistr;
    ifstream fin("strings.txt", ios_base::in | ios_base::binary);
    if (!fin.is_open())
    {
        cerr << "Could not open the file for input.\n";
        exit(EXIT_FAILURE);
    }

    GetStrs(fin, vistr);
    cout << "\nHere are the strings read from the file:\n";
    for_each(vistr.begin(), vistr.end(), ShowStr);

    return 0;
}
```

第 18 章　探讨 C++新标准

本章知识点总结

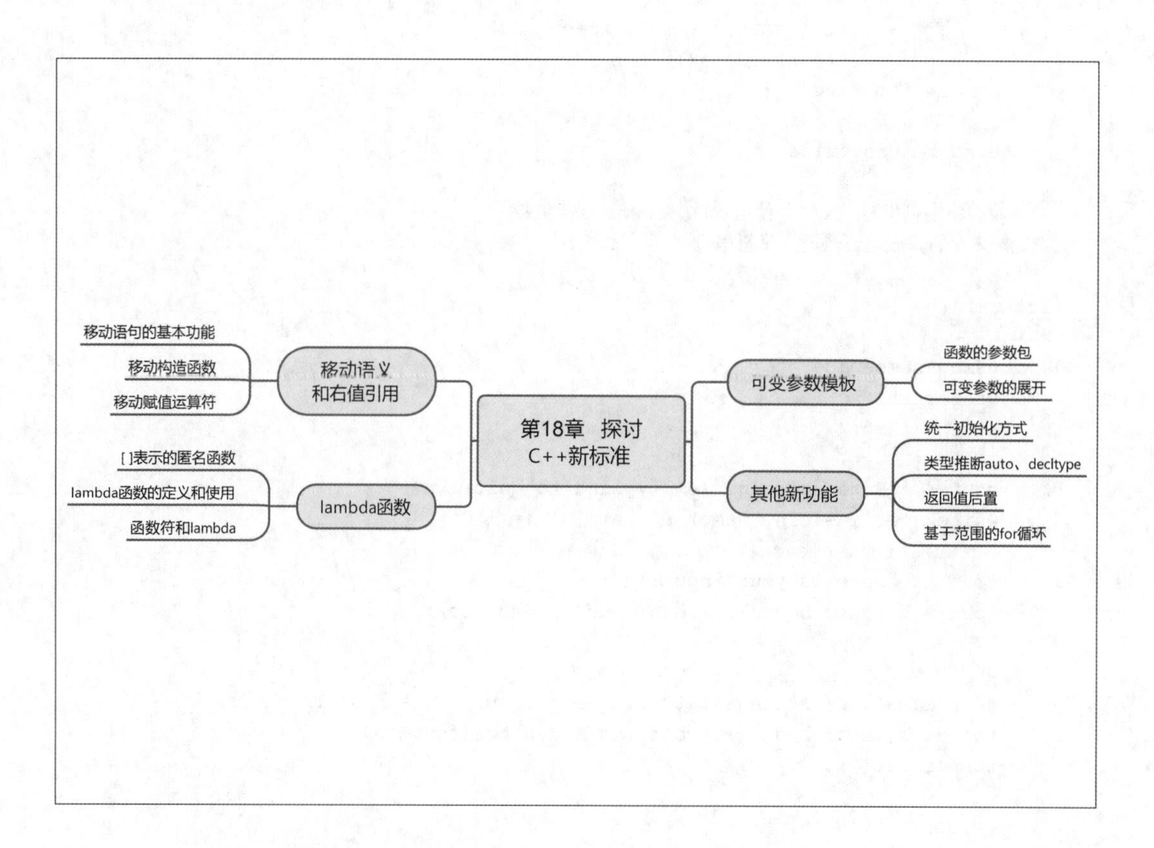

C++11 在原有标准的基础上更新了很多的功能，本章主要介绍移动语义、lambda 匿名函数和可变参数模板。除此之外，应用花括号统一初始化表达式，以及 decltype 和 auto 类型推断，基于范围的 for 循环等内容，都需要在编程实践中练习使用。

18.1　移动语义和右值引用

在 C++语言中，在数据复制操作中通常会使用临时变量暂时存储数据，待运算完成后再销毁这些临时数据。最典型的例子就是函数返回值。函数在运算中会建立一个临时变量以存储返回值，在函数返回时再将临时变量通过数据副本的形式赋值给主调函数的变量，随后销毁被调函数中的临时数据对象，这样的操作在数据量较大时运行效率低下。主要由于临时数据对象的复制和销毁在很多情况下是不必要的，因此 C++通过移动语义的形式来实现数据的快速传递，即保持临时数据对象的存储单元不变，只修改临时数据的引用指向，这样就能够快速实现数据复制和临时数据销毁的操作。总之，移动语义实际上避免了移动原始数据，而只修改了数据存储的引用记录。

C++语言中的左值是指能够取得地址的表达式；右值是指不能取得地址的表达式，如函数的非引用返回值或者字面量。要实现移动语义就需要使用右值引用，因为通常情况下右值是以临时变量的形式进行数据运算和存储的，针对右值的引用会使编译器了解当前需要移动语义。右值引用通常使用&&表示，例如，以下语句中的字面值是典型的右值。

```
int && n = 13;
```

因此可以使用右值引用使变量 n 指向字面量 13。

C++11 在类的设计中新增了两个特殊的成员函数——移动构造函数和移动赋值运算符。编译器会自动提供默认的移动构造函数和移动赋值运算符，函数进行逐成员初始化并复制内置类型。因此用户可以通过自定义一个使用右值引用的构造函数来创建移动构造函数。当使用右值对象初始化对象时，系统将使用移动构造函数。即通常情况下如果要实现移动语义，则需要定义两个构造函数：一个是常规复制构造函数，它使用 const 左值引用作为参数，这个引用关联到左值实参；一个是移动构造函数，它使用右值引用作为参数，该引用关联到右值实参。复制构造函数可执行深拷贝，而移动构造函数只调整引用记录。某些情况下为了使用移动赋值运算符，可以通过运算符 static_cast<>将对象的类型强制转换为 Type &&，或者使用 C++11 提供的函数 std::move(Object)。

18.2　lambda 函数和可变参数模板

lambda 函数是一种匿名函数，它主要应用于 STL 中的算法信息传递，可以用来替代函数指针进行参数传递，其基本语法规则如下。

```
[](Type_Name Var){return expression;}
```

其中，以方括号替代函数名，标识其为 lambda 匿名函数，圆括号内为其参数。如果匿名函数没有返回值，那么系统将会通过 decltyp 推断返回值类型，或者可以使用返回值后置的语法格式进行表示，格式如下。

```
[] (Type_Name Var) ->Retune_Type {return expression;}
```

应用 lambda 函数能够避免使用函数指针，简化代码并且提高运行效率。此外，匿名函数可以通过方括号内设置的变量名直接访问和使用作用域内的任何动态变量，这样能够更加灵活地实现特定的运算。

通过可变参数模板，可以创建接受可变数量的参数的模板函数和模板类。C++11 提供了一个用省略号（...）表示的元运算符（meta-operator），表示模板参数包的标识符。模板参数包基本上是一个类型列表。同样，它还能够用来声明表示函数参数包的标识符，而函数参数包就是一个值列表。其语法如下。

```
template<typename...Args>
void vf(Args...args){}
```

其中，Args 是一个模板参数包，而 args 是一个函数参数包。可将这些参数包的名称指定为任何符合 C++标识符规则的名称。在调用过程中可以直接代入不定量个参数，例如：

```
vf('a', 13, 23, 45);
```

可变参数模板函数在调用中会将参数包展开，其展开方式是使用递归方式进行函数的调用展开。具体实现方法如下所示：

```
void vf(){}
template<typename Ttypename...Args>
void vf(T value, Args...args){
 vf{ args...};
}
```

18.3　复习题

1．使用以大括号括起来的初始化列表重写下述代码。重写后的代码不应使用数组 ar。

```
class Z200 {
private:
    int j;
    char ch;
    double z;
public:
    Z200(int jv, char chv, double zv) : j(jv), ch(chv), z(zv) {}
    ...
};
    double x = 8.8;
    std::string s = "What a bracing effect!";
    int k(99);
    Z200 zip(200,'Z',0.675);
    std::vector<int> ai(5);
    int ar[5] = {3, 9, 4, 7, 1};
    for (auto pt = ai.begin(), int i = 0; pt != ai.end(); ++pt, ++i)
        *pt = ai[i];
```

习题解析：

C++11 标准扩大了用大括号括起来的列表（初始化列表）的适用范围，使其可用于所有内置类型和用户定义的类型。在使用初始化列表时，可添加等号（=），也可不添加。代码首先定义了类 Z200，并通过多种方式初始化了浮点型变量、整型变量、Z200 对象以及数组和 vector 对象 ai 等。为了使用大括号进行数据初始化，修改如下。

```
class Z200 {
private:
    int j;
    char ch;
    double 2;
public:
    Z200(int jv, char chv, double zv) : j(jv), ch(chv), z(zv) {}
    ...
};
    double x = {8.8};
    std::string s {"What a bracing effect!"};
```

```
    int k {99};
    Z200 zip {200,'Z',0.675};
    std::vector<int> ai {3, 9, 4, 7, 1};
```

2. 在下述简短的程序中，哪些函数调用不合法？为什么？对于合法的函数调用，指出其引用参数指向的是什么。

```
include <iostream>
using namespace std;
double up(double x) { return 2.0*x;}
void r1(const double &rx) {cout << rx << endl;}
void r2(double &rx) {cout << rx << endl;}
void r3(double &&rx) {cout << rx << endl;}

int main()
{
    double w = 10.0;
    r1(w);
    r1 (w+1);
    r1(up(w));
    r2(w) ;
    r2 (w+1);
    r2(up(w) );
    r3 (w) ;
    r3 (w+1) ;
    r3(up(w));
    return 0;
}
```

习题解析：

函数 r1()、r2()、r3()分别使用 const 引用参数、引用参数和右值引用参数，因此调用效果分别如下。

- r1(w)合法，形参 rx 指向 w。
- r1(w+1)合法，形参 rx 指向一个临时变量，这个变量被初始化为 w+1。
- r1(up(w))合法，形参 rx 指向一个临时变量，这个变量被初始化为 up(w)的返回值。一般而言，当将左值传递给 const 左值引用参数时，参数将被初始化为左值。当将右值传递给函数时，const 左值引用参数将指向右值的临时副本。
- r2(w)合法，形参 rx 指向 w。
- r2(w+1)非法，因为 w+1 是一个右值。
- r2(up(w))非法，因为 up(w)的返回值是一个右值。一般而言，当将左值传递给非 const 左值引用参数时，参数将被初始化为左值，但非 const 左值形参不能接受右值实参。
- r3(w)非法，因为右值引用不能指向左值（如 w）。
- r3(w+1)合法，rx 指向表达式 w+1 的临时副本。
- r3(up(w))合法，rx 指向 up(w)的临时返回值。

3. a. 下述简短的程序显示什么？为什么？

```
#include <iostream>
using namespace std;
double up(double x) { return 2.0*x;}
void r1(const double &rx) {cout <<"const double & rx\n";}
void r1(double &rx) {cout <<"double & rx\n";}
int main()
{
    double w = 10.0;
    r1(w) ;
    r1 (w+1);
    r1(up(w) ) ;
    return 0;
}
```

b. 下述简短的程序显示什么？为什么？

```
#include <iostream>
using namespace std;
double up(double x) { return 2.0*x;}
void r1(double &rx) {cout << "double & rx\n";}
void r1(double &&rx) {cout << "double && rx\n";}
int main()
{
    double w = 10.0;
    r1(w) ;
    r1(w+1);
    r1(up(w) );
    return 0;
}
```

c. 下述简短的程序显示什么？为什么？

```
#include <iostream>
using namespace std;
double up(double x) {return 2.0*x;}
void r1(const double &rx) {cout << "const double & rx\n";}
void r1(double &&rx) {cout << "double && rx\n";}
int main()
{
    double w = 10.0;
    r1(w) ;
    r1 (w+1);
    r1(up(w));
    return 0;
}
```

习题解析：

a. 本小题中的代码重载了函数 r1()，通过参数的特征值可以确定：

r1(w)的参数 w 不是 const 值，因此调用以(double & rx)为参数的函数；w+1 和 up(w)均是右值，因此匹配 const 值，调用以(const double & rx)为参数的函数，最后的输出如下。

```
double & rx
const double & rx
const double & rx
```

b. 本小题的关键在于第 2 个重载使用了右值引用参数，因此可以确定：

以 w 作为变量调用第 1 个左值引用函数；w+1 和 up(w) 均是右值，因此使用第 2 个右值引用参数，输出如下。

```
double & rx
double && rx
double && rx
```

c. 本小题中两个函数分别使用了 const 左值引用和右值引用参数。调用中 w 作为非 const 变量进行类型转换，匹配(const double & rx)参数，w+1 和 up(w)均是右值，因此使用第 2 个右值引用参数，输出如下。

```
const double & rx
double && rx
double && rx
```

4．哪些成员函数是特殊的成员函数？它们特殊的原因是什么？

习题解析：

C++语言中特殊的成员函数主要包括默认构造函数、复制构造函数、移动构造函数、析构函数、复制赋值运算符和移动赋值运算符。这些函数之所以特殊，是因为编译器将根据情况自动提供它们的默认版本。

5. 假设 Fizzle 类只有如下所示的数据成员，为什么不适合给这个类定义移动构造函数？

```
class Fizzle
{
private:
        double bubbles [4000] ;
}
```

要使这个类适合定义移动构造函数，应如何修改 4 000 个 double 值的存储方式?

习题解析：

在通过右值引用实现的移动语义中，主要目的是实现临时数据的存储空间的所有权转让和再利用，但对于标准数组没有转让其所有权的机制，因此无法使用移动语义。但是如果 Fizzle 使用指针和动态内存分配，则可将数据的地址赋给新指针，以转让其所有权。

6．修改下述简短的程序，使其使用 lambda 表达式而不是 f1()。请不要修改 show2()。

```
#include <iostream>
template<typename T>
void show2(double x, T& fp) {std::cout << x << "-> " «<< fp(x) << '\n';}
double f1(double x) { return 1.8*x + 32;}
int main()
{
        show2(18.0, £1);
        return 0;
}
```

习题解析：

代码中 f1()函数的调用在 show()函数内，因此可以使用 lambda 函数的形式实现。修改方法是直接将 f1()函数的调用修改为 lambda 函数及其实现。代码如下。

```
#include <iostream>
#include <algorithm>
template<typename T>
void show2(double x, T fp) {std::cout << x << " -> " << fp(x) << '\n';}
int main()
{
    show2(18.0, [] (double x){return 1.8*x + 32;});
    return 0;
}
```

7. 修改下述简短而混乱的程序，使其使用 lambda 表达式而不是函数符 Adder。请不要修改 sum()。

```
#include <iostream>
#include <array>
const int Size = 5;
template<typename T>
void sum(std::array<double,Size> a, T& fp);
class Adder {
    double tot;
public:
    Adder(double q = 0) : tot(q) {}
    void operator()(double w) { tot += w; }
    double tot_v() const { return tot; };
};
int main()
{
    double total = 0.0;
    Adder ad(total) ;
    std::array<double, Size> temp_c = {32.1, 34.3, 37.8, 35.2, 34.7};
    sum(temp_c, ad) ;
    total = ad.tot_v();
    std::cout << "total: " << ad.tot_v() << '\n';
    return 0;
}
template<typename T>
void sum(std::array<double,Size> a, T& fp)
{
    for(auto pt = a.begin(); pt != a.end(); ++pt)
    {
        Ep (*pt) ;
    }
}
```

习题解析：

Adder 函数符的主要功能在运算符 operator()(double w) 部分实现。函数符在 sum()函数的调用中作为参数调用，因此将 Adder 修改为 lambda 函数可以简化整个程序。完整代码如下。

```
#include <iostream>
#include <array>
#include <algorithm>
const int Size = 5;
```

```
template<typename T>
void sum(std::array<double,Size> a, T& fp);
int main()
{
    double total = 0.0;
    std::array<double, Size> temp c = {32.1, 34.3, 37.8, 35.2, 34.7};
    sum(temp_c, [&total] (double w) {total += w;});
    std::cout << "total: " << total << '\n';
    std::cin.get();
    return 0;
}
template<typename T>
void sum(std::array<double,Size> a, T& fp)
{
    for(auto pt = a.begin(); pt != a.end(); ++pt)
    {
        fp(*pt) ;
    }
}
```

18.4　编程练习

1．下面是一个简短程序的一部分。

```
int main()
{
    using namespace std;

    auto q = average_list({15.4, 10.7, 9.0});
    cout << q << endl;

    cout << average_list({20, 30, 19, 17, 45, 38} ) << endl;

    auto ad = average_list<double>({'A', 70, 65.33});
    cout << ad << endl;
    return 0;
}
```

请提供函数 average_list()，使该程序变得完整。它应该是一个模板函数，其中的类型参数指定了用作函数参数的 initilize_list 模板的类型以及函数的返回类型。

编程分析：

题目要求定义模板函数 average_list()，计算 double 类型数据的平均值。average_list()函数需要使用 initilize_list 模板作为函数的参数。其中 initilize_list 的基本使用方法可以参考程序清单 16.22。模板函数的声明如下。

```
template<typename T>
T average_list(const std::initializer_list<T> il )
```

完整代码如下。

```
/*第 18 章的编程练习 1*/
#include <iostream>
```

```
#include <initializer_list>

using namespace std;

template<typename T>
T average_list(const std::initializer_list<T> il )
{
    int count = 0;
    T sum=0;
    for (auto p = il.begin(); p <= il.end(); p++)
    {
        count++;
        sum += *p;
    }
    return sum / count;
}
int main()
{
    auto q = average_list({ 15.4,10.7,9.0 });
    cout << q << endl;
    cout << average_list({ 20,30,19,17,45,38 }) << endl;
    auto ad = average_list<double>({ 'A',70,65,33 });
    cout << ad << endl;
    return 0;
}
```

2. 下面是类 Cpmv 的声明。

```
class Cpmv
{
public:
    struct Info
    {
       std::string qcode;
       std::string zcode;
    };
private:
    Info *pi;
public:
    Cpmv () ;
    Cpmv (std::string q, std::string z);
    Cpmv(const Cpmv & cp);
    Cpmv(Cpmv && mv) ;
    ~Cpmv () ;
    Cpmv & operator=(const Cpmv & cp);
    Cpmv & operator=(Cpmv && mv) ;
    Cpmv operator+(const Cpmv & obj) const;
    void Display() const;
};
```

函数 operator+ ()应创建一个对象，其成员 qcode 和 zcode 由操作数的相应成员拼接而成。请提供为移动构造函数和移动赋值运算符实现移动语义的代码。编写一个使用所有这些方法的程序。为方便测试，使各个方法都显示特定的内容，以便知道它们被调用了。

编程分析：

题目要求为 Cpmv 类定义移动构造函数和移动赋值运算符。题目中已经提供类 Cpmv 的声明，构造函数和析构函数的定义中需要注意指针数据成员 pi 的处理。完整代码如下。

```
/*第 18 章的编程练习 2*/
#include<iostream>
#include<string>
using std::cout;
using std::endl;
class Cpmv
{
public:
    struct Info
    {
        std::string qcode;
        std::string zcode;
    };
private:
    Info *pi;
public:
    Cpmv() { pi = nullptr; cout<<"Default Constructor\n"; Display();};
    Cpmv(std::string q, std::string z);
    Cpmv(const Cpmv &cp);
    Cpmv(Cpmv &&mv);
    ~Cpmv();
    Cpmv &operator=(const Cpmv&cp);
    Cpmv &operator=(Cpmv&&mv);
    Cpmv operator+(const Cpmv &obj)const;
    void Display()const;
};

Cpmv::Cpmv(std::string q, std::string z)
{
    pi = new Info;
    pi->qcode = q;
    pi->zcode = z;
    cout<<"Constructor with args."<<endl;
    Display();
}
Cpmv::Cpmv(const Cpmv &cp)
{
    pi = new Info;
    pi->qcode = cp.pi->qcode;
    pi->zcode = cp.pi->zcode;
    cout<<"Constructor Copy."<<endl;
    Display();
}
Cpmv::Cpmv(Cpmv &&mv)
{
    cout<<"\nMove Constructor."<<endl;
    pi = mv.pi;
```

```
    mv.pi = nullptr;
}
Cpmv::~Cpmv()
{
    delete pi;
    cout<<"Deconstructor."<<endl;

}
Cpmv& Cpmv::operator=(const Cpmv&cp)
{
    if (this == &cp)
        return *this;
    delete pi;
    pi = new Info;
    pi->qcode = cp.pi->qcode;
    pi->zcode = cp.pi->zcode;
    cout<<"Assinment Normal."<<endl;
    return *this;
}
Cpmv& Cpmv::operator=(Cpmv&&mv)
{
    if (this == &mv)
        return *this;
    delete pi;
    pi = mv.pi;
    mv.pi = nullptr;
    cout<<"\nAssinment R-values."<<endl;
    Display();
    return *this;
}

Cpmv Cpmv::operator+(const Cpmv &obj) const
{
    cout<<"operator + ()."<<endl;
    return Cpmv((pi->qcode + obj.pi->qcode),(pi->zcode + obj.pi->zcode));
}

void Cpmv::Display()const
{
    cout << "Display Info: ";
    if (pi == nullptr)
        cout << "pi is null.\n";
    else
    {
        cout<<"address: "<<pi<< " qcode: " << pi->qcode
            << " zcode: " << pi->zcode << endl;
    }
}

int main()
{
    using namespace std;
```

```
    Cpmv cp1("C ","++ ");
    Cpmv cp2("Computer ", "Science ");
    Cpmv cp3 = cp2;
    cp2 = cp1;
    cp1 = cp2 + cp3;
    Cpmv cp4 (cp1 + cp2);
    return 0;
}
```

3．编写并测试可变参数模板函数 sum_value()，它接受任意长度的参数列表（其中包含数值，但可以是任何类型），并以 long double 的方式返回这些数值的和。

编程分析：

定义可变参数模板函数的关键在于函数的参数包以及函数的递归展开，其中 Args 是模板参数包，而 args 是函数的参数包。函数递归展开本质上就是利用函数的参数包不断递归调用该函数的一个过程，当函数包展开到最后时，将调用不含任何参数的函数。因此，需要首先定义最后的函数，并确保该函数正确的返回值和返回值类型。程序的完整代码如下。

```
/*第 18 章的编程练习 3*/
#include<iostream>

long double sum_value() {return 0;};
/*定义可变参数函数最后的展开函数*/
template<typename T, typename...Args>
long double sum_value(T value, Args...args)
{
    long double sum = (long double)value + (long double) sum_value(args...);
    return sum;
}
/*可变参数模板函数*/
int main()
{
    using namespace std;
    cout << sum_value(52, 34, 98, 101)<<endl;
    cout << sum_value('x', 'y', 95, 74, 'Z')<<endl;
    cout << sum_value(0.2, 1e2, 54, 'M','\t')<<endl;
    /*简单测试可变参数函数，计算参数的和*/
    return 0;
}
```

4．使用 lambda 函数重写程序清单 16.15。具体地说，使用一个有名称的 lambda 函数替换函数 outint()，并将函数符替换为两个匿名 lambda 表达式。

编程分析：

程序清单 16.15 使用函数符 TooBig 实现了数据筛选功能。函数符实质上是通过重载运算符 operator()实现的，因此将函数符替换成匿名函数需要保持该重载运算符的功能完整。完整代码如下。

```
/*第 18 章的编程练习 4*/
//functor.cpp —— 使用函数符
#include <iostream>
#include <list>
#include <iterator>
#include <algorithm>

template<class T>  //定义 operator()()的函数符类
class TooBig
{
private:
    T cutoff;
public:
    TooBig(const T & t) : cutoff(t) {}
    bool operator()(const T & v) { return v > cutoff; }
};
/*定义函数符*/
//void outint(int n) {std::cout << n << " ";}
auto Loutint = [](int n){std::cout << n << " ";};
/*命名匿名 lambda 函数 Loutin*/
int main()
{
    using std::list;
    using std::cout;
    using std::endl;
    using std::for_each;
    using std::remove_if;

    TooBig<int> f100(100); //limit = 100
    int vals[10] = {50, 100, 90, 180, 60, 210, 415, 88, 188, 201};
    list<int> yadayada(vals, vals + 10); //范围构造函数
    list<int> etcetera(vals, vals + 10);

    //C++0x 可以使用以下语句
// list<int> yadayada = {50, 100, 90, 180, 60, 210, 415, 88, 188, 201};
// list<int> etcetera {50, 100, 90, 180, 60, 210, 415, 88, 188, 201};

    cout << "Original lists:\n";
    for_each(yadayada.begin(), yadayada.end(), Loutint);
    /*调用匿名 lambda 函数*/
    cout << endl;
    for_each(etcetera.begin(), etcetera.end(), Loutint);
    cout << endl;
    yadayada.remove_if([](int n)->bool{ return n > 100;});  //使用命名的函数对象
    etcetera.remove_if([](int n)->bool{ return n > 200;});  //构造一个函数对象
    cout <<"Trimmed lists:\n";
    for_each(yadayada.begin(), yadayada.end(), Loutint);
    cout << endl;
    for_each(etcetera.begin(), etcetera.end(), Loutint);
    cout << endl;
    //std::cin.get();
    return 0;
}
```